TRACKING AND KALMAN FILTERING MADE EASY

TRACKING AND KALMAN FILTERING MADE EASY

ELI BROOKNER
Consulting Scientist
Raytheon Comp.
Sudbury, MA

A Wiley-Interscience Publication
JOHN WILEY & SONS, INC
New York · Chichester · Weinheim · Brisbane · Singapore · Toronto

Copyright © 1998 by John Wiley & Sons. All rights reserved.

Published simultaneously in Canada.

Library of Congress Cataloging-in-Publication Data:

Brookner, Eli, 1931–
  Tracking and Kalman filtering made easy/Eli Brookner.
    p.   cm.
  "A Wiley-Interscience publication."
  Includes bibliographical references (p.   ) and index.
  ISBN 0-471-18407-1 (cloth : alk. paper)
    1. Electric filters—Mathematics.   2. Tracking radar—Mathematics.
  3. Kalman filtering.   I. Title.
  TK7872.F5B76 1998
  621.3848—dc21   97-33122

Printed in the United States of America

10 9 8 7 6 5

To Larry and Vera,
Richard and Connie,
and Daniel, the little miracle

CONTENTS

PREFACE

At last a book that hopefully will take the mystery and drudgery out of the g–h, α–β, g–h–k, α–β–γ and Kalman filters and makes them a joy. Many books written in the past on this subject have been either geared to the tracking filter specialist or difficult to read. This book covers these filters from very simple physical and geometric approaches. Extensive, simple and useful design equations, procedures, and curves are presented. These should permit the reader to very quickly and simply design tracking filters and determine their performance with even just a pocket calculator. Many examples are presented to give the reader insight into the design and performance of these filters. Extensive homework problems and their solutions are given. These problems form an integral instructional part of the book through extensive numerical design examples and through the derivation of very key results stated without proof in the text, such as the derivation of the equations for the estimation of the accuracies of the various filters [see Note (1) on page 388]. Covered also in simple terms is the least-squares filtering problem and the orthonormal transformation procedures for doing least-squares filtering.

The book is intended for those not familiar with tracking at all as well as for those familiar with certain areas who could benefit from the physical insight derived from learning how the various filters are related, and for those who are specialists in one area of filtering but not familiar with other areas covered. For example, the book covers in extremely simple physical and geometric terms the Gram–Schmidt, Givens, and Householder orthonormal transformation procedures for doing the filtering and least-square estimation problem. How these procedures reduce sensitivity to computer round-off errors is presented. A simple explanation of both the classical and modified Gram–Schmidt procedures is given. Why the latter is less sensitive to round-off errors is explained in physical terms. For the first time the discrete-time orthogonal Legendre polynomial (DOLP) procedure is related to the voltage-processing procedures.

Important real-world issues such as how to cope with clutter returns, elimination of redundant target detections (observation-merging or clustering), editing for inconsistent data, track-start and track-drop rules, and data association (e.g., the nearest-neighbor approach and track before detection) are covered in clear terms. The problem of tracking with the very commonly used chirp waveform (a linear-frequency-modulated waveform) is explained simply with useful design curves given. Also explained is the important moving-target detector (MTD) technique for canceling clutter.

The Appendix gives a comparison of the Kalman filter (1960) with the Swerling filter (1959). This Appendix is written by Peter Swerling. It is time for him to receive due credit for his contribution to the "Kalman–Swerling" filter.

The book is intended for home study by the practicing engineer as well as for use in a course on the subject. The author has successfully taught such a course using the notes that led to this book. The book is also intended as a design reference book on tracking and estimation due to its extensive design curves, tables, and useful equations.

It is hoped that engineers, scientists, and mathematicians from a broad range of disciplines will find the book very useful. In addition to covering and relating the $g–h$, $\alpha–\beta$, $g–h–k$, $\alpha–\beta–\gamma$, Kalman filters, and the voltage-processing methods for filtering and least-squares estimation, the use of the voltage-processing methods for sidelobe canceling and adaptive-array processing are explained and shown to be the same mathematically as the tracking and estimated problems. The massively parallel systolic array sidelobe canceler processor is explained in simple terms. Those engineers, scientists, and mathematicians who come from a mathematical background should get a good feel for how the least-squares estimation techniques apply to practical systems like radars. Explained to them are matched filtering, chirp waveforms, methods for dealing with clutter, the issue of data association, and the MTD clutter rejection technique. Those with an understanding from the radar point of view should find the explanation of the usually very mathematical Gram–Schmidt, Givens, and Householder voltage-processing (also called square-root) techniques very easy to understand. Introduced to them are the important concepts of ill-conditioning and computational accuracy issues. The classical Gram–Schmidt and modified Gram–Schmidt procedures are covered also, as well as why one gives much more accurate results. Hopefully those engineers, scientists, and mathematicians who like to read things for their beauty will find it in the results and relationships given here. The book is primarily intended to be light reading and to be enjoyed. It is a book for those who need or want to learn about filtering and estimation but prefer not to plow through difficult esoteric material and who would rather enjoy the experience. We could have called it "The Joy of Filtering."

The first part of the text develops the $g–h$, $g–h–k$, $\alpha–\beta$, $\alpha–\beta–\gamma$, and Kalman filters. Chapter 1 starts with a very easy heuristic development of $g–h$

filters for a simple constant-velocity target in "lineland" (one-dimensional space, in contrast to the more complicated two-dimensional "flatland"). Section 1.2.5 gives the g–h filter, which minimizes the transient error resulting from a step change in the target velocity. This is the well-known Benedict–Bordner filter. Section 1.2.6 develops the g–h filter from a completely different, common-sense, physical point of view, that of least-squares fitting a straight line to a set of range measurements. This leads to the critically damped (also called discounted least-squares and fading-memory) filter. Next, several example designs are given. The author believes that the best way to learn a subject is through examples, and so numerous examples are given in Section 1.2.7 and in the homework problems at the end of the book.

Section 1.2.9 gives the conditions (on g and h) for a g–h filter to be stable (these conditions are derived in problem 1.2.9-1). How to initiate tracking with a g–h filter is covered in Section 1.2.10. A filter (the g–h–k filter) for tracking a target having a constant acceleration is covered in Section 1.3. Coordinate selection is covered in Section 1.5.

The Kalman filter is introduced in Chapter 2 and related to the Benedict–Bordner filter, whose equations are derived from the Kalman filter in Problem 2.4-1. Reasons for using the Kalman filter are discussed in Section 2.2, while Section 2.3 gives a physical feel for how the Kalman filter works in an optimum way on the data to give us a best estimate. The Kalman filter is put in matrix form in Section 2.4, not to impress, but because in this form the Kalman filter applies way beyond lineland—to multidimensional space.

Section 2.6 gives a very simple derivation of the Kalman filter. It requires differentiation of a matrix equation. But even if you have never done differentiation of a matrix equation, you will be able to follow this derivation. In fact, you will learn how to do matrix differentiation in the process! If you had this derivation back in 1958 and told the world, it would be your name filter instead of the Kalman filter. You would have gotten the IEEE Medal of Honor and \$20,000 tax-free and the \$340,000 Kyoto Prize, equivalent to the Nobel Prize but also given to engineers. You would be world famous.

In Section 2.9 the Singer g–h–k Kalman filter is explained and derived. Extremely useful g–h–k filter design curves are presented in Section 2.10 together with an example in the text and many more in Problems 2.10-1 through 2.10-17. The issues of the selection of the type of g–h filter is covered in Section 2.11.

Chapter 3 covers the real-world problem of tracking in clutter. The use of the track-before-detect retrospective detector is described (Section 3.1.1). Also covered is the important MTD clutter suppression technique (Section 3.1.2.1). Issues of eliminating redundant detections by observation merging or clustering are covered (Section 3.1.2.2) as well as techniques for editing out inconsistent data (Section 3.1.3), combining clutter suppression with track initiation (Section 3.1.4), track-start and track-drop rules (Section 3.2), data association (Section 3.3), and track-while-scan systems (Section 3.4).

In Section 3.5 a tutorial is given on matched filtering and the very commonly used chirp waveform. This is followed by a discussion of the range bias error problem associated with using this waveform and how this bias can be used to advantage by choosing a chirp waveform that predicts the future—a fortune-telling radar.

The second part of the book covers least-squares filtering, its power and voltage-processing approaches. Also, the solution of the least-squares filtering problem via the use of the DOLP technique is covered and related to voltage-processing approaches. Another simple derivation of the Kalman filter is presented and additional properties of the Kalman filter given. Finally, how to handle nonlinear measurement equations and nonlinear equations of motion are discussed (the extended Kalman filter).

Chapter 4 starts with a simple formulation of the least-squares estimation problem and gives its power method solution, which is derived both by simple differentiation (Section 4.1) and by simple geometry considerations (Section 4.2). This is followed by a very simple explanation of the Gram–Schmidt voltage-processing (square-root) method for solving the least-squares problem (Section 4.3). The voltage-processing approach has the advantage of being much less sensitive to computer round-off errors, with about half as many bits being required to achieve the same accuracy. The voltage-processing approach has the advantage of not requiring a matrix inverse, as does the power method.

In Section 4.4, it is shown that the mathematics for the solution of the tracking least-squares problem is identical to that for the radar and communications sidelobe canceling and adaptive nulling problems. Furthermore, it is shown how the Gram–Schmidt voltage-processing approach can be used for the sidelobe canceling and adaptive nulling problem.

Often the accuracy of the measurements of a tracker varies from one time to another. For this case, in fitting a trajectory to the measurements, one would like to make the trajectory fit closer to the accurate data. The minimum-variance least-squares estimate procedure presented in Section 4.5 does this. The more accurate the measurement, the closer the curve fit is to the measurement.

The fixed-memory polynomial filter is covered in Chapter 5. In Section 5.3 the DOLP approach is applied to the tracking and least-squares problem for the important cases where the target trajectory or data points (of which there are a fixed number $L + 1$) are approximated by a polynomial fit of some degree m. This method also has the advantage of not requiring a matrix inversion (as does the power method of Section 4.1). Also, its solution is much less sensitive to computer round-off errors, half as many bits being required by the computer.

The convenient and useful representation of the polynomial fit of degree m in terms of the target equation motion derivatives (first m derivatives) is given in Section 5.4. A useful general solution to the DOLP least-squares estimate for a polynomial fit that is easily solved on a computer is given in Section 5.5. Sections 5.6 through 5.10 present the variance and bias errors for the least-squares solution and discusses how to balance these errors. The important method of trend removal to lower the variance and bias errors is discussed in Section 5.11.

In Chapter 5, the least-squares solution is based on the assumption of a fixed number $L + 1$ of measurements. In this case, when a new measurement is made, the oldest measurement is dropped in order to keep the number of measurements on which the trajectory estimate is based equal to the fixed number $L + 1$. In Chapter 6 we consider the case when a new measurement is made, we no longer throw away the oldest data. Such a filter is called a growing-memory filter. Specifically, an mth-degree polynomial is fitted to the data set, which now grows with time, that is, L increases with time. This filter is shown to lead to the easy-to-use recursive growing-memory g–h filter used for track initiation in Section 1.2.10. The recursive g–h–k ($m = 2$) and g–h–k–l ($m = 3$) versions of this filter are also presented. The issues of stability, track initiation, root-mean-square (rms) error, and bias errors are discussed.

In Chapter 7 the least-squares polynomial fit to the data is given for the case where the error of the fit is allowed to grow the older the data. In effect, we pay less and less attention to the data the older it is. This type of filter is called a fading-memory filter or discounted least-squares filter. This filter is shown to lead to the useful recursive fading-memory g–h filter of Section 1.2.6 when the polynomial being fitted to is degree $m = 1$. Recursive versions of this filter that apply to the case when the polynomial being fitted has degree $m = 2, 3, 4$ are also given. The issues of stability, rms error, track initiation, and equivalence to the growing-memory filters are also covered.

In Chapter 8 the polynomial description of the target dynamics is given in terms of a linear vector differential equation. This equation is shown to be very useful for obtaining the transition matrix for the target dynamics by either numerical integration or a power series in terms of the matrix coefficient of the differential equation.

In Chapter 9 the Bayes filter is derived (Problem 9.4-1) and in turn from it the Kalman filter is again derived (Problem 9.3-1). In Chapters 10 through 14 the voltage least-squares algorithms are revisited. The issues of sensitivity to computer round-off error in obtaining the inverse of a matrix are elaborated in Section 10.1. Section 10.2 explains physically why the voltage least-squares algorithm (square-root processing) reduces the sensitivity to computer round-off errors. Chapter 11 describes the Givens orthonormal transformation voltage algorithm. The massively parallel systolic array implementation of the Givens algorithm is detailed in Section 11.3. This implementation makes use of the CORDIC algorithm used in the Hewlett-Packard hand calculators for trigonometric computations.

The Householder orthonormal transformation voltage algorithm is described in Chapter 12. The Gram–Schmidt orthonormal transformation voltage algorithm is revisited in Chapter 13, with classical and modified versions explained in simple terms. These different voltage least-squares algorithms are compared in Section 14.1 and to QR decomposition in Section 14.2. A recursive version is developed in Section 14.3. Section 14.4 relates these voltage-processing orthonormal transformation methods to the DOLP approach used in Section 5.3 for obtaining a polynomial fit to data. The two methods are shown to be essentially identical. The square-root Kalman filter, which is less sensitive to round-off errors, is discussed in Section 14.5.

Up until now the deterministic part of the target model was assumed to be time invariant. For example, if a polynomial fit of degree m was used for the target dynamics, the coefficients of this polynomial fit are constant with time. Chapter 15 treats the case of time-varying target dynamics.

The Kalman and Bayes filters developed up until now depend on the observation scheme being linear. This is not always the situation. For example, if we are measuring the target range R and azimuth angle θ but keep track of the target using the east-north x, y coordinates of the target with a Kalman filter, then errors in the measurement of R and θ are not linearly related to the resulting error in x and y because

$$x = R \cos \theta \tag{1}$$

and

$$y = R \sin \theta \tag{2}$$

where θ is the target angle measured relative to the x axis. Section 16.2 shows how to simply handle this situation. Basically what is done is to linearize Eqs. (1) and (2) by using the first terms of a Taylor expansion of the inverse equations to (1) and (2) which are

$$R = \sqrt{x^2 + y^2} \tag{3}$$

$$\theta = \tan \frac{y}{x} \tag{4}$$

Similarly the equations of motion have to be linear to apply the Kalman–Bayes filters. Section 16.3 describes how a nonlinear equation of motion can be linearized, again by using the first term of a Taylor expansion of the nonlinear equations of motion. The important example of linearization of the nonlinear observation equations obtained when observing a target in spherical coordinates (R, θ, ϕ) while tracking it in rectangular (x, y, z) coordinates is given. The example of the linearization of the nonlinear target dynamics equations obtained when tracking a projectile in the atmosphere is detailed. Atmospheric drag on the projectile is factored in.

In Chapter 17 the technique for linearizing the nonlinear observation equations and dynamics target equations in order to apply the recursive Kalman and Bayes filters is detailed. The application of these linearizations to a nonlinear problem in order to handle the Kalman filter is called the extended Kalman filter. It is also the filter Swerling originally developed (without the target process noise). The Chapter 16 application of the tracking of a ballistic projectile through the atmosphere is again used as an example.

The form of the Kalman filter given in Kalman's original paper is different from the forms given up until now. In Chapter 18 the form given until now is related to the form given by Kalman. In addition, some of the fundamental results given in Kalman's original paper are summarized here.

ELI BROOKNER

Sudbury, MA
January 1998

ACKNOWLEDGMENT

I would like to thank Fred Daum (Raytheon Company), who first educated me on the Kalman filter and encouraged me throughout this endeavor. I also would like to thank Erwin Taenzer (formally of the Raytheon Company), from whose many memos on the g–k and g–h–k filters I first learned about these filters. I am indebted to Barbara Rolinski (formerly of the Raytheon Company), who helped give birth to this book by typing a good part of its first draft, including its many complicated equations. This she did with great enthusiasm and professionalism. She would have finished it were it not that she had to leave to give birth to her second child. I would also like to thank Barbara Rolinski for educating her replacement on the typing of the text with its complicated equations using the VAX Mass-11 Word Processor. I would like to thank Lisa Cirillo (formerly of the Raytheon Company), Barbara Rolinski's first replacement, for typing the remainder of the first draft of the book. I am most appreciative of the help of Richard P. Arcand, Jr. (Raytheon Company), who helped Barbara Rolinski and Lisa Cirillo on difficult points relative to the use of the VAX Mass-11 for typing the manuscript and for educating Ann Marie Quinn (Raytheon Company) on the use of the Mass-11 Word Processor. Richard Arcand, Jr. also meticulously made the second-draft corrections for the first part of the book. I am most grateful to Ann Marie Quinn for retyping some of the sections of the book and making the legion of corrections for the many successive drafts of the book. Thanks are also due the Office Automation Services (Raytheon Company) for helping to type the second draft of the second part of the book. Sheryl Evans (Raytheon Company) prepared many of the figures and tables for the book and for that I am grateful. I am grateful to Richard Arcand, Jr. for converting the text to Microsoft Word on the MAC. I am extremely grateful to Joanne Roche, who completed the horrendous task of retyping the equations into Microsoft Word and for correcting some of the figures. I am grateful to Joyce Horne for typing some of the problems and solutions and some of the tables and to Jayne C. Stokes for doing the final typing. Thanks are due to Peter Maloney (Raytheon Company) for helping to convert the manuscript from the MAC to the PC Microsoft Word format. Thanks are due Margaret M. Pappas, Filomena Didiano, Tom Blacquier, and Robert C. Moore of the Raytheon library for helping obtain many of the references used in preparing this book. I would like to thank Tom Mahoney and Robert E. Francois for providing some of the secretarial support needed for typing the book. Thanks are also due Jack Williamson and Sally Lampi (both of Raytheon Company) for their support.

Thanks are due to Robert Fitzgerald (Raytheon Company) for permitting me to extract from two of his excellent tracking papers and for his helpful proofreading of the text. Fritz Dworshak, Morgan Creighton, James Howell, Joseph E. Kearns, Jr., Charles W. Jim, Donald R. Muratori, Stavros Kanaracus, and Gregg Ouderkirk (all of Raytheon Company), Janice Onanian McMahon and Peter Costa (both formerly of Raytheon Company), and Dr. H. Kerr (TeK Associates) provided useful comments. Thanks are due Allan O. Steinhardt (DARPA) and Charles M. Rader (Lincoln Laboratory, MIT) for initially educating me on the Givens transformation. Special thanks is due Norman Morrison (University of Cape Town) for allowing me to draw freely from his book *Introduction to Sequential Smoothing and Prediction* [5]. His material formed the basis for Chapters 5 to 9, and 15 to 18.

Finally I would like to thank my wife, Ethel, for her continued encouragements in this endeavor. Her support made it possible.

E. B.

TRACKING AND KALMAN FILTERING MADE EASY

PART I

TRACKING, PREDICTION, AND SMOOTHING BASICS

In the first part of this book the mystery is taken out of the Kalman, α–β, g–h, α–β–γ and g–h–k filters. We will analyze these filters in simple physical terms rather than in difficult mathematical terms. First the α–β and g–h filters are introduced, followed by track initiation, the α–β–γ and g–h–k filters, coordinate selection, the Kalman filter, ill-conditioning, a summary of examples from the literature, tracking in clutter, track-start and track-drop rules, data association, and finally tracking with chirp waveforms.

1

g–h AND g–h–k FILTERS

1.1 WHY TRACKING AND PREDICTION ARE NEEDED IN A RADAR

Let us first start by indicating why tracking and prediction are needed in a radar. Assume a fan-beam surveillance radar such as shown in Figure 1.1-1. For such a radar the fan beam rotates continually through $360°$, typically with a period of 10 sec. Such a radar provides two-dimensional information about a target. The first dimension is the target range (i.e., the time it takes for a transmitted pulse to go from the transmitter to the target and back); the second dimension is the azimuth of the target, which is determined from the azimuth angle (see Figure 1.1-1) the fan beam is pointing at when the target is detected [1]. Figures 1.1-2 through 1.1-6 show examples of fan-beam radars.

Assume that at time $t = t_1$ the radar is pointing at scan angle θ and two targets are detected at ranges R_1 and R_2; see Figure 1.1-7. Assume that on the next scan at time $t = t_1 + T$ (i.e., $t_1 + 10$ see), again two targets are detected; see Figure 1.1-7. The question arises as to whether these two targets detected on the second scan are the same two targets or two new targets. The answer to this question is important for civilian air traffic control radars and for military radars. In the case of the air traffic control radar, correct knowledge of the number of targets present is important in preventing target collisions. In the case of the military radar it is important for properly assessing the number of targets in a threat and for target interception.

Assume two echoes are detected on the second scan. Let us assume we correctly determine these two echoes are from the same two targets as observed on the first scan. The question then arises as to how to achieve the proper association of the echo from target 1 on the second scan with the echo from

3

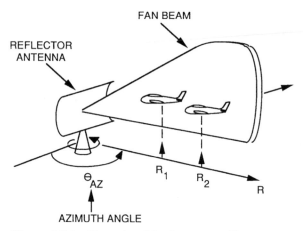

Figure 1.1-1 Example of fan-beam surveillance radar.

Figure 1.1-2 New combined Department of Defense (DOD) and Federal Aviation Administration (FAA) S Band fan-beam track-while-scan Digital Airport Surveillance Radar (DASR) ASR-11. This primary system uses a 17-kW peak-power solid-state "bottle" transmitter. Mounted on top of ASR-11 primary radar antenna is L-band open-array rectangular antenna of colocated Monopulse Secondary Surveillance Radar (MSSR). Up to 200 of these systems to be emplaced around the United States. (Photo courtesy of Raytheon Company.)

Figure 1.1-3   Fan-beam track-while-scan S-band and X-band radar antennas emplaced on tower at Prince William Sound Alaska (S-band antenna on left). These radars are part of the Valdez shore-based Vessel Traffic System (VTS). (Photo courtesy of Raytheon Company.)

target 1 on the first scan and correspondingly the echo of target 2 on the second scan with that of target 2 on the first scan.

If an incorrect association is made, then an incorrect velocity is attached to a given target. For example, if the echo from target 1 on the second scan is associated with the echo from target 2 of the first scan, then target 2 is concluded to have a much faster velocity than it actually has. For the air traffic control radar this error in the target's speed could possibly lead to an aircraft collision; for a military radar, a missed target interception could occur.

The chances of incorrect association could be greatly reduced if we could accurately predict ahead of time where the echoes of targets 1 and 2 are to be expected on the second scan. Such a prediction is easily made if we had an estimate of the velocity and position of targets 1 and 2 at the time of the first scan. Then we could predict the distance target 1 would move during the scan-to-scan period and as a result have an estimate of the target's future position. Assume this prediction was done for target 1 and the position at which target 1 is expected at scan 2 is indicated by the vertical dashed line in Figure 1.1-7. Because the exact velocity and position of the target are not known at the time of the first scan, this prediction is not exact. If the inaccuracy of this prediction is known, we can set up a $\pm 3\sigma$ (or $\pm 2\sigma$) window about the expected value, where σ is the root-mean-square (rms), or equivalently, the standard deviation of the sum of the prediction plus the rms of the range measurement. This window is defined by the pair of vertical solid lines straddling the expected position. If an echo is detected in this window for target 1 on the second scan,

Figure 1.1-4 Fan-beam track-while-scan shipboard AN/SPS-49 radar [3]. Two hundred ten radars have been manufactured. (Photo courtesy of Raytheon Company.)

then with high probability it will be the echo from target 1. Similarly, a $\pm 3\sigma$ window is set for target 2 at the time of the second scan; see Figure 1.1-7.

For simplicity assume we have a one-dimensional world. In contrast to a term you may have already heard, "flatland", this is called "linland". We assume a target moving radially away or toward the radar, with x_n representing the slant range to the target at time n. In addition, for further simplicity we assume the target's velocity is constant; then the prediction of the target position (range) and velocity at the second scan can be made using the following simple target equations of motion:

$$x_{n+1} = x_n + T\dot{x}_n \qquad (1.1\text{-}1a)$$

$$\dot{x}_{n+1} = \dot{x}_n \qquad (1.1\text{-}1b)$$

where x_n is the target range at scan n, $\dot{x}_n$ is the target velocity at scan n, and T the scan-to-scan period. These equations of motion are called the *system dynamic model*. We shall see later, once we understand the above simple case,

Figure 1.1-5 L-band fan-beam track-while-scan Pulse Acquisition Radar of HAWK system, which is used by 17 U.S. allied countries and was successfully used during Desert Storm. Over 300 Hawk systems have been manufactured. (Photo courtesy of Raytheon Company.)

Figure 1.1-6 New fan-beam track-while-scan L-band airport surveillance radar ASR-23SS consisting of dual-beam cosecant squared antenna shown being enclosed inside 50-ft radome in Salahah, Oman. This primary radar uses a 25-kW peak-power solid-state "bottle" transmitter. Mounted on top of primary radar antenna is open-array rectangular antenna of colocated MSSR. This system is also being deployed in Hong Kong, India, The People's Republic of China, Brazil, Taiwan, and Australia.

FIRST SCAN, t=t ᵢ

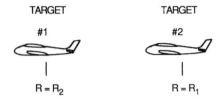

SECOND SCAN, $t = t_2 = t_1 + T = t_1 + 10$ SEC

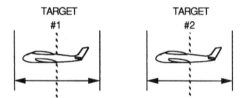

Figure 1.1-7   Tracking problem.

that we can easily extend our results to the real, multidimensional world where we have changing velocity targets.

The α–β, α–β–γ, and Kalman tracking algorithms described in this book are used to obtain running estimates of x_n and $\dot{x}_n$, which in turn allows us to do the association described above. In addition, the prediction capabilities of these filters are used to prevent collisions in commercial and military air traffic control applications. Such filter predictions also aid in intercepting targets in defensive military situations.

The fan-beam ASR-11 Airport Surveillance Radar (ASR) in Figure 1.1-2 is an example of a commercial air traffic control radar. The fan-beam marine radar of Figure 1.1-3 is used for tracking ships and for collision avoidance. These two fan-beam radars and those of the AN/SPS-49, HAWK Pulse Acquisition Radar (PAR), and ASR-23SS radars of Figures 1.1-4 to 1.1-6 are all examples of radars that do target tracking while the radar antenna rotates at a constant rate doing target search [1]. These are called track-while-scan (TWS) radars. The tracking algorithms are also used for precision guidance of aircraft onto the runway during final approach (such guidance especially needed during bad weather). An example of such a radar is the GPS-22 High Performance Precision Approach Radar (HiPAR) of Figure 1.1-8 [1–4]. This radar uses electronic scanning of the radar beam over a limited angle (20° in azimuth,

Figure 1.1-8   Limited-scan, electronically scanned phased-array AN/GPS-22 HiPAR. Used for guiding aircraft during landing under conditions of poor visibility [1–3]. Sixty systems deployed around the world [137]. (Photo courtesy of Raytheon Company.)

Figure 1.1-9   Multifunction PATRIOT electronically scanned phased-array radar used to do dedicated track on many targets while doing search on time-shared basis [1–3]. One hundred seventy-three systems built each with about 5000 radiating elements for front and back faces for a total of about 1.7 million elements [137]. (Photo courtesy of Raytheon Company.)

Figure 1.1-10 Multifunction shipboard AEGIS electronically scanned phased-array radar used to track many targets while also doing search on a time-shared basis. [1, 3]. Two hundred thirty-four array faces built each with about 4000 radiating elements and phase shifters [137]. (Photo courtesy of Raytheon Company.)

$8°$ in elevation) instead of mechanical scanning [1–4]. An example of a wide-angle electronically scanned beam radar used for air defense and enemy target intercept is the PATRIOT radar of Figure 1.1-9 used successfully during Desert Storm for the intercept of SCUD missiles. Another example of such a radar is the AEGIS wide-angle electronically scanned radar of Figure 1.1-10.

The Kalman tracking algorithms discussed in this book are used to accurately predict where ballistic targets such as intercontinental ballistic missiles (ICBMs) will impact and also for determining their launch sites (what country and silo field). Examples of such radars are the upgraded wide-angle electronically steered Ballistic Missile Early Warning System (BMEWS) and the Cobra Dane radars of Figures 1.1-11 and 1.1-12 [1–3]. Another such wide-angle electronically steered radar is the tactical ground based 25,000-element X-band solid state active array radar system called Theater High Altitude Area

Figure 1.1-11 Upgrade electronically steered phased-array BMEWS in Thule, Greenland [1]. (Photo courtesy of Raytheon Company.)

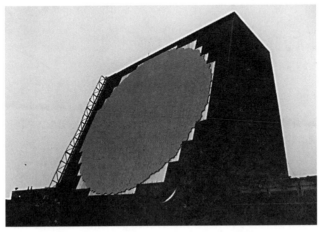

Figure 1.1-12 Multifunction electronically steered Cobra Dane phased-array radar (in Shemya, Alaska). Used to track many targets while doing search on a time-shared basis [1, 3]. (Photo by Eli Brookner.)

Figure 1.1-13 A 25,000-element X-band MMIC (monolithic microwave integrated circuit) array for Theater High Altitude Area Defense (THAAD; formerly GBR) [136, 137]. (Photo courtesy of Raytheon Company.)

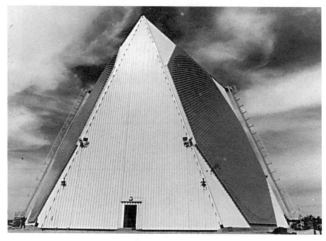

Figure 1.1-14 Multifunction electronically steered two-faced Pave Paws solid-state, phase-steered, phased-array radar [1–3]. (Photo by Eli Brookner.)

Defense (THAAD; formerly called GBR) system used to detect, track, and intercept, at longer ranges than the PATRIOT, missiles like the SCUD; see Figure 1.1-13 [136, 137]. Still another is the Pave Paws radar used to track satellites and to warn of an attack by submarine-launched ballistic missiles; see Figure 1.1-14 [1–3].

Figure 1.1-15   Long-range limited electronically scanned (phase–phase), phased-array artillery locating Firefinder AN/TPQ-37 radar [1]. One hundred two have been built and it is still in production [137]. (Photo courtesy of Hughes Aircraft.)

Two limited-scan electronically steered arrays that use the algorithms discussed in this book for the determination of artillery and mortar launch sites are the Firefinder AN/TPQ-36 and AN/TPQ-37 radars of Figures 1.1-15 and 1.1-16 [1]. An air and surface–ship surveillance radar that scans its beam electronically in only the azimuth direction to locate and track targets is the Relocatable Over-the-Horizon Radar (ROTHR) of Figure 1.1-17 [1].

All of the above radars do target search while doing target track. Some radars do dedicated target track. An example of such a radar is the TARTAR AN/SPG-51 dish antenna of Figure 1.1-18, which mechanically slews a pencil beam dedicated to tracking one enemy target at a time for missile interception. Two other examples of dedicated pencil beam trackers are the HAWK and NATO SEASPARROW tracker-illuminators; see Figures 1.1-19 and 1.1-20.

Figure 1.1-16 Short-range, limited-scan, electronically scanned (phase-frequency) phased-array artillery locating Firefinder AN/TPQ-36 radar [1]. Two hundred forty-three have been built. (Photo courtesy of Hughes Co.)

1.2 *g–h* FILTERS

1.2.1 Simple Heuristic Derivation of *g–h* Tracking and Prediction Equations

Equation (1.1-1) enables us to predict forward from time n to time $n + 1$. We still need to show how to improve our estimate of the target position and velocity after an observation is made of the target position at some time n and at successive times. Assume for the moment we have an estimate for the target position and velocity at time $n - 1$. (Later we shall show how we get our initial estimates for the target position and velocity.)

Assume the target is estimated to have a velocity at time $n - 1$ of 200 ft/sec. Let the scan-to-scan period T for the radar be 10 sec. Using (1.1-a) we estimate the target to be (200 ft/sec) (10 sec) = 2000 ft further away at time n than it was at time $n - 1$. This is the position x_n indicated in Figure 1.2-1. Here we are assuming the aircraft target is flying away from the radar, corresponding to the situation where perhaps enemy aircraft have attacked us and are now leaving

(a)

(b)

Figure 1.1-17 Very long range (over 1000 nmi) one-dimensional electronically scanned (in azimuth direction) phased-array ROTHR: (a) transmit antenna; (b) receive antenna [1]. (Photos courtesy of Raytheon Company.)

Figure 1.1-18 AN/SPG-51 TARTAR dedicated shipboard tracking C-band radar using offset parabolic reflector antenna [3]. Eighty-two have been manufactured. (Photo courtesy of Raytheon Company.)

Figure 1.1-19 Hawk tracker-illuminator incorporating phase 3 product improvement kit, which consisted of improved digital computer and the Low Altitude Simultaneous HAWK Engagement (LASHE) antenna (small vertically oriented antenna to the left of main transmit antenna, which in turn is to the left of main receive antenna). Plans are underway to use this system with the AMRAAM missile. (Photo courtesy of Raytheon Company.)

Figure 1.1-20 NATO SEASPARROW shipborne dedicated tracker-illuminator antenna [3]. One hundred twenty-three have been built. (Photo courtesy of Raytheon Company.)

and we want to make sure they are still leaving. Assume, however, at time n the radar observes the target to be at position y_n instead, a distance 60 ft further away; see Figure 1.2-1. What can we conclude as to where the target really is? Is it at x_n, at y_n, or somewhere in between? Let us initially assume the radar range measurements at time $n - 1$ and n are very accurate; they having been made with a precise laser radar that can have a much more accurate range measurement than a microwave radar. Assume the laser radar has a range measurement accuracy of 0.1 ft. In this case the observation of the target at a distance 60 ft further out than predicted implies the target is going faster than we originally estimated at time $n - 1$, traveling 60 ft further in 10 sec, or (60 ft)/(10 sec)=6 ft/sec faster than we thought. Thus the updated target velocity should be

$$\text{Updated velocity } = 200\,\text{ft/sec} + \left(\frac{60\,\text{ft}}{10\,\text{sec}}\right) = 206\,\text{ft/sec} \qquad (1.2\text{-}1)$$

This is all right for a very accurate laser radar. However, generally we will have an ordinary microwave radar. What do we do then? Assume its range accuracy is only 50 ft, 1σ. Then the 60-ft deviation from the expected target location at time n could be due to the measurement error of the radar alone and not due to the target's higher velocity. However, the target could really be going faster than we anticipated, so we would like to allow for this possibility. We do this by not giving the target the full benefit of the 6-ft/sec apparent increase in velocity but instead a fraction of this increase. Let us use the fraction $\frac{1}{10}$th of the

FIRST SCAN, t=t₁ :

SECOND SCAN, t=t +T :

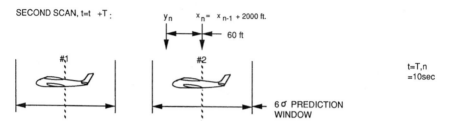

Figure 1.2-1 Target predicted and measured position, x_n and y_n, respectively, on nth scan.

6-ft/sec apparent velocity increase. (How we choose the fraction $\frac{1}{10}$ will be indicated later.) The updated velocity now becomes

$$\text{Updated velocity} = 200\,\text{ft/s} + \frac{1}{10}\left(\frac{60\,\text{ft}}{10\,\text{sec}}\right)$$
$$= 200\,\text{ft/sec} + 0.60\,\text{ft/sec} = 200.6\,\text{ft/sec} \qquad (1.2\text{-}2)$$

In this way we do not increase the velocity of the target by the full amount. If the target is actually going faster, then on successive observations the observed position of the target will on the average tend to be biased further in range than the predicted positions for the target. If on successive scans the target velocity is increased by 0.6 ft/sec on average, then after 10 scans the target velocity will be increased by 6 ft/sec and we would have the correct velocity. On the other hand, if the target velocity were really 200 ft/sec, then on successive observations the measured position of the target would be equally likely to be in front or behind the predicted position so that on average the target velocity would not be changed from its initial estimated value of 200 ft/sec.

Putting (1.2-2) in parametric form yields

$$\dot{x}_n = \dot{x}_n + h_n\left(\frac{y_n - x_n}{T}\right) \qquad (1.2\text{-}3)$$

The fraction $\frac{1}{10}$ is here represented by the parameter h_n. The subscript n is used to indicate that in general the parameter h will depend on time. The above equation has a problem: The symbol for the updated velocity estimate after the

FIRST SCAN, t=t₁ :

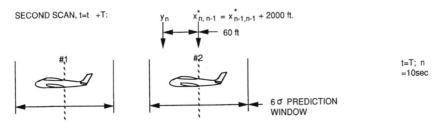

Figure 1.2-2   Target predicted, filtered, and measured positions using new notation.

measurement at time n is the same as the symbol for the velocity estimate at time n just before the measurement was made, both using the variable $\dot{x}_n$. To distinguish these two estimates, a second subscript is added. This second subscript indicates the time at which the last measurement was made for use in estimating the target velocity. Thus (1.2-3) becomes

$$\dot{x}^*_{n,n} = \dot{x}^*_{n,n-1} + h_n \left(\frac{y_n - x^*_{n,n-1}}{T} \right) \tag{1.2-4}$$

The second subscript, $n-1$, for the velocity estimate $\dot{x}^*_{n,n-1}$ indicates an estimate of the velocity of the target at time n based on measurement made at time $n-1$ and before.[†] The second subscript n for the velocity estimate $\dot{x}^*_{n,n}$ given before the equal sign above indicates that this velocity estimate uses the range measurement made at time n, that is, the range measurement y_n. The superscript asterisk is used to indicate that the parameter is an estimate. Without the asterisk the parameters represent the true values of the velocity and position of the target. Figure 1.2-2 gives Figure 1.2-1 with the new notation.

We now have the desired equation for updating the target velocity, (1.2-4). Next we desire the equation for updating the target position. As before assume that at time $n-1$ the target is at a range of 10 nautical miles (nmi) and at time n, $T = 10$ sec later, the target with a radial velocity of 200 ft/sec is at a range 2000 ft further out. As before, assume that at time n the target is actually

---

[†] This is the notation of reference 5. Often, as shall be discussed shortly, in the literature [6, 7] a caret over the variable is used to indicate an estimate.

observed to be 60 ft further downrange from where predicted; see Figure 1.2-2. Again we ask where the target actually is. At $x^*_{n,n-1}$, at y_n, or somewhere in between? As before initially assume a very accurate laser radar is being used for the measurements at time $n - 1$ and n. It can then be concluded that the target is at the range it is observed to be at time n by the laser radar, that is, 60 ft further downrange than predicted. Thus

$$\text{Updated position} = 10\,\text{nmi} + 2000\,\text{ft} + 60\,\text{ft} \qquad (1.2\text{-}5)$$

If, however, we assume that we have an ordinary microwave radar with a 1σ of 50 ft, then the target could appear to be 60 ft further downrange than expected just due to the measurement error of the radar. In this case we cannot reasonably assume the target is actually at the measured range y_n, at time n. On the other hand, to assume the target is at the predicted position is equally unreasonable. To allow for the possibility that the target could actually be a little downrange from the predicted position, we put the target at a range further down than predicted by a fraction of the 60 ft. Specifically, we will assume the target is $\frac{1}{6}$ of 60 ft, or 10 ft, further down in range. (How the fraction $\frac{1}{6}$ is chosen will be indicated later.) Then the updated range position after the measurement at time n is given by

$$\text{Updated position} = 10\,\text{nmi} + 2000\,\text{ft} + \tfrac{1}{6}(60\,\text{ft}) \qquad (1.2\text{-}6)$$

Putting (1.2-6) in parametric form yields

$$x^*_{n,n} = x^*_{n,n-1} + g_n(y_n - x^*_{n,n-1}) \qquad (1.2\text{-}7)$$

where the fraction $\frac{1}{6}$ is represented by the parameter g_n, which can be dependent on n. Equation (1.2-7) represents the desired equation for updating the target position.

Equations (1.2-4) and (1.2-7) together give us the equations for updating the target velocity and position at time n *after* the measurement of the target range y_n has been made. It is convenient to write these equations together here as the present position and velocity *g–h track update (filtering) equations:*

$$\dot{x}^*_{n,n} = \dot{x}^*_{n,n-1} + h_n\left(\frac{y_n - x^*_{n,n-1}}{T}\right) \qquad (1.2\text{-}8a)$$

$$x^*_{n,n} = x^*_{n,n-1} + g_n(y_n - x^*_{n,n-1}) \qquad (1.2\text{-}8b)$$

These equations provide an updated estimate of the *present target* velocity and position based on the present measurement of target range y_n as well as on prior measurements. These equations are called the *filtering equations.* The estimate $x^*_{n,n}$ is called the *filtered estimate* an estimate of x_n at the present time based on the use of the present measurement y_n as well as the past measurements. This estimate is in contrast to the *prediction estimate* $x_{n,n-1}$, which is an estimate of x_n based on past measurements. The term *smoothed* is used sometimes in place of the term *filtered* [8]. "Smoothed" is also used [7] to indicate an estimate of the position or velocity of the target at some past time between the first and last measurement, for example, the estimate $x^*_{h,n}$, where $n_0 < h < n$, n_0 being the time of the first measurement and n the time of the last measurement. In this book, we will use the latter definition for smoothed.

Often in the literature [6, 7] a caret is used over the variable x to indicate that x is the predicted estimate $x^*_{n,n-1}$ while a bar over the x is used to indicate that x is the filtered estimate $x^*_{n,n}$. Then *g–h track update equations* of (1.2-8a) and (1.2-8b) become respectively

$$\bar{\dot{x}}_n = \hat{\dot{x}}_n + h_n \left(\frac{y_n - \hat{x}_n}{T} \right) \tag{1.2-9a}$$

$$\bar{x}_n = \hat{x}_n + g_n(y_n - \hat{x}_n) \tag{1.2-9b}$$

It is now possible by the use of (1.1-1) to predict what the target position and velocity will be at time $n + 1$ and to repeat the entire velocity and position update process at time $n + 1$ after the measurement y_{n+1} at time $n + 1$ has been made. For this purpose (1.1-1) is rewritten using the new notation as the *g–h transition equations* or *prediction equations*:

$$\dot{x}^*_{n+1,n} = \dot{x}^*_{n,n} \tag{1.2-10a}$$

$$x^*_{n+1,n} = x^*_{n,n} + T\dot{x}^*_{n+1,n} \tag{1.2-10b}$$

These equations allow us to transition from the velocity and position at time n to the velocity and position at time $n + 1$ and they are called the *transition equations*. We note in (1.2-10a) the estimated velocity at time $n + 1$, $\dot{x}^*_{n+1,n}$, is equal to the value $\dot{x}^*_{n,n}$ at time n, because a constant-velocity target model is assumed.

Equations (1.2-8) together with (1.2-10) allow us to keep track of a target. In a tracking radar, generally one is not interested in the present target position x_n but rather in the predicted target position x_{n+1} to set up the range prediction windows. In this case (1.2-8) and (1.2-10) can be combined to give us just two equations for doing the track update. We do this by substituting (1.2-8) into (1.2-10) to yield the following prediction update equations:

$$\dot{x}^*_{n+1,n} = \dot{x}^*_{n,n-1} + \frac{h_n}{T}(y_n - x^*_{n,n-1}) \tag{1.2-11a}$$

$$x^*_{n+1,n} = x^*_{n,n-1} + T\dot{x}^*_{n+1,n} + g_n(y_n - x^*_{n,n-1}) \tag{1.2-11b}$$

Equations (1.2-11) represent the well-known *g–h* tracking-filter equations. These *g–h* tracking-filter equations are used extensively in radar systems [5, 6, 8–10]. In contrast to the filtering equation of (1.2-8), those of (1.2-11) are called prediction equations because they predict the target position and velocity at the next scan time. An important class of *g–h* filters are those for which *g* and *h* are fixed. For this case the computations required by the radar tracker are very simple. Specifically, for each target update only four adds and three multiplies are required. The memory requirements are very small. Specifically, for each target only two storage bins are required, one for the latest predicted target velocity and one for the latest predicted target position, past measurements and past predicted values not being needed for future predictions.

We have developed the filtering and prediction equations (1.2-8) and (1.2-10) above through a simple heuristic development. Later (Section 1.2.6, Chapter 2 and the second part of the book) we shall provide more rigorous developments. In the meantime we will give further commonsense insight into why the equations are optimal.

In Figure 1.2-2 we have two estimates for the position of the target at time n, $x^*_{n,n-1}$ and y_n. The estimate y_n is actually the radar measurement at time n. The estimate $x^*_{n,n-1}$ is based on the measurement made at time $n-1$ and all preceding times. What we want to do is somehow combine these two estimates to obtain a new best estimate of the present target position. This is the filtering problem. We have the estimates y_n and $x^*_{n,n-1}$ and we would like to find a combined estimate $x^*_{n,n}$, as illustrated in Figure 1.2-3. The problem we face is how to combine these two estimates to obtain the combined estimate $x^*_{n,n}$. If y_n and $x^*_{n,n-1}$ were equally accurate, then we would place $x^*_{n,n-1}$ exactly in the middle between y_n and $x^*_{n,n-1}$. For example, assume you weigh yourself on two scales that are equally accurate with the weight on one scale being 110 lb and that on the other scale being 120 lb. Then you would estimate your weight based on these two measurements to be 115 lb. If on the other hand one scale were more accurate than the other, then we would want the combined estimate of the weight to be closer to that of the more accurate scale. The more accurate the scale, the closer we would place our combined estimate to it. This is just what the filtering equation (1.2-8b) does. To see this, rewrite (1.2-8b) as a *position filtering equation*:

$$x^*_{n,n} = x^*_{n,n-1}(1 - g_n) + y_n g_n \qquad (1.2\text{-}12)$$

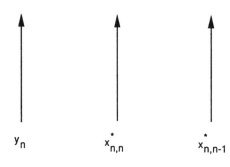

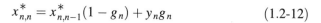

y_n $x^*_{n,n}$ $x^*_{n,n-1}$

Figure 1.2-3 Filtering problem. Estimate of $x^*_{n,n}$ based on measurement y_n and prediction $x^*_{n,n-1}$.

The above equation gives the updated estimate as a weighted sum of the two estimates. The selection of the fraction g_n determines whether we put the combined estimate closer to y_n or to $x^*_{n,n-1}$. For example, if y_n and $x^*_{n,n-1}$ are equally accurate, then we will set g_n equal to $\frac{1}{2}$. In this case the combined estimate $x^*_{n,n}$ is exactly in the middle between y_n and $x^*_{n,n-1}$. If on the other hand $x^*_{n,n-1}$ is much more accurate than y_n (perhaps because the former is based on many more measurements), then we will want to have the coefficient associated with y_n much smaller than that associated with $x^*_{n,n-1}$. For example, in this case we might want to pick $g_n = \frac{1}{5}$, in which case the combined estimate is much closer to $x^*_{n,n-1}$. How we select g_n shall be shown later.

1.2.2 α–β Filter

Now that we have developed the *g–h* filter, we are in a position to develop the α–β filter. To obtain the α–β filter, we just take (1.2-11) and replace g with α and h with β—we now have the α–β filter. Now you know twice as much, knowing the α–β filter as well as the *g–h* filter.

1.2.3   Other Special Types of Filters

In this section we will increase our knowledge 22-fold because we will cover 11 new tracking filters. Table 1.2-1 gives a list of 11 new tracking filters. The equations for all of these filters are given are given by (1.2-11). Consequently, all 11 are *g–h* filters. Hence all 11 are α–β filters. Thus we have increased our tracking-filter knowledge 22 fold! You are a fast learner! How do these filters differ? They differ in the selection of the weighting coefficients g and h as shall be seen later. (Some actually are identical). For some of these filters g and h depend on n. This is the case for the Kalman filter. It is worthwhile emphasizing that (1.2-11a) and (1.2-11b) are indeed the Kalman filter prediction equations, albeit for the special case where only the target velocity and position are being tracked in one dimension. Later we will give the Kalman filter for the multi-

---

TABLE 1.2-1. Special Types of Filters

1.   Wiener filter
2.   Fading-memory polynomial filter
3.   Expanding-memory (or growing-memory) polynomial filter
4.   Kalman filter
5.   Bayes filter
6.   Least-squares filter
7.   Benedict–Bordner filter
8.   Lumped filter
9.   Discounted least-squares *g–h* filter
10. Critically damped *g–h* filter
11. Growing-memory filter

dimensional situation involving multiple states. For many of the *g–h* tracking filters of Table 1.2-1, *g* and *h* are related.

1.2.4   Important Properties of *g–h* Tracking Filters

1.2.4.1   *Steady-State Performance for Constant-Velocity Target*
Assume a target with a constant-velocity trajectory and an errorless radar range measurement. Then in steady state the *g–h* will perfectly track the target without any errors; see Figure 1.2-4. This is not surprising since the equations were developed to track a constant-velocity target. In the steady state the estimate of the target velocity obtained with the tracking equations of (1.2-11) will provide a perfect estimate of the target velocity if there is no range error present in the radar. For this case in steady state g_n and h_n become zero. The prediction equations given by (1.2-11) then become the transition equations or target equation of motion as given by (1.2-10).

1.2.4.2   *For What Conditions is the Constant-Velocity Assumption Reasonable*
Assume a target having a general arbitrary one-dimensional trajectory as a function of time *t* given by $x(t)$. Expressing $x(t)$ in terms of its Taylor expansion yields

$$x(t) = x(t_n) + \Delta t \quad \dot{x}(t_n) + \frac{(\Delta t)^2}{2!} \ddot{x}(t_n)$$
$$+ \frac{(\Delta t^3)}{3!} \dddot{x}(t_n) + \cdots \qquad (1.2\text{-}13)$$

For

$$\frac{(\Delta t)^2}{2!} \ddot{x}(t_n)$$

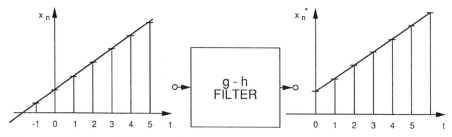

Figure 1.2-4 The *g–h* filter predicts position of constant-velocity target perfectly in steady state if there are no measurement errors.

small

$$x(t_{n+1}) = x(t_n) + T\dot{x}(t_n) \tag{1.2-14}$$

where we replaced Δt by T. Equation (1.2-14) is the equation for a target having a constant velocity. Thus the assumption of a constant-velocity target is a reasonable one as long as the time between observations T is small or the target acceleration $\ddot{x}$ is small or their combination is small.

1.2.4.3 Steady-State Response for Target with Constant Acceleration

Assume we are using the prediction equations given by (1.2-11) that were developed for a target having a constant velocity but in fact the target has a constant acceleration given by $\ddot{x}$. We ask: How well does the tracking filter do? It turns out that in steady state the constant *g–h* tracking filter will have a constant prediction error given by b^* for the target position that is expressed in terms of the target acceleration and scan-to-scan period by [5, 12]

$$b^* \equiv b^*_{n+1,n} = \frac{-\ddot{x}T^2}{h} \tag{1.2-15}$$

Figure 1.2-5 illustrates the constant lag error prediction resulting when the tracking equations of (1.2-11) are used for a constant accelerating target.

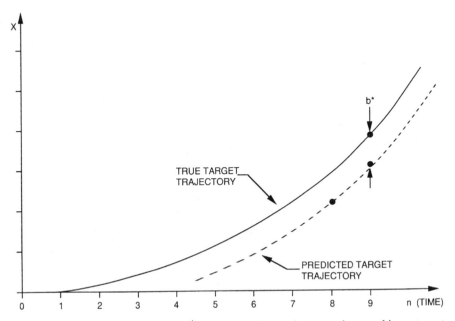

Figure 1.2-5 Constant lag error b^* that results in steady state when tracking a target having a constant acceleration with a constant *g–h* filter.

Equation (1.2-15) is not surprising. The acceleration error is proportional to $\frac{1}{2}$ $(\ddot{x}T^2)$ and we see that correspondingly b^* is proportional to $\frac{1}{2}(\ddot{x}T^2)$. Equation (1.2-15) also indicates that the steady-state error b^*, called the lag error, is also inversely proportional to the parameter h of (1.2-11a). This is not unreasonable. Assume as in Figure 1.2-2 that y_n is 60 ft further downrange than predicted and interpret the additional distance as due to the target having an increased speed of 6 ft/sec by making $h_n = 1$. In this case our tracking filter responds immediately to a possible increase in the target velocity, increasing the velocity immediately by 6 ft/sec. If on the other hand we thought that the location of the target 60 ft further downrange could be primarily due to inaccuracies in the radar range measurement, then, by only allotting a fraction of this 6 ft/sec to the target velocity update by setting $h_n = \frac{1}{10}$, the tracking filter will not respond as quickly to a change in velocity if one actually did occur. In this latter case several scans will be necessary before the target velocity will have increased by 6 ft/sec, if indeed the target actually did increase by 6 ft/sec. Thus the larger is h, the faster the tracking filter responds to a change in target velocity. Alternatively, the smaller is h, the more sluggish is the filter. Thus quite reasonably the lag error for the filter is inversely proportional to h.

When tracking a constant-accelerating target with a g–h filter, there will also be in steady state constant lag errors for the filtered target position $x_{n,n}^*$ and the velocity $\dot{x}_{n,n}^*$ given respectively by [12]

$$b_{n,n}^* = -\ddot{x}\,T^2\,\frac{1-g}{h} \tag{1.2-16a}$$

$$\dot{b}_{n,n}^* = -\ddot{x}\,T\,\frac{2g-h}{2h} \tag{1.2-16b}$$

Unless indicated otherwise, b^* without the subscript shall be the one-step prediction bias error $b_{n+1,n}^*$.

As in many other fields, knowledge of the terms used is a major part of the battle. Otherwise one will be snowed by the tracking specialist. An example is *lag error* which goes by other names that should be known as well. (see Table 1.2-2). Another term given, in the table for lag error is dynamic error. This is not surprising since lag error is not an error due to random effects but one due to target dynamics. Lag error is also called systematic error, quite reasonably,

TABLE 1.2-2. Common Names for Steady-State Prediction Error b^* Due to Constant Acceleration

Lag error
Dynamic error
Systematic error
Bias error
Truncation error

since it is a systematic error rather than a random error, again caused by the target motion. Appropriately lag error is called a bias error since the error is a fixed deviation from the true value in steady state. Finally, lag error is called a truncation error, since the error results from the truncation of the acceleration term in the Taylor expansion given by (1.2-13).

1.2.4.4 Tracking Errors due to Random Range Measurement Error

The radar range measurement y_n can be expressed as

$$y_n = x_n + \nu_n \tag{1.2-17}$$

where x_n without the asterisk is the true target position and ν_n is the range measurement error for the nth observation. Assume that ν_n is a random zero mean variable with an rms of σ_ν that is the same for all n. Because σ_ν represents the rms of the range x measurement error, we shall replace it by σ_x from here on. The variance of the prediction $x^*_{n+1,n}$ is defined as

$$\mathrm{VAR}(x^*_{n+1,n}) = E[\{x^*_{n+1,n} - E(x^*_{n+1,n})\}^2] \tag{1.2-18}$$

where $E[\cdot]$ stands for "expected value of". We would like to express $\mathrm{VAR}(x^*_{n+1,n})$ in terms of σ_x and the tracking-filter parameters. In the literature expressions are given for a normalized $\mathrm{VAR}(x^*_{n+1,n})$. Specifically it is given normalized with respect to σ_x^2, that is, it is given for $[\mathrm{VAR}(x_{n+1,n})]/\sigma_x^2$. This normalized variance is called the variance reduction factor (VRF). Using the tracking prediction equations of (1.2-11), in steady state the VRF for $x^*_{n+1,n}$ for a constant g–h filter is given by [12; see also problems 1.2.4.4-1 and 1.2.6-2]

$$\mathrm{VRF}\,(x^*_{n+1,n}) = \frac{\mathrm{VAR}(x^*_{n+1,n})}{\sigma_x^2} = \frac{2g^2 + 2h + gh}{g(4 - 2g - h)} \tag{1.2-19}$$

The corresponding VRFs for $x^*_{n,n}$ and $\dot{x}^*_{n+1,n}$ are given respectively by [12]

$$\mathrm{VRF}(x^*_{n,n}) = \frac{\mathrm{VAR}(x^*_{n,n})}{\sigma_x^2} = \frac{2g^2 + 2h - 3gh}{g(4 - 2g - h)} \tag{1.2-20}$$

$$\mathrm{VRF}(\dot{x}^*_{n+1,n}) = \frac{\mathrm{VAR}(\dot{x}^*_{n+1,n})}{\sigma_x^2} = \frac{1}{T^2} \frac{2h^2}{g(4 - 2g - h)} \tag{1.2-21}$$

Thus the steady-state normalized prediction error is given simply in terms of g and h. Other names for the VRF are given in Table 1.2-3.

1.2.4.5 Balancing of Lag Error and rms Prediction Error

Equation (1.2-19) allows us to specify the filter prediction error $\mathrm{VAR}(x_{n+1,n})$ in terms of g and h. The non random lag prediction error b^* is given in turn by

TABLE 1.2-3. Common Names for Normalized Var $(x^*_{n+1,n})$ of Prediction Error

Variance reduction factor [5]
Variance reduction ratio [12]
Noise ratio [72]
Noise amplification factor [12]

(1.2-15) in terms of $\ddot{x}$, T, and h. For convenience let the rms prediction error (i.e., the rms of $x^*_{n+1,n}$) be designated as $\sigma_{n+1,n}$. In designing a tracking filter there is a degree of freedom in choice of the magnitude of $\sigma_{n+1,n}$ relative to b^*. They could be made about equal or one could be made much smaller than the other. Generally, making one much smaller than the other does not pay because the total prediction error is the sum $b^* + \sigma_{n+1,n}$. Fixing one of these errors and making the other error much smaller does not appreciably reduce the total prediction error. We would be paying a high price to reduce one of the prediction errors without significantly reducing the total prediction error. Intuitively a good approach would be a balanced system where the random and non random error components are about the same order of magnitude. One way to do this is to make b^* equal to three times the 1σ rms prediction error, that is,

$$3\sqrt{\text{VAR}(x^*_{n+1,n})} = b^* \qquad (1.2\text{-}22)$$

or equivalently

$$3\sigma_{n+1,n} = b^* \qquad (1.2\text{-}23)$$

where

$$\sigma_{n+1,n} = \sigma(x^*_{n+1,n}) = \sqrt{\text{VAR}(x^*_{n+1,n})} \qquad (1.2\text{-}23a)$$

where b^* is the lag error obtained for the maximum expected target acceleration. The choice of the factor 3 above is somewhat arbitrary. One could just as well have picked a factor 2, 1.5, or 1, as will be done later in Section 5.10. Using (1.2-22) determines b^* if $\sigma_{n+1,n}$ is known. Equation (1.2-15) can then in turn be used to determine the track filter update period T for a given maximum target acceleration $\ddot{x}$. This design procedure will shortly be illustrated by an example.

Another approach to designing a *g–h* filter is to choose $\sigma_{n+1,n}$ and b^* such that their total error

$$E_T = 3\sigma_{n+1,n} + b^* \qquad (1.2\text{-}24)$$

is minimized [11]. What is done is E_T is plotted versus g for a given T and $\ddot{x}$ and the minimum found. Using this procedure yielded that the total error could be reduced as much as 7% for $1 \geq g \geq 0.9$ and by as much as 15% for $0 \geq g \leq 0.9$ [11]. The improvement tended to be smaller for larger g because the minimum E_{TN} point tended to be broader for larger g. These results were obtained in reference 11 for the critically damped and Benedict–Bordner filters and will be detailed in Sections 1.2.6 and 1.2.5, where these filters will be introduced.

It is worth recalling that the larger is h, the smaller is the dynamic error, however; in turn, the larger $\mathrm{VAR}(x^*_{n+1,n})$ will be and vice versa; see (1.2-15) and (1.2-19). Typically, the smaller are g and h, the larger is the dynamic error and the smaller are the errors of (1.2-19) to (1.2-21) due to the measurement noise. In fact for the critically damped and Benedict–Bordner g–h filters introduced in Sections 1.2.5 and 1.2.6, $\sigma_{n+1,n}$ decreases monotonically with decreasing g while the prediction bias $b^*_{n+1,n}$ increases monotinically with decreasing g [11]. Similarly, for the critically damped filter and steady-state Kalman g–h–k filter (often called optimum g–h–k filter) introduced in Sections 1.3 and 2.4, $\sigma_{n+1,n}$ decreases monotonically with decreasing h while the prediction bias $b^*_{n+1,n}$ increases monotonically with decreasing h [11]. Here the bias is calculated for a constant jerk, that is, a constant third derivative of x, ; see Section 1.3. The subject of balancing the errors $\sigma_{n+1,n}$ and b^* as well as minimization of their total is revisited in Sections 5.10 and 5.11.

1.2.5 Minimization of Transient Error (Benedict–Bordner Filter)

Assume the g–h tracking filter is tracking a constant-velocity target and at time zero the target velocity takes a step function jump to a new constant-velocity value. This is illustrated in Figure 1.2-6 where the target position versus time is plotted as the g–h filter input. Initially the target is at zero velocity, jumping at time zero to a nonzero constant-velocity value. Initially the g–h tracking filter will not perfectly follow the target, the filter's sluggishness to a change in target

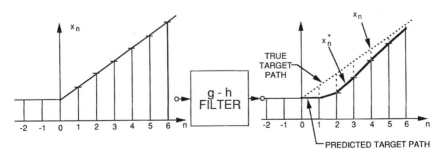

Figure 1.2-6 Transient error resulting from step jump in velocity for target being tracked with g–h filter.

velocity being the culprit. The filter output in Figure 1.2-6 illustrates this. As a result, the tracking filter has an initial error in its prediction, the difference between the true trajectory and the predicted trajectory, as indicated in the figure. A measure of the total transient error is the sum of the squares of these differences, that is,

$$D_{x_{n+1,n}^*} = \sum_{n=0}^{\infty} (x_{n+1,n}^* - x_{n+1})^2 \tag{1.2-25}$$

One would like to minimize this total transient error. In the literature such a minimization was carried out for the weighted sum of the total transient error plus the variance of the prediction error due to measurement noise errors, that is, the minimization of

$$E = \text{VAR}(x_{n,n+1}^*) + \lambda D x_{n+1,n}^* \tag{1.2-26}$$

This minimization was done for a step jump in velocity. Rather than concentrating merely on the *g–h* tracking filters, which minimize the total error given in (1.2-26), what was done in the literature was to find a filter among a much wider class of tracking filters minimizing the total error given by (1.2-26). From this much wider class of filters the filter minimizing the total error of (1.2-26) is (surprise) the constant *g–h* tracking filter of (1.2-11) with *g* and *h* related by

$$h = \frac{g^2}{2 - g} \tag{1.2-27}$$

Table 1.2-4 gives a list of values for $\sigma_{n+1,n}/\sigma_x, g$, and *h* for the Benedict–Bordner filter. The parameter θ in the *g–h* filter will be explained shortly, at the end of Section 1.2.6. Basically, for the present think of it as just a convenient index for the tabulation in Table 1.2-4. In the literature the Benedict–Bordner filter has also been referred to as an optimum *g–h* filter [11, 12].

The transient error in $x_{n+1,n}^*$ defined by (1.2-25) for any constant *g–h* filter is given by [6, 12]

$$D_{x_{n,n+1}^*} = \frac{\Delta v^2 \, T^2 (2 - g)}{gh(4 - 2g - h)} \tag{1.2-28}$$

where Δv is the step change in velocity that produces the filter transient error. The corresponding transient errors for $x_{n,n}^*$ and $\dot{x}_{n,n}^*$ are given respectively by [12]

$$D_{x_{n,n}^*} = \frac{\Delta v^2 T^2 (2 - g)(1 - g)^2}{gh(4 - 2g - h)} \tag{1.2-29}$$

$$D_{\dot{x}_{n+1,n}^*} = \frac{\Delta v^2 [g^2 (2 - g) + 2h(1 - g)]}{gh(4 - 2g - h)} \tag{1.2-30}$$

TABLE 1.2-4. Smoothing Constant for Benedict–Bordner *g–h* Filter

θ	$\sigma_{n+1,n}/\sigma_x$	g	h
.00	2.236	1.000	1.000
.05	2.066	.975	.929
.10	1.913	.949	.856
.15	1.773	.920	.783
.20	1.644	.889	.711
.25	1.526	.855	.639
.30	1.416	.819	.568
.35	1.313	.781	.500
.40	1.216	.739	.434
.45	1.124	.695	.371
.50	1.036	.649	.311
.55	.952	.599	.256
.60	.871	.546	.205
.65	.791	.490	.159
.70	.711	.431	.118
.75	.632	.368	.083
.80	.551	.302	.054
.85	.465	.232	.030
.90	.370	.159	.014
.95	.256	.081	.003

Figure 1.2-7 gives convenient design curves of three times the normalized prediction error $3\sigma_{n+1,n}/\sigma_x$ and normalized b^* for the Benedict–Bordner filter versus g. Using these curves the design point at which $3\sigma_{n+1,n} = b^*$ can easily be found. Figure 1.2-8 plots the sum of these normalized errors

$$
E_{TN} = \frac{E_T}{\sigma_x} = \frac{3\sigma_{n+1,n}}{\sigma_x} + \frac{b^*}{\sigma_x}
$$
$$
= \frac{3\sigma_{n+1,n} + b^*}{\sigma_x} \tag{1.2-31}
$$

versus g for different values of the normalized maximum acceleration given by $A_N = T^2 \ddot{x}_{\max}/\sigma_x$. These curves allow us to obtain the Benedict–Bordner filters, which minimizes the total error given by (1.2-31). These minimum total error designs are plotted in Figure 12-9. Problems 1.2.5-1 and 1.2.5-2 compare Benedict–Bordner filter designs obtained using (1.2-23) and obtained for when (1.2-31) is minimized.

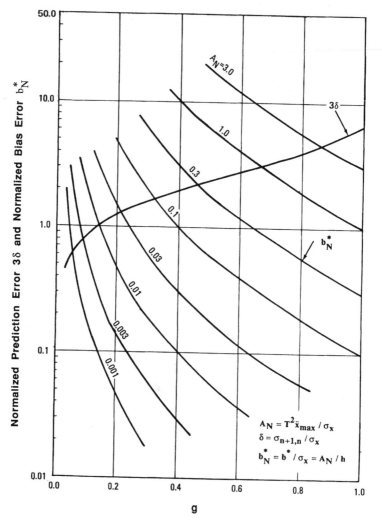

Figure 1.2-7 Normalized bias error $b_N^* = b^*/\sigma_x$ and prediction error $\delta = \sigma_{n+1,n}/\sigma_x$ versus weight g for Benedict–Bordner g–h filter. (After Asquith and Woods [11].)

1.2.6 New Approach to Finding Optimum g–h Tracking Filters (The Critically Damped Filter)

To obtain further insight into the development of tracking equations, let us start afresh with a new point of view. We take an approach that is optimal in some other sense. We shall use a very simple commonsense approach.

Let $y_0, y_1, \ldots, y_6$ be the first seven range measurements made of the target to be tracked. These observations could be plotted as shown in Figure 1.2-10. With these measurements we would like to now predict where the target will most

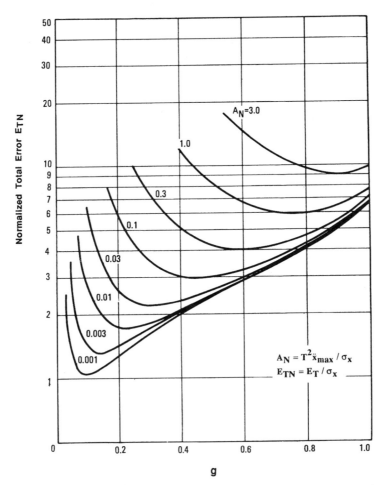

Figure 1.2-8 Normalized total error E_{TN} versus weight g for Benedict–Bordner $g–h$ filter. (After Asquith and Woods [11].)

likely be at time $n = 7$. A simple approach would be to eyeball a best-fitting straight line to the points, as shown in Figure 1.2-10, this best fitting straight line providing an estimate of the true target trajectory. The target position at time $n = 7$ will then be estimated to be the position on the straight line of the target at time $n = 7$. In the figure v_0^* is the slope of the best-fitting line and x_0^* its ordinate intercept. These two parameters define the line.

Instead of using the above "eyeball fit" approach, one could use an "optimum" approach for finding the best-fitting line. Specifically, one could find the least-squares fitting line [5, 6]. The least-squares fitting line is the line that gives the minimum error between the line and the measurements by minimizing the sum of the squares of the difference between the measurements and the straight-line fit. We now formulate this approach mathematically. The

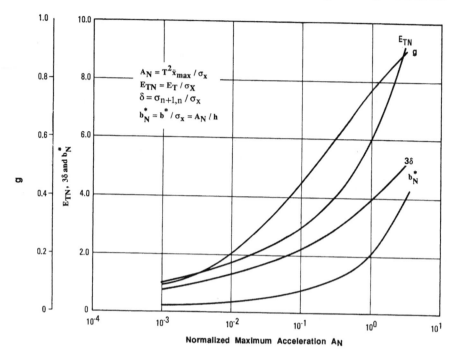

Figure 1.2-9 Minimum normalized total error E_{IN} versus normalized maximum acceleration A_N for Benedict–Bordner *g–h* filter. (After Asquith and Woods [11].)

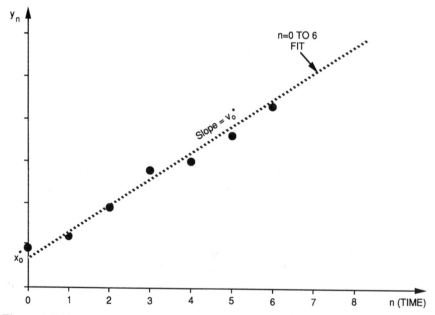

Figure 1.2-10 "Eyeball best fitting" straight line to data points $y_0, y_1, \ldots, y_6$; least-squares fit.

deviation between the measurement and the straight line at time n is given by

$$\varepsilon_n = x_n^* - y_n \qquad (1.2\text{-}32)$$

where x_n^* is the target range indicated by the straight-line fit at at time n. Then the least-squares fitting line is that line that minimizes the total sum of squares of differences given by

$$e_T = \sum_{n=0}^{N} \varepsilon_n^2 = \sum_{n=0}^{N} (x_n^* - y_n)^2 \qquad (1.2\text{-}33)$$

where, for Figure 1.2-10, $N = 6$. At time $N + 1$ the new range measurement y_{N+1} is obtained and the whole process is repeated. This is illustrated in Figure 1.2-11, where the fit is now made to eight points, $y_0, \ldots, y_7$.

We thus now have an optimum procedure for tracking the target. The eyeball-fit procedure, your mind providing a pretty good least-squares fit.

There is one major problem with the above approach. It weights the old data equally as importantly as the latest data when calculating the least-squares fitting line that predicts where the target would be at the next observation. If the target should turn, then the old data would incorrectly influence the target predicted path. To eliminate this problem, old data must somehow have a reduced influence when predicting the future path of the target. We do this by

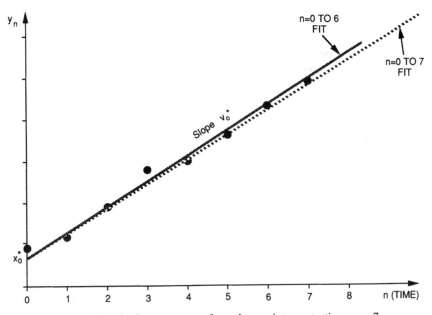

Figure 1.2-11 Least-squares fit to data points up to time $n = 7$.

giving the older errors ε_n^2 in (1.2-33) lesser significance when forming the total error. This is done by weighting the older errors by a factor less than 1 when forming the total error sum. We do this by weighting the most recent error by unity, the next most recent error by factor θ (where $\theta < 1$), the next oldest error by θ^2, the next oldest error by θ^3, and so on. The total error now is given by

$$e_D = \sum_{r=0}^{N} \theta^r \varepsilon_{n-r}^2 \qquad (1.2\text{-}34)$$

where

$$0 \le \theta \le 1 \qquad (1.2\text{-}34a)$$

The straight-line fit that minimizes the total error given by (1.2-34) is called the *discounted least-squares fit*, because this line fit minimizes the sum of the weighted errors with the weighting decreasing as the data get older; that is, the older the error, the more it is discounted.

If we could do the computation in real time for the above discounted least-squares filter tracker, it would represent a perfectly good usable filter. However, it does have some disadvantages relative to the simple *g–h* filter of (1.2-11). First, all of the old data must be stored. (Actually it is possible to drop the very old data for which $\theta^r \ll 1$.) In contrast, the filter of (1.2-11) only has to store two numbers, the predicted target position and velocity from the last observations, that is, $x_{n,n-1}^*$ and $\dot{x}_{n,n-1}^*$. Second, a complicated discounted least-squares computation must be carried out at each update point whereas the tracking filter given by (1.2-11) requires just a simple recursive equation calculation for the case of a constant *g* and *h*.

Thus we might conclude that although the discounted least-squares approach is mathematically optimal, we might be better using the less optimal, heuristically derived filter given by (1.2-11) since the implementation is easier. Wrong! If one is clever enough and works hard enough, the optimum discounted least-squares tracking filter described above can be put in the form of the constant *g–h* filter given by (1.2-11) where parameters *g* and *h* are related to θ by

$$g = 1 - \theta^2 \qquad (1.2\text{-}35a)$$
$$h = (1 - \theta)^2 \qquad (1.2\text{-}35b)$$

Thus, the two filters, amazingly, are identical!

A truly remarkable result! We do not really have to store all of the past data to do the optimal discounted least-squares tracking filter! *We only need to store the last predicted target position and velocity estimates* $x_{n,n-1}^*$ *and* $\dot{x}_{n,n-1}^*$. These two numbers contain all the information we need about the past. These two numbers thus are sufficient statistics (even in the rigorous statistical sense

[8, 9, 100] as well as in the commonsense use of the word). No other past measurement information need be stored. Furthermore, we need only do a simple recursive computation as given by (1.2-11) instead of the complicated discounted least-squares fit computation at each update point.

The constant *g–h* filter with *g* and *h* related to θ by (1.2-35) is called a discounted least-squares *g–h* filter, the ninth filter of Table 1.2-1. This filter also goes by other names. For example, it is called a fading-memory polynomial filter of degree 1 (the second filter of Table 1.2-1) because it has a fading memory. It is designated as having a degree 1 because it is for a constant-velocity target, in which case its position is a linear function of time, that is, it is given by a constant plus a constant times *t* to degree 1. It is called a critically damped *g–h* filter (the tenth filter of Table 1.2-1) because it is critically damped and not overdamped or underdamped. To see this, one obtains the *z*-transform of (1.2-11) with *g* and *h* given by (1.2-35a) and (1.2-35b). In doing this, one finds that the poles of the filter are equal and real and given by $z_1 = z_1 = \theta$. Hence this double pole lies inside the unit circle on the *z*-plane as long as $\theta < 1$; see Figure 1.2-12. Actually in finding the poles of the filter of (1.2-11) one finds that critically damped conditions are realized when *h* is related to *g* by [12]

$$h = 2 - g \pm 2\sqrt{1 - g} \qquad (1.2\text{-}36)$$

Substituting (1.2-35a) into (1.2-36) yields (1.2-35b) if the minus sign is used and yields $h = (1 + \theta)^2$ if the plus sign in used. As shall be indicated in Section 1.2.9, the minus sign solution is the useful solution. The plus solution gives a larger $\sigma_{n+1,n}$ for a given transient response [12].

Using (1.2-35a) and (1.2-35b) the convenient Table 1.2-5 of *g* and *h* (versus θ) is obtained for the critically damped constant *g–h* filter. The table also gives *g*, *h*, and *k* versus θ for the critically damped *g–h–k* filter to be discussed later in Section 1.3 and Chapter 7. We are now in a position to define the indexing parameter θ of Table 1.2-4. Physically, for a given $\sigma_{n+1,n}/\sigma_x$ for the Benedict–Bordner design summarized in Table 1.2-4, θ is that of a critically damped filter

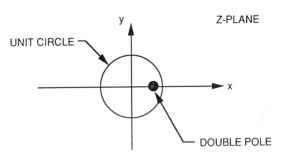

Figure 1.2-12 Poles of critically damped *g–h* filter; double pole inside unit circle and on real axis in *z* plane.

TABLE 1.2-5. Smoothing Constants for Critically Damped *g–h* and *g–h–k* Digital Filters

	g–h Filter		*g–h–k* Filter		
θ	*g*	*h*	*g*	*h*	*k*
.00	1.000	1.000	1.000	1.500	.500
.01	1.000	.980	1.000	1.485	.485
.02	1.000	.960	1.000	1.469	.471
.03	.999	.941	1.000	1.454	.456
.04	.998	.922	1.000	1.438	.442
.05	.998	.903	1.000	1.421	.429
.06	.996	.884	1.000	1.405	.415
.07	.995	.865	1.000	1.388	.402
.08	.994	.846	.999	1.371	.389
.09	.992	.828	.999	1.354	.377
.10	.990	.810	.999	1.337	.365
.11	.988	.792	.999	1.319	.352
.12	.986	.774	.998	1.301	.341
.13	.983	.757	.998	1.283	.329
.14	.980	.740	.997	1.265	.318
.15	.978	.723	.997	1.246	.307
.16	.974	.706	.996	1.228	.296
.17	.971	.689	.995	1.209	.286
.18	.968	.672	.994	1.190	.276
.19	.964	.656	.993	1.171	.266
.20	.960	.640	.992	1.152	.256
.21	.956	.624	.991	1.133	.247
.22	.952	.608	.989	1.113	.237
.23	.947	.593	.988	1.094	.228
.24	.942	.578	.986	1.074	.219
.25	.938	.563	.984	1.055	.211
.26	.932	.548	.982	1.035	.203
.27	.927	.533	.980	1.015	.195
.28	.922	.513	.978	.995	.187
.29	.916	.504	.976	.975	.179
.30	.910	.490	.973	.956	.172
.31	.904	.476	.970	.936	.164
.32	.898	.462	.967	.916	.157
.33	.891	.449	.964	.896	.150
.34	.884	.436	.961	.876	.144
.35	.878	.423	.957	.856	.137

TABLE 1.2-5. (*Continued*)

	g–h Filter			*g–h–k* Filter	
θ	*g*	*h*	*g*	*h*	*k*
.36	.870	.410	.953	.836	.131
.37	.863	.397	.949	.816	.125
.38	.856	.384	.945	.796	.119
.39	.848	.372	.941	.776	.113
.40	.840	.360	.936	.756	.108
.41	.832	.348	.931	.736	.103
.42	.824	.336	.926	.717	.098
.43	.815	.325	.920	.697	.093
.44	.806	.314	.915	.677	.088
.45	.798	.303	.909	.658	.083
.46	.788	.292	.903	.639	.079
.47	.779	.281	.896	.619	.074
.48	.770	.270	.889	.600	.070
.49	.760	.260	.882	.581	.066
.50	.750	.250	.875	.563	.063
.51	.740	.240	.867	.544	.059
.52	.730	.230	.859	.525	.055
.53	.719	.221	.851	.507	.052
.54	.708	.212	.843	.489	.049
.55	.698	.203	.834	.471	.046
.56	.686	.194	.824	.453	.043
.57	.675	.185	.815	.435	.040
.58	.664	.176	.805	.418	.037
.59	.652	.168	.795	.401	.034
.60	.640	.160	.784	.384	.032
.61	.628	.152	.773	.367	.030
.62	.616	.144	.762	.351	.027
.63	.603	.137	.750	.335	.025
.64	.590	.130	.738	.319	.023
.65	.578	.123	.725	.303	.021
.66	.564	.116	.713	.288	.020
.67	.551	.109	.699	.273	.018
.68	.538	.102	.686	.258	.016
.69	.524	.096	.671	.244	.015
.70	.510	.090	.657	.230	.014
.71	.496	.084	.642	.216	.012
.72	.482	.078	.627	.202	.011
.73	.467	.073	.611	.189	.010
.74	.452	.068	.595	.176	.009
.75	.438	.063	.578	.164	.008

TABLE 1.2-5. (*Continued*)

	g–h Filter		*g–h–k* Filter		
θ	*g*	*h*	*g*	*h*	*k*
.76	.422	.058	.561	.152	.007
.77	.407	.053	.543	.140	.006
.78	.392	.048	.525	.129	.005
.79	.376	.044	.507	.118	.005
.80	.360	.040	.488	.108	.004
.81	.344	.036	.469	.098	.003
.82	.328	.032	.449	.088	.003
.83	.311	.029	.428	.079	.002
.84	.294	.026	.407	.071	.002
.85	.278	.023	.386	.062	.002
.86	.260	.020	.364	.055	.001
.87	.243	.017	.341	.047	.001
.88	.226	.014	.319	.041	.001
.89	.208	.012	.295	.034	.001
.90	.190	.010	.271	.029	.001
.91	.172	.008	.246	.023	.000
.92	.154	.006	.221	.018	.000
.93	.135	.005	.196	.014	.000
.94	.116	.004	.169	.010	.000
.95	.098	.003	.143	.007	.000
.96	.078	.002	.115	.005	.000
.97	.059	.001	.087	.003	.000
.98	.040	.000	.059	.001	.000
.99	.020	.000	.030	.000	.000
1.00	.000	.000	.000	.000	.000

that produces the same $\sigma_{n+1,n}/\sigma_x$. Figure 1.2-13 to 1.2-15 gives, for the critically damped filter, corresponding useful design curves to Figures 1.2-7 to 1.2-9 for the Benedict–Bordner filter. Problems 1.2.6-6 to 1.2.6-8 compare critically damped filter designs obtained using (1.2-23) and when the total error of (1.2-31) is minimized.

1.2.7 *g–h* Filter Examples

Example 1: Critically Damped g–h Filter   Assume that the radar rms range measurement error $\sigma_x = 50$ ft; the maximum acceleration expected for the target is $5\,g$; the desired standard deviation of the filter prediction error $\sigma_{n+1,n} = 31.6$ ft. What we now want to do is design a suitable critically damped

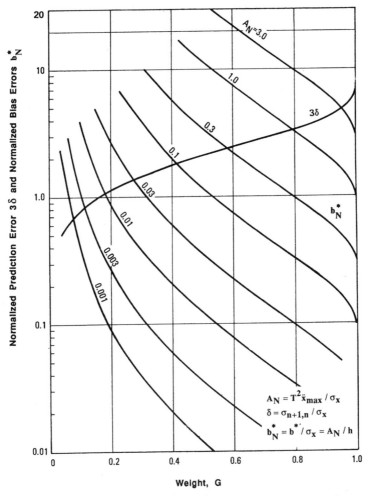

Figure 1.2-13 Normalized bias error $b_N^* = b^*/\sigma_x$ and prediction error $\delta = \sigma_{n+1,n}/\sigma_x$ versus weight *g* for critically damped *g–h* filter. (After Asquith and Woods [11].)

g–h filter using these assumptions. From (1.2-19)

$$
\text{VRF}(x_{n+1,n}^*) = \frac{\text{VAR}(x_{n+1,n}^*)}{\sigma_x^2} = \frac{\sigma_{n+1,n}^2}{\sigma_x^2}
$$

$$
= \frac{2g^2 + 2h + gh}{g(4 - 2g - h)}
$$

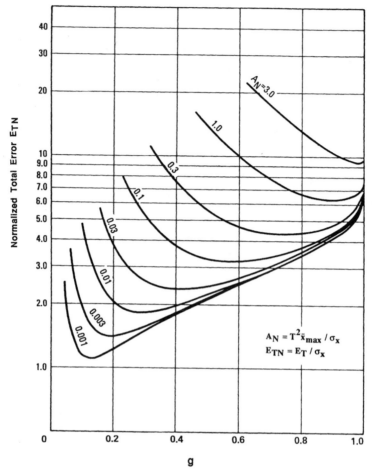

Figure 1.2-14 Normalized total error E_{TN} versus weight g for critically damped g–h filter. (After Asquith and Woods [11].)

and from (1.2-35)

$$g = 1 - \theta^2$$
$$h = (1 - \theta)^2$$

Substituting (1.2-35) into (1.2-19) yields an expression with only one unknown, that of θ, because σ_x and $\sigma_{n+1,n}$ are given. Solving yields $\theta = 0.75$. From (1.2-35) it follows in turn that $g = 0.4375$ and $h = 0.0625$. These values are close to those used in the heuristic development presented in Section 1.2.1. The reason is that we knew the answers for this example and chose values for g and h close to these in Section 1.2.1. The above gives a procedure for finding the constants g and h of (1.2-11).

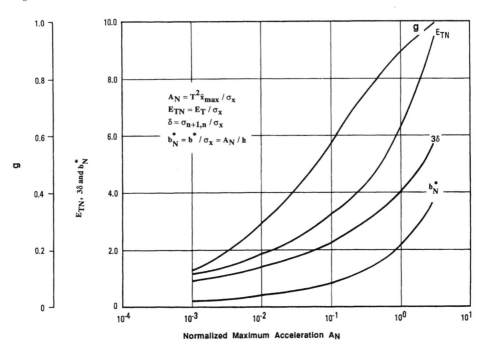

Figure 1.2-15 Minimum normalized total error E_{TN} versus normalized maximum acceleration A_N for critically damped *g–h* filter. (After Asquith and Woods [11].)

We still need to select the tracking–filter update period T. To do this, we make use of (1.2-23) to determine b^*. It yields $b^* = 94.8$ ft. With the latter we now use (1.2-15) to find the update period T. The acceleration (1.2-15) is set equal to the maximum expected $5g$ yielding $T = 0.1924$ sec. This is the tracking-filter update period needed to achieve the specified lag error $b^* = 94.8$ ft. Let us now determine the transient error $D_{x^*_{n+1,n}}$ given by (1.2-28). In (1.2-28), D can be normalized with respect to $(\Delta v)^2$. We solve for this normalised transient error D. From (1.2-28) it follows that the normalized transient error equals 0.691 sec^2 for our critically damped filter design.

Example 2: Benedict-Bordner g–h Filter　For the same assumptions as used in Example 1 we want to design a Benedict–Bordner *g–h* filter. Substituting (1.2-27) into (1.2-19) gives an equation with only one unknown, that of g. Solving yields $g = 0.368$. From (1.2-27) it in turn follows that $h = 0.083$. We can now proceed to solve for the tracking-filter update period, as was done for the critically damped *g–h* filter through the use of (1.2-23) and (1.2-15). We shall do this later. Right now we shall design the Benedict–Bordner filter using the same track update period as obtained for the critically damped filter, thereby permitting us to compare the performance of the Benedict–Bordner filter with that of the critically damped filter for the case where they have the same update period.

Using in (1.2-15) the update period $T = 0.1924$ sec obtained for the critically damped filter yields a lag error of 71.3 ft versus the 94.8 ft obtained for the critically damped filter. From (1.2-28) it follows that the normalized transient error for the Benedict–Bordner filter is 0.622 sec^2, as compared to the value of 0.691 sec^2 obtained for the critically damped filter. So, the Benedict–Bordner filter has a smaller transient error and a smaller lag error when the two filters are specified to have the same track update period, an expected result because the Benedict–Bordner filter is designed to minimize the transient error. As a result the Benedict–Bordner filter responds faster to a change in the target velocity and will also have a smaller lag error. The Benedict–Bordner filter is slightly underdamped. That is, it has slight overshoot in response to a step function change in the target velocity.

Columns 2 and 3 of Table 1.2-6 summarize the parameters obtained for respectively the critically damped and Benedict–Bordner filters obtained for Examples 1 and 2 so far. The fourth column (Example 2a) gives an alternate Benedict–Bordner filter design. This alternate design was obtained using a procedure paralleling that used for the critically damped filter. Specifically, (1.2-23) was used to obtain b^*, and in turn (1.2-15) was used to obtain the tracking-filter update period. This alternate design results in a lag error identical to that obtained for the critically damped filter but with an update period larger than that for the critically damped filter and also a transient error that is larger. The design obtained in column 3 (Example 2) is more in the spirit of what is intended to be obtained with a Benedict-Bordner filter, specifically a filter that minimizes the transient error. The filter in the last column does have the advantage of giving the same lag error with a larger track update period.

TABLE 1.2-6. Comparison of Critically Damped and Benedict–Bordner *g–h* Filters

Parameter	Critically Damped (Example 1)	Benedict–Bordner	
		(Example 2)	(Example 2a)
σ_v, ft	50	50	50
$\sqrt{\mathrm{VAR}\left(x^*_{n+1,n}\right)}$, ft	31.6	31.6	31.6
b^*, ft	94.8	71.3	94.8
D/v^2, ft^2	0.691	0.622	0.828
g	0.4375	0.368	0.368
h	0.0625	0.083	0.083
θ	0.75	—	—
T, sec	0.1924	0.1924	0.222

Additional Examples: Comparing Dependence of Critically Damped Filter Design on Prediction Error and Target Maximum Acceleration   Table 1.2-7 compares four critically damped *g–h* filter designs. The second column summarizes again the design obtained in Example 1 above. It is our reference design with which the new designs of columns 3 to 5 are to be compared. The second design in column 3 (Example 1a) gives the design obtained when the 1σ tracking-filter prediction error is decreased to 18.52 ft, all other input assumptions remaining unchanged. This reduction in the required prediction error by almost a factor of 2 leads to a tracking-filter update period reduced by over a factor of 3 to 0.0589 sec. For the third design (Example 1b) given in column 4 of Table 1.2-7 the 18.52-ft 1σ prediction error of Example 1a was still used; however, the maximum target acceleration expected was reduced from 5 g to 3 g to see if this would relax the update period required for the filter to a more reasonable level. The table shows an update period increased by only a small amount, from 0.0589 to 0.0761 sec. For the fourth design (Example 1c) in the last column of Table 1.2-7, the maximum target acceleration was increased back to the value of 5 g and the predicted 1σ rms error was increased to 88.6 ft. Increasing the predicted error by this factor of almost 3 from the original value of Example 1 increases the tracking-filter update period by almost a factor of 6 to 1.096 sec (from 0.1924 sec). The update period of 0.1924 sec could not be achieved with a typical TWS radar, requiring instead a dedicated dish tracking radar, tracking only one target. With a dedicated dish tracker the track update rate could be very high, the rate not limited by the mechanical scan limitation of the radar. Alternately, a phase–phase electronically steered array [1–3] could be used for tracking multiple targets, a radar like the Raytheon PATRIOT radar (see Figure 1.1-9), which is a phased array designed to track multiple targets capable of high-*g* maneuvers with high update rates. The track update period of 1.096 sec could be done with a TWS radar possessing a high rotation rate of the

TABLE 1.2-7. Comparison of Three Critically Damped *g–h* Filters

Parameter	Example 1	Example 1a	Example 1b	Example 1c
σ_R, ft	50	50	50	50
$\sigma_{n+1,n}$, ft	31.6	18.52	18.52	88.6
b^*, ft	94.8	55.6	55.6	266
D/v^2, sec$^{2\,a}$	0.691	0.916	1.528	1.314
g	0.0625	0.190	0.010	0.9775
h	0.0625	0.010	0.010	0.7225
θ	0.75	0.90	0.90	0.15
T, sec	0.1924	0.0589	0.0761	1.096
$\dot{x}_{max}$, ft/sec^2	$5g = 160$	$5\,g$	$3\,g$	$5\,g$

$$^a\ \frac{D}{\Delta v} = \frac{T^2(2 - g)}{gh(4 - 2g - h)}$$

order of 60 rpm. However, a high scan rate does not permit a long enough dwell time per scan to coherently suppress ground and rain clutter.

Note that Example 1c of Table 1.2-7 has very little memory, θ being very small at 0.15. Hence this filter should have a rapid response to a change in target velocity. The design value for the 1σ prediction error $\sigma_{n+1,n}$ and measurement error σ_x will depend on the target density. The 1σ error for the prediction window, $\sigma_w = \sqrt{\sigma_{n+1,n}^2 + \sigma_x^2}$, should be small enough that the 6σ prediction window ($6\sigma_w$) of Figure 1.2-2 does not generally contain a second target in addition to the return from the target being tracked. If a second target return were in this window, the association problem could become difficult. Specifically, we would have a problem in determining which of these two returns should be associated with the target in track. There are ways of coping with this problem [6, 8, 9; see Section 3.3]; however, when it occurs, frequently there is an increased probability of losing target track. It is clear from this discussion that the denser the target environment, the smaller one would desire to have $\sigma_{n+1,n}$.

1.2.8 Circuit Diagram of General *g–h* Filter

For those engineers who prefer seeing a filter in circuit form we give Figure 1.2-16. This circuit applies for the general *g–h* tracking filter of (1.2-11).

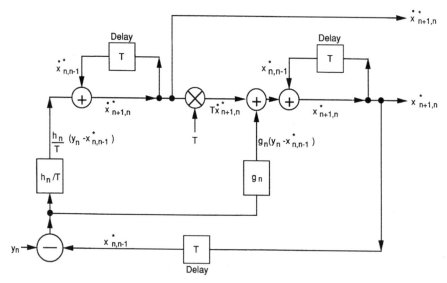

Figure 1.2-16 Equivalent circuit diagram of *g–h* filter.

1.2.9 Stability of Constant *g–h* Filter

Stability is often a concern in designing a circuit for a filter. The literature shows [12, 74] the constant *g–h* filter is stable when the filter parameters *g* and *h* obey the following conditions [12; see also problem 1.2.9-1]:

$$g > 0 \tag{1.2-37a}$$

$$h > 0 \tag{1.2-37b}$$

$$(4 - 2g - h) > 0 \tag{1.2-37c}$$

The above conditions define the triangular region of Figure 1.2-17. Because, generally, $g \leq 1$ and $h < 2$, it follows that normally a constant *g–h* filter is stable. Also plotted in Figure 1.2-17 is the curve relating *h* and *g* for the critically damped filter [i.e., (1.2-36)]. As can be seen, the critically damped *g–h* filter is always stable. The part of the *g–h* curve for $h \leq 1$ corresponds to the minus sign solution of (1.2-36) and is the useful part of the solution.

1.2.10 Track Initiation

So far when discussing the *g–h* filter we did not say how we initiated track. We implicitly assumed we were in track and we wanted to update the estimate of the next target position and velocity, that is, update $x^*_{n+1,n}$ and $\dot{x}^*_{n+1,n}$. (The Kalman filter to be discussed in Chapter 2 actually automatically provides track initiation.) If we use the constant *g–h* filters like the critically damped and Benedict–Bordner filters for track initiation, tracking will be poor. The problem arises because the constant weights *g* and *h*, such as those obtained above (in Examples 1 and 2) for steady-state filtering, are poor choices for track initiation. They overweigh the old estimate for the target position at time *n* given by $x^*_{n,n-1}$ and underweigh the newly acquired measurement y_n. As a result the updated estimate is placed closer to $x^*_{n,n-1}$ than it should be; see Figure 1.2-3. Initially the weights *g* and *h* should be larger than those provided by the steady-state

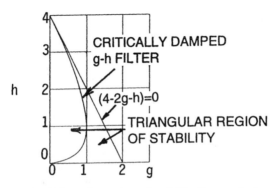

Figure 1.2-17 Triangular region of stability for constant *g–h* filter.

critically damped and Benedict–Bordner filters because initially the estimates $x_{n,n-1}^*$ and $\dot{x}_{n,n-1}^*$ are not accurate at the start of the track. The g–h filters with constant g–h weights are meant to be used after the tracking has reached steady-state conditions, that is, after track initiation. The use of the steady-state constant g–h filters for track initiation could result in loss of track.

For track initiation the least-squares fit procedure (with uniform weighting, that is, without the discounting so that $\theta = 1$) outlined in Section 1.2.6 is used. Figure 1.2-10 illustrates the least-squares fit line (constant-velocity trajectory) for the case of seven measurements, the first seven, that is, $n = 0, \ldots, 6$. The least-squares fitted line is that line that minimizes the sum of the squares of differences between the line and the measurements as given by (1.2-33). The position of the target at time $n = 7$ is then estimated by the position on the least-squares fitting line at time $n = 7$. Knowing this position, a $\pm 3\sigma$ range window is set up about it and the next target echo looked for inside this window. If it is detected, then the whole process is repeated. This procedure represents the least-squares tracking filter. It is not the discounted least-squares filter discussed in Section 1.2.6 because there is no discounting of old data. All the past measurements are weighted with equal importance. (Again, here θ of Section 1.2.6 equals 1.) Note that the slope v_0^* of the least-squares fitting line represents the least-squares estimate of the target velocity. The intercept x_0^* of the line with the ordinate represents a least-squares estimate of the position of the target at time $n = 0$.

Let us step back a little. When $n = 0$, we do not have enough measurements to form a least-squares fitting line. What do we do then? Generally we will have an estimate of the maximum approaching and receding velocities of the target. If we know the maximum approaching and receding velocities the target could have, we can determine the range interval into which the target would fall on the next observation, observation $n = 1$. If the target is observed at time $n = 1$ in this expected range window, we can then obtain an estimate of the target's velocity by drawing a straight line through the two observation points at time $n = 0$ and $n = 1$ using the line to predict the target's position for time $n = 2$ and to determine the $\pm 3\sigma$ window in which the echo is to be looked for at time $n = 2$. If an observation y_2 is obtained in this window, then we would like to obtain a new estimate of the target velocity when predicting where the target is expected to be at the next observation at $n = 3$. That is, one would determine $x_{3,2}^*$. One cannot draw a straight line through these three points, because the measurement noise forces the three points to not fall on a straight line. The dotted line representing the target trajectory is "over-determined" in this case. At this point the least-squares filter procedure discussed above starts.

The least-squares filtering outlined above is a complex one to implement. However, just as the discounted least-square filter of Figures 1.2-10 and 1.2-11 is identical to a constant g–h filter, so too can the least-squares filter be represented by a g–h filter, but with g and h not constant. The weights are given by

$$h_n = \frac{6}{(n+2)(n+1)} \tag{1.2-38a}$$

$$g_n = \frac{2(2n+1)}{(n+2)(n+1)} \tag{1.2-38b}$$

For the above equations, the first measurement is assumed to start at $n = 0$. Here the weights g and h vary with n as indicated. The above equations indicate the weights decrease with increasing n. This is expected. Initially one has a very poor estimate $x^*_{n,n-1}$ so it should not be given a heavy weighting with respect to y_n. As time goes on, with estimate $x_{n,n-1}$ improving, the weighting on it should increase or equivalently the weighting for y_n should decrease, implying a corresponding decrease in the constants g and h. Note than $n + 1$ represents the number of target measurements. The weight h_n is expressed in terms of g_n by the equation [22]

$$h_n = 4 - 2g_n - \sqrt{4(g_n - 2)^2 - 3g_n^2} \tag{1.2-39}$$

The least-squares *g–h* filter just developed (line 6 of Table 1.2-1) is also called an expanding-memory (growing-memory) polynomial filter (line 3 of Table 1.2-1), because its memory increases linearly with time. The term *polynomial* refers to the equation used to model the target motion. For a constant-velocity target, one uses a polynomial of time of degree 1 (i.e., a

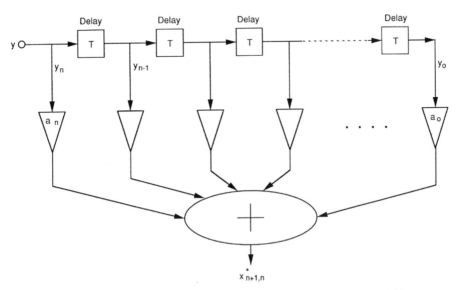

Figure 1.2-18 Equivalent circuit for expanding-memory polynomial filter.

constant plus a constant times t). Figure 1.2-18 gives an equivalent circuit representation for the expanding-memory polynomial filter.

Poor results will be obtained if the track initiation expanding-memory filter is used too long since the weights g and h decrease as n increases, becoming essentially zero for large n; see (1.2-38). So, for large n the tracking filter in effect will not pay any attention to new data. The filter has tracked the target for a long time, knows its trajectory, and does not need new observations to determine where the target is going. Indeed, this would be the case if the target truly were going in a straight line, but in the real world the target will turn or maneuver; hence one does not want the weights g and h to go to zero. For this reason one switches at some point from the expanding-memory polynomial filter to a steady-state constant $g–h$ filter.

The question now to be addressed is when should this transition take place? The answer is that we switch from the track initiation filter to a steady-state $g–h$ filter (such as the critically damped or Benedict–Bordner filter) at that n for which the variance of the target's predicted position $x^*_{n+1,n}$ for the track initiation filter equals the variance of the target's predicted position for the steady-state filter [as given by (1.2-19)]; that is, when

$$\text{VAR } (x^*_{n+1,n}) \text{ for expanding-memory polynomial filter}$$

$$= \text{steady-state VAR } (x^*_{n+1,n}) \text{ for steady-state } g–h \text{ filter} \qquad (1.2\text{-}40)$$

Table 1.2-8 summarizes when the transition should occur.

We shall now illustrate this transition assuming the use of a steady-state critically damped filter. For the critically damped filter one finds, on substituting (1.2-35) into (1.2-19), that

$$\text{Steady-state VAR } (x^*_{n+1,n}) \text{ for critically damped } g–h \text{ filter}$$

$$= \frac{1-\theta}{(1+\theta)^3}(5+4\theta+\theta^2)\sigma_x^2 \qquad (1.2\text{-}41)$$

For the expanding-memory polynomial filter [5; see also Section 6]

$$\text{VAR } (x^*_{n+1,n}) \text{ for expanding-memory polynomial filter} = \frac{2(2n+3)}{(n+1)n}\sigma_x^2$$

$$(1.2\text{-}42)$$

TABLE 1.2-8. Track Initiation for *g–h* Filter

Procedure

1. Start with expanding-memory polynomial filter for constant-velocity trajectory.
2. Switch to steady-state $g–h$ filter at $n = n_0$ for which

$$\text{VAR}(x^*_{n+1,n}) \text{ for expanding-memory polynomial filter}$$

$$= \text{steady-state VAR } (x^*_{n+1,n}) \text{ for steady-state } g–h \text{ filter} \qquad (1.2\text{-}40)$$

We shall use the same assumptions as given in Example 1 of Section 1.2.7. There $\theta = 0.75$ and $\sigma_x = 50$ ft. Substituting (1.2-41) and (1.2-42) into (1.2-40) and solving for n yield that $n = n_0 = 10.46 = 11$. Hence, one switches over to the critically damped filter steady state weights of $g_n = g = 0.4375$ and $h_n = h = 0.0625$ on the 12th observation. The expanding-memory (growing-memory) filter is self-starting; specifically, it properly initiates track independent of the initial conditions assumed; see Section 6.5 and problems 6.5-1 and 6.5-2. The expression (1.2-42) for the accuracy of a growing-memory least-squares filter is a very important one. It tells us that for the normalised one-stop prediction $\sigma_{n+1,n}/\sigma_x = 2/\sqrt{n}$ for large n independent of the track time.

1.3 *g–h–k* FILTER

So far we have developed the filter for tracking a target modeled as on a constant-velocity trajectory. Now we will consider the case of a target having a constant acceleration. The target equations of motion given by (1.1-1) now become, for the target having a constant acceleration,

$$x_{n+1} = x_n + \dot{x}_n T + \ddot{x}_n \frac{T^2}{2} \tag{1.3-1a}$$

$$\dot{x}_{n+1} = \dot{x}_n + \ddot{x}_n T \tag{1.3-1b}$$

$$\ddot{x}_{n+1} = \ddot{x}_n \tag{1.3-1c}$$

Following the heuristic procedure used to develop (1.2-11) for the constant-velocity target, it is a straightforward matter to develop the tracking equations needed for updating the prediction estimates of position, velocity, and acceleration for the constant-accelerating target model. The *g–h–k* track update equations [corresponding to (1.2-8) for the *g–h* filter] become

$$\ddot{x}^*_{n,n} = \ddot{x}^*_{n,n-1} + \frac{2k}{T^2}(y_n - x^*_{n,n-1}) \tag{1.3-2a}$$

$$\dot{x}^*_{n,n} = \dot{x}^*_{n,n-1} + \frac{h}{T}(y_n - x^*_{n,n-1}) \tag{1.3-2b}$$

$$x^*_{n,n} = x^*_{n,n-1} + g(y_n - x^*_{n,n-1}) \tag{1.3-2c}$$

The *g–h–k* prediction equations or transition equations [corresponding to (1.2-10) for the *g–h* filter] become

$$\ddot{x}^*_{n+1,n} = \ddot{x}^*_{n,n} \tag{1.3-3a}$$

$$\dot{x}^*_{n+1,n} = \dot{x}^*_{n,n} + \ddot{x}^*_{n,n} T \tag{1.3-3b}$$

$$x^*_{n+1,n} = x^*_{n,n} + \dot{x}^*_{n,n} T + \ddot{x}^*_{n,n} \frac{T^2}{2} \tag{1.3-3c}$$

The above is for the well-known *g–h–k* filter, also referred to as the α–β–γ filter when *g*, *h*, and *k* are respectively replaced by α, β, and $\gamma/2$. This filter has the advantage that it can track a constant-accelerating target with zero lag error in the steady state. It will have a constant lag error for a target having a constantly changing acceleration with time, that is, for a target having a constant jerk. It is a three state filter—tracking, position, velocity, and acceleration.

The VRF and bias error for $x^*_{n+1,n}$ of the general *g–h–k* filter are given by [11]

$$
VRF(x^*_{n+1,n}) = \frac{VAR(x^*_{n+1,n})}{\sigma^2_x}
$$

$$
= \frac{gk(2g + h - 4) + h[g(2g + h) + 2h]}{[2k - g(h + k)][2g + h - 4]} \tag{1.3-4}
$$

$$
b^* = b^*_{n+1,n} = -\frac{T^3 \dddot{x}}{2k} \tag{1.3-5}
$$

where $\dddot{x}$ is the third derivative of *x* with respect to time (the jerk). The VRFs for $x^*_{n,n}$, $\dot{x}^*_{n,n}$ and $\ddot{x}^*_{n,n}$ are given by respectively [8]

$$
VRF(x^*_{n,n}) = \frac{VAR(x_{n,n})}{\sigma^2_x}
$$

$$
= \frac{2h(2g^2 + 2h - 3gh) - 2gk(4 - 2g - h)}{2(4 - 2g - h)(gh + gk - 2k)} \tag{1.3-6}
$$

$$
VRF(\dot{x}^*_{n,n}) = \frac{VAR(\dot{x}^*_{n,n})}{\sigma^2_x}
$$

$$
= \frac{2h^3 - 4h^2 k + 4k^2(2 - g)}{T^2(4 - 2g - h)(gh + gk - 2k)} \tag{1.3-7}
$$

$$
VRF(\ddot{x}^*_{n,n}) = \frac{VAR(\ddot{x}^*_{n,n})}{\sigma^2_x} = \frac{8hk^2}{T^4(4 - 2g - h)(gh + gk - 2k)}
$$

For the *g–h–k* filters with constants *g*, *h*, and *k* there are equivalent filters to the *g–h* filters, such as the critically damped (fading-memory), Benedict–Bordner, and expanding-memory polynomial filters. The critically damped *g–h–k* filter [5] has three real roots and represents the filter minimizing the discounted least-squares error for a constantly accelerating target. It is shown in Chapter 7 that for the critically damped filter

$$
g = 1 - \theta^3 \tag{1.3-8a}
$$

$$
h = 1.5(1 - \theta^2)(1 - \theta) \tag{1.3-8b}
$$

$$
k = 0.5(1 - \theta)^3 \tag{1.3-8c}
$$

where θ is the discounting factor used for the critically damped *g–h* filter of Section 1.2.6. Chapter 7 also gives the equations for the critically damped *g–h–k–i* filter, that is, the filter designed to track a constant-jerk target (a target whose derivative of acceleration with respect to time is constant), which is a four-state filter. The variance reduction factors for position, velocity, and acceleration for the critically damped *g–h–k* and *g–h–k–i* filters are also given in Chapter 7. Figures 1.3-1 to 1.3-3 give, for the *g–h–k* fading-memory filter, the corresponding useful design curves to Figures 1.2-13 to 1.2-15 for calculating and balancing the bias error b^* and prediction error $\sigma_{n+1,n}$; see problems 2.10-6 and 2.10-7, to be done after reading Section 2.10

The Benedict–Bordner *g–h–k* filter, also called the Simpson filter, minimizes the transient error [13]. The relationship between *g*, *h*, and *k* for the Simpson filter is given as [13]

$$2h - g(g + h + k) = 0 \tag{1.3-9}$$

This filter minimizes the transient due to a unit step in acceleration for a given $\sigma_{n,n}$ and the transient due to a unit step in velocity. For $k = 0$, (1.3-9) yields the relationship between *g* and *h* for the Benedict–Bordner filter given by (1.2-27).

A form of the Benedict–Bordner *g–h–k* filter can also be derived as a steady-state Kalman filter for a target having a constant-accelerating trajectory with a

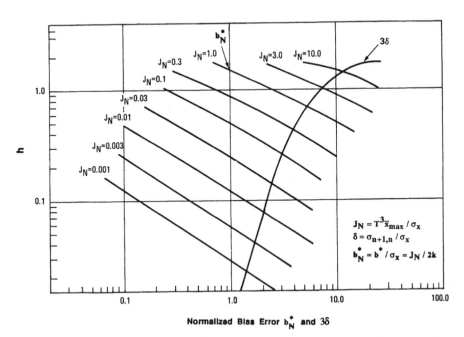

Figure 1.3-1 Normalized bias error $b_N^* = b^*/\sigma_x$ and prediction error $\delta = \sigma_{n+1,n}/\sigma_x$ versus weight *h* for critically damped *g–h–k* filter. (After Asquith and Woods [11].)

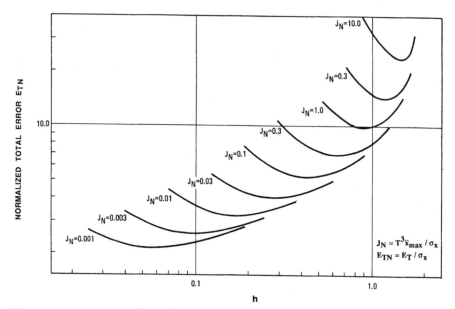

Figure 1.3-2   Normalized total error E_{TN} versus weight h for critically damped g–h–k filter. (After Asquith and Woods [11].)

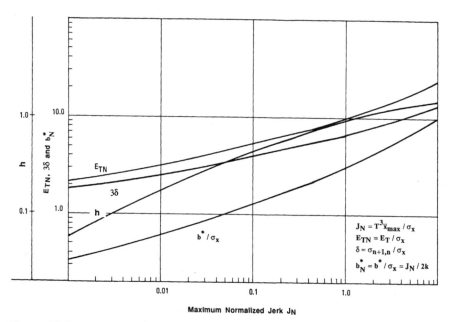

Figure 1.3-3   The minimum total error, E_{TN}, and 3δ, b_N^*, and h versus normalized jerk for critically damped g–h–k filter. (After Asquith and Woods [11].)

random-acceleration component [given by w_n of (2.4-11) and by (2.4-12)]. For this case [14; see also 11 and 15]

$$g = 0.5(-h + \sqrt{8h}) \tag{1.3-10a}$$

or

$$h = 2(2 - g) - 4\sqrt{1 - g} \tag{1.3-10b}$$

and

$$k = \frac{h^2}{4g} \tag{1.3-10c}$$

One can easily show that (1.3-10a) to (1.3-10c) satisfy (1.3-9); see problem (1.3-1). The version of the Benedict–Bordner filter given above by (1.3-10a) to (1.3-10c) is often called the optimum *g–h–k* filter [14, 15]. Figures 1.3-4 to 1.3-6 give, for the optimum *g–h–k* filter, the corresponding useful design curves to Figures 1.2-7 to 1.2-9 for calculating and balancing the bias error b^* and

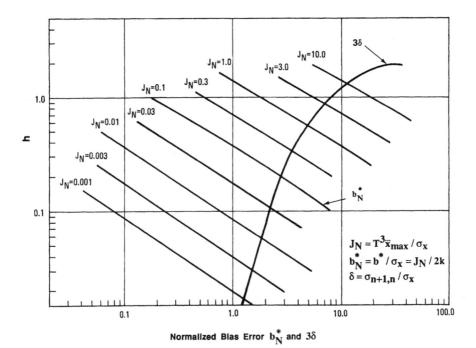

Figure 1.3-4 Normalized bias error $b_N^* = b^*/\sigma_x$ and prediction error $\delta = \sigma_{n+1,n}/\sigma_x$ versus weight h for optimum *g–h–k* filter. (After Asquith and Woods [11].)

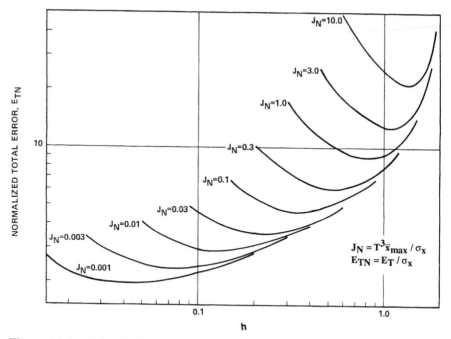

Figure 1.3-5 Normalized total error versus weight h for optimum *g–h–k* filter. (After Asquith and Woods [11].)

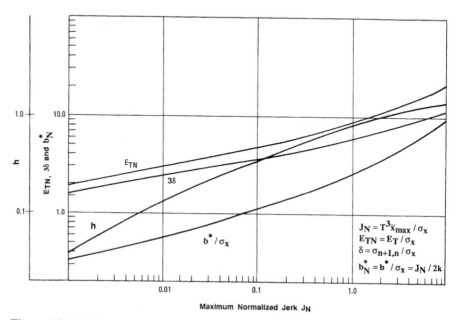

Figure 1.3-6 The minimum total error, E_{TN}, and 3δ, b_N^*, and h versus normalized jerk for optimum *g–h–k* filter. (After Asquith and Woods [11].)

prediction error $\sigma_{n+1,n}$; see problems 2.10-8 and 2.10-9, to be done after reading Section 2.10.

   The *g–h–k* expanding-memory polynomial filter is described in Chapter 6. For a constantly accelerating target the target motion is expressed by a polynomial of time of degree 2. Chapter 6 discusses the expanding-memory polynomial filter for degrees ranging from 0 to 3. The *g–h–k* expanding memory polynomial filter is the least-squares filter for a target having a constant acceleration. These expanding memory polinomial filters are used for track initiating three-state steady-state fading-memory filters having the same number of states as done in Section 1.2.10 for a two-state *g–h* filter and to be further discussed for higher order filters in Section 7.6. Thus a *g–h–k* steady-state filter would use a *g–h–k* growing-memory filter for track initiation. Figures 1.3-7 and 13-8 show the starting transient observed when using the critically damped *g–h–k* filter for track initiation versus using the expanding-memory polynomial filter for track initiation, the steady-state critically damped *g–h–k* filter being used for tracking in steady state. For the latter case of Figure 1.3-8, the switching from the expanding-memory polynomial to the steady-state filter occurs when (1.2-40) holds for the *g–h–k* filters involved.

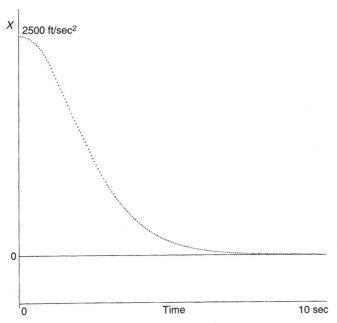

Figure 1.3-7   Starting transient error for critically damped *g–h–k* filter when critically damped filter itself is used for track initiation with first three data points fitted to second-degree polynomial in order to initialize this filter. Parameters: $\theta = 0.942, T = 0.05$ sec, and $\sigma_x = 5$ ft. (From Morrison [5, p. 539].)

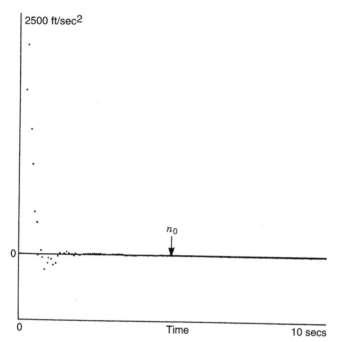

Figure 1.3-8 Starting transient error when expanding-memory polynomial filter is used for track initiation. Switch from expanding-memory polynomial filter to *g–h–k* critically damped filter occurs after 76th observation, that is, $n_0 = 75$. Parameters: $\theta = 0.942$, $T = 0.05$ sec, and $\sigma_x = 5$ ft. (From Morrison [5, p. 540].)

When track initiating with the *g–h–k* critically damped filter without using the expanding-memory filter, the track initiation actually starts by fitting a polynomial of degree 2 to the first data points observed at observation times $n = 0, 1, 2$. This polynomial fit provides the initial estimate of the target position, velocity, and acceleration. This procedure is actually equivalent to using the expanding-memory polynomial filter for track initiation, with only the first three data points used. In contrast, when using the expanding-memory polynomial filter for track initiation until (1.2-40) is satisfied, the first 76 data points are used as indicated in Figure 1.3-8. Figure 1.3-7 shows the large transient error obtained when using the steady-state fading-memory polynomial filter for track initiation as compared to the small transient error obtained when using the expanding-memory polynomial filter; see Figure 1.3-8. As indicated in Section 1.2.10, the growing-memory filter is self-starting; see also Section 6.5 and Problems 6.5-1 and 6.5-2.

A set of extremely useful normalized design curves for constant *g–h–k* filters are given in Section 2.10.

1.4   TRACKING IN MULTIDIMENSIONS

A two-dimensional radar tracks a target in slant range R and azimuth θ. To do this, typically a g–h filter is used to track the target in slant range and a separate, independent g–h filter is used to track the target in azimuth. These filters would generally have different weights g–h. The range state variables R and $\dot{R}$ and azimuth state variables θ and $\dot{\theta}$ generally are correlated, but for simplicity they are often assumed to be independent. When this is done, they are called uncoupled.

Define the four-state vector X by the transpose of the row matrix $[R, \dot{R}, \theta, \dot{\theta}]$, that is, $X = [R, \dot{R}, \theta, \dot{\theta}]^T$, where the superscript T stands for matrix transpose. (Recall that the transpose of an arbitrary $m \times n$ matrix Z is obtained by having the columns of Z become the rows of the $n \times m$ matrix $A = Z^T$.) The column matrix X will be used later to indicate the quantities being estimated by the tracking filter or filters. In this example, these quantities are R, $\dot{R}$, θ, and $\dot{\theta}$. The covariance of X is defined by

$$\text{COV}(X) = E[X^T X] \qquad (1.4\text{-}1)$$

where, as before, $E[\cdot]$ stands for "expected value of". This definition applies as long as the entries of the state vector X have zero mean. Otherwise, X has to be replaced by $X - E[X]$ in (1.4-1). When the range and azimuth variates are assumed to be uncoupled, the covariance of X takes the form

$$\text{COV}(X) = S = \begin{bmatrix} \sigma^2_{RR} & \sigma^2_{R\dot{R}} & 0 & 0 \\ \sigma^2_{\dot{R}R} & \sigma^2_{\dot{R}\dot{R}} & 0 & 0 \\ \hline 0 & 0 & \sigma^2_{\theta\theta} & \sigma^2_{\theta\dot{\theta}} \\ 0 & 0 & \sigma^2_{\dot{\theta}\theta} & \sigma^2_{\dot{\theta}\dot{\theta}} \end{bmatrix} \qquad (1.4\text{-}2)$$

This permits the use of independent g–h filters for the range and azimuth variates.

Often the range variate R is tracked using a g–h–k filter while the azimuth variate θ is tracked using a g–h filter. This is because range is measured much more precisely (to feet whereas angle typically is measured to miles). As a result the target trajectory in slant range is more sensitive to small accelerations, which results in the requirement for a three-state filter that estimates the target acceleration.

For a three-dimensional radar the target is tracked in slant range R, azimuth θ, and elevation ϕ. In this case a third independent (decoupled) g–h filter that tracks the target elevation angle variable ϕ is used in parallel with the g–h–k (or g–h) filter for slant range R and g–h filter for azimuth θ.

1.5 SELECTION OF COORDINATES FOR TRACKING FILTER

For the two dimensional radar the natural coordinates for tracking are the slant range R and azimuth θ coordinates. It is the one generally used. However, this coordinate system has an important disadvantage. Specifically, when a target going in a straight line with a constant velocity flies by the radar at close range, a large geometry-induced acceleleration is seen for the slant range even though the target itself has no acceleration. This acceleration is sometimes referred to as the pseudoacceleration of the target. It is illustrated in Figure 1.5-1. The closer the target flies by the radar, the larger is the maximum geometry-induced acceleration seen for the slant range coordinate. The maximum value for this acceleration is given by

$$a_{max} = \frac{v^2}{R_c} \tag{1.5-1}$$

where v is the target velocity and R_c is the closest approach range of the target to the radar. Thus a 520-knot target having a closest approach of 2 nmi induces a 2g pseudoacceleration onto the slant range measurement even though the target is actually not accelerating. The presence of the pseudoacceleration causes tracking to be more difficult. It could necessitate the need for a higher order tracking filter in the slant range coordinate in order to maintain track on targets passing by at a close range. The pseudoacceleration problem can be eliminated by tracking the target in rectangular coordinates instead of polar coordinates. To do this, a transformation is made from the polar $R–\theta$ coordinates, in which the radar measurements are made, to the rectangular $x–y$ coordinates, in which the tracking is to be done. The predicted $x–y$ coordinates of the target are then transformed into $R–\theta$ coordinates for locating the (n+1)st measurement windows.

Often in practice what is actually done is to track the target in the radar measurement polar $R–\theta$ coordinates when it is at far range, switching to rectangular coordinates when the target is at close range where the pseudoacceleration problem exists. This procedure was used for the terminal ASR-7 air surveillance radar used at Burlington, Vermont, when evaluating the Lincoln Laboratory MTD processor [16]. This was also done for the Applied Physics Laboratory Integrated Automatic Detection and Tracking (IADT) shipboard tracker [17]. Moreover, for the latter case three-dimensional spherical coordinate R, θ, ϕ tracking information is available on the target [see Figure 1.5-2], and the tracking was done at close range using a fully coupled $x–y–z$ rectangular coordinate system tracking filter.

For the ASR-7 system the transformation to rectangular coordinates was made when the target had a slant range than 6 nmi. For simplicity, independent (uncoupled) *g–h* tracking filters were used for the x and y coordinates. Tracking in the rectangular $x–y$ coordinates is not done for all ranges because of the increased computer computation required to do the tracking in these

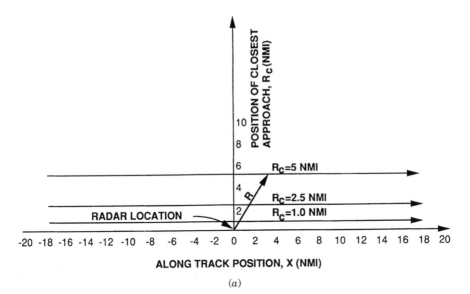

Figure 1.5-1 (*a*) Geometry of constant-velocity target passing by radar to generate pseudoacceleration. (*b*) Pseudoacceleration generated for different points of closest approach; target velocity = 600 knots.

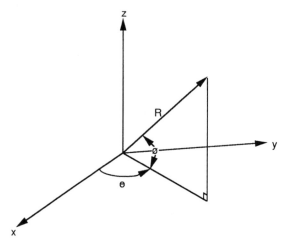

Figure 1.5-2 Spherical coordinates typically used for three-dimensional radar measurements.

coordinates. First, the x and y tracking filters should be coupled because ignoring the coupling degrades tracking accuracy when the parameters have different measurement accuracies, as is the case for range and azimuth measurements. Second, one has to do the transformation from polar to rectangular coordinates and back for each track update. Specifically, for each new target measurements one has to make the transformation from the polar coordinates in which the measurements are made to the rectangular coordinates in which the tracking filters operate and then back to the polar coordinates after each new prediction ahead is made in order to set up the windows in measurement space for the next look.

For the case of spherical and rectangular coordinates these transformations are

$$x = R \cos \phi \cos \theta \tag{1.5-2a}$$
$$y = R \cos \phi \sin \theta \tag{1.5-2b}$$
$$z = R \sin \phi \tag{1.5-2c}$$

for spherical-to-rectangular coordinates and

$$R = (x^2 + y^2 + z^2)^{1/2} \tag{1.5-3a}$$
$$\theta = \tan\left(\frac{y}{x}\right) \tag{1.5-3b}$$
$$\phi = \tan^{-1}\left[\frac{z}{(x^2 + y^2)^{1/2}}\right] \tag{1.5-3c}$$

for rectangular-to-spherical coordinates.

Reference 17 has proposed the use of a dual coordinate system (DCS) to reduce the effects of the pseudoacceleration. In the DCS system the target filtering is done with uncoupled range and azimuth filters but the track prediction is done in cartesian coordinates. Reference 17 indicates that simulation has verified that use of the DCS filter produces the same accuracy as obtained with the coupled Cartesian filter except when one has extremely high angular rates (such as, $20°/\sec$) where both filters fail to perform well. However, such rates are not of practical interest.

The literature is rich on the subject of the selection of the tracking-filter coordinate system. The reader is referred to references 6, 8, and 18 for further detailed discussions on this subject. Extensive discussions are given in reference 8 on the selection of coordinates for sensors on moving platforms and in reference 18 for ground-based intercontinental ballistic missile (ICBM) systems.

2

KALMAN FILTER

2.1 TWO-STATE KALMAN FILTER

Up to now we have used a *deterministic* description for the target motion. Specifically, we have assumed a target having a constant-velocity motion as given by

$$x_{n+1} = x_n + T\dot{x}_n \qquad (1.1\text{-}1a)$$

$$\dot{x}_{n+1} = \dot{x}_n \qquad (1.1\text{-}1b)$$

In the real world the target *will not have a constant velocity for all time*. There is actually uncertainty in the target trajectory, the target accelerating or turning at any given time. Kalman allowed for this uncertainty in the target motion by adding a random component to the target dynamics [19, 20]. For example, a random component u_n could be added to the target velocity as indicated by the following equations for the target dynamics:

$$x_{n+1} = x_n + T\dot{x}_n \qquad (2.1\text{-}1a)$$

$$\dot{x}_{n+1} = \dot{x}_n + u_n \qquad (2.1\text{-}1b)$$

where u_n is a random change in velocity from time n to time $n + 1$. We assume u_n is independent from n to $n + 1$ for all n and that it has a variance σ_u^2. Physically u_n represents a random-velocity jump occurring just prior to the $n + 1$ observation.

We now have a system dynamics model with some randomness. This model is called the constant-velocity trajectory model with a random-walk velocity.

The random-velocity component u_n is sized to account for a possible target acceleration or unexpected target turn. The random dynamics model component u_n in the literature goes by the names process noise [6, 30], plant noise [8, 29, 30], driving noise [5, 8], dynamics noise [119], model noise, and system noise (see Appendix).

Let x_{n+1} represent the true location of the target at time $n + 1$. Let $x^*_{n+1,n}$ represent an estimated predicted position of the target at time $n + 1$ based on the measurements made up to and including time n. Kalman addressed the question of finding the optimum estimate among the class of all linear and nonlinear estimates that minimizes the mean square error

$$\left(x^*_{n+1,n} - x_{n+1} \right)^2 \tag{2.1-2}$$

After much effort Kalman found that the optimum filter is given by the equations

$$\dot{x}^*_{n+1,n} = \dot{x}^*_{n,n-1} + \frac{h_n}{T}\left(y_n - x^*_{n,n-1} \right) \tag{2.1-3a}$$

$$x^*_{n+1,n} = x^*_{n,n-1} + T\dot{x}^*_{n+1,n} + g_n\left(y_n - x^*_{n,n-1} \right) \tag{2.1-3b}$$

But these are identical to the g–h filter given previously, specifically (1.2-11). For the Kalman filter the weights g_n and h_n depend on n. Furthermore, as shall be seen later, g_n and h_n are functions of the variance of the radar position measurement, that is, the variance of ν_n; see (1.2-17). These filter constants are also a function of the accuracy to which we know the position and velocity before any measurements are made, that is, of our prior knowledge of the target trajectory, as given by the a priori variance of the target position and velocity. (Such information might be available when the track is being started after handoff from another sensor.) In the steady state the filter constants g_n and h_n are given by [12]

$$h = \frac{g^2}{2 - g} \tag{2.1-4}$$

This equation is identical to that for the Benedict–Bordner filter given by Eq. (1.2-27). Thus the steady-state Kalman filter is identical to the Benedict–Bordner g–h filter. The more general Kalman filter was developed before the Benedict–Bordner filter. The Kalman filter was first published in 1960 [19] while the Benedict–Bordner filter was published in 1962 [10]. In the literature, when the g–h Benedict–Bordner filter is derived as a steady-state g–h Kalman filter as described above, it is usually referred to as the optimum g–h filter [14, 15].

The Kalman filter has the advantage that it does not require the use of the ad hoc equation relating the rms target prediction error to the g–h bias error as given by (1.2-22) and the bias equation as given by (1.2-15) to determine the tracking-filter update period T as done in Example 1 Section 1.2.7. Instead for the Kalman filter the update period is obtained using the equation [12]

$$T^2 \frac{\sigma_u^2}{\sigma_x^2} = \frac{g^4}{(2-g)^2(1-g)} = \frac{h^2}{1-g} \tag{2.1-5}$$

which relates the target update period to the variance of the target dynamics σ_u^2 as well as to the noise measurement error and the filter parameter g. [See problem 2.4-1 for derivation of (2.1-4) and (2.1-5).] Figure 1.2-7 to 1.2-9 were actually developed for optimum $g-h$ filter in Reference 11. However, in developing these figures, (2.1-5) is not used, only (2.1-4).

It remains to determine how to specify the variance of the target dynamics σ_u^2. Let the maximum expected target acceleration be $\ddot{x}_{max}$. Then the greatest change in velocity in one sample period T is $T\ddot{x}_{max}$ and σ_u is chosen to be given by [12]

$$\sigma_u = \frac{T\ddot{x}_{max}}{B} \tag{2.1-6}$$

where B is a constant. A good value for B is 1 when tracking ballistic targets [12]. With this choice of B the errors $3\sigma_{n+1,n}$ and b^* will be about equal. For maneuvering targets a larger B may be better [12]. If the maneuver were independent from sample to sample, then $B = 3$ would be suitable [12].

Using (1.2-15), (2.1-6), and (2.1-5), the following expression for the normalized bias error b^* is obtained for the $g-h$ Kalman filter in steady state [12]:

$$\frac{b^*}{\sigma_x} = \frac{B}{\sqrt{1-g}} \tag{2.1-7}$$

2.2 REASONS FOR USING THE KALMAN FILTER

Since the steady-state Kalman filter is identical to the Benedict–Bordner filter, the question arises as to why we should use the Kalman filter. The benefits accrued by using the Kalman filter are summarized in Table 2.2-1. First, while in the process of computing the filter weights g_n and h_n for the Kalman filter, calculations of the accuracy of the Kalman filter predictions are made. This prediction information is needed for a weapon delivery system to determine if the predicted position of the target is known accurately enough for a target kill. It is also needed to accurately predict where a detected SCUD, intermediate-

TABLE 2.2-1. Benefits of Kalman Filter

Provides running measure of accuracy of predicted position needed for weapon
   kill probability calculations; impact point prediction calculation
Permits optimum handling of measurements of accuracy that varies with n;
   missed measurements; nonequal times between measurements
Allows optimum use of a priori information if available
Permits target dynamics to be used directly to optimize filter parameters
Addition of random-velocity variable, which forces Kalman filter to be always stable

---

range ballistic missile (IRBM), or intercontinental ballistic missile (ICBM) would land. It makes a difference whether the SCUD, IRBM, or ICBM is landing in neutral territory such as the ocean or in a major city. It is also needed for determining where an artillery shell, mortar shell, SCUD, IRBM, or ICBM was launched. In the cases of the mortar shell, artillery shell, and SCUD this information is needed in order to destroy the launcher or canon. In the case of the IRBM and ICBM this information is needed to determine who is firing so that the appropriate action can be taken. One does not want to take action against country A when it is country B that is firing. The Kalman filter allows one to make estimates of the launcher location and estimate the accuracy of these estimates.

The Kalman filter makes optimal use of the target measurments by adjusting the filter weights g_n and h_n to take into account the accuracy of the nth measurement. For example, if on the nth measurement the signal-to-noise ratio (SNR) is very good so that a very accurate target position measurement is obtained, then g_n and h_n are automatically adjusted to take this into account. Specifically, they are made larger so as to give more weight to this more accurate measurement. If the target had been missed on the $(n-1)$st look, then g_n and h_n are optimally adjusted to account for this. The filter parameters g_n and h_n are also adjusted to allow for nonequal times between measurements.

The Kalman filter optimally makes use of a priori information. Such a priori information could come from another radar that had been previously tracking the target and from which the target is being handed over—such as handover from a search to tracking radar or from one air route surveillance radar (ARSR) to another. The data from the other radar can be used to optimally set up the Kalman g–h filter for the new radar. The Kalman filter automatically chooses weights that start with large g and h values as needed for optimum track initiation. The weights slowly transition to a set of small constant g's and h's after track initiation. The target dynamics model incorporated by the Kalman filter allows direct determination of the filter update rate by the use of (2.1-5). Finally the addition of the random-velocity variable u_n forces the Kalman filter to always be stable.

2.3 PROPERTIES OF KALMAN FILTER

We will now give some physical feel for why the Kalman filter is optimum. Let us go back to our discussion in Section 1.2. Recall that for our two-state g–h tracking we have at time n two estimates of the target position. The first is y_n, based on the measurement made at time n (see Figure 1.2-3). The second is the prediction $x^*_{n,n-1}$, based on past measurements. The Kalman filter combines these two estimates to provide a filtered estimate $x^*_{n,n}$ for the position of the target at time n. The Kalman filter combines these two estimates so as to obtain an estimate that has a minimum variance, that is, the best accuracy. The estimate $x^*_{n,n}$ will have a minimum variance if it is given by [5–7]

$$x^*_{n,n} = \left[\frac{x^*_{n,n-1}}{\text{VAR}(x^*_{n,n-1})} + \frac{y_n}{\text{VAR}(Y_n)} \right] \frac{1}{1/\text{VAR}(x^*_{n,n-1}) + 1/\text{VAR}(y_n)} \quad (2.3\text{-}1)$$

That (2.3-1) provides a good combined estimate can be seen by examining some special cases. First consider the case where y_n and $x^*_{n,n-1}$ have equal accuracy. To make this example closer to what we are familiar with, we use the example we used before; that is, we assume that y_n and $x^*_{n,n-1}$ represent two independent estimates of your weight obtained from two scales having equal accuracy (the example of Section 1.2.1). If one scale gives a weight estimate of 110 lb and the other 120 lb, what would you use for the best combined-weight estimate? You would take the average of the two weight estimates to obtain 115 lb. This is just what (2.3-1) does. If the variances of the two estimates are equal (say to σ^2), then (2.3-1) becomes

$$x^*_{n,n} = \left(\frac{x^*_{n,n-1}}{\sigma^2} + \frac{y_n}{\sigma^2} \right) \frac{1}{1/\sigma^2 + 1/\sigma^2} = \frac{x^*_{n,n-1} + y_n}{2} \quad (2.3\text{-}2)$$

Thus in Figure 1.2-3 the combined estimate $x^*_{n,n}$ is placed exactly in the middle between the two estimates y_n and $x^*_{n,n-1}$.

Now consider the case where $x^*_{n,n-1}$ is much more accurate than the estimate y_n. For this case $\text{VAR}(x^*_{n,n-1}) \ll \text{VAR}(y_n)$ or equivalently $1/\text{VAR}(x^*_{n,n-1}) \gg 1/\text{VAR}(y_n)$. As a result, (2.3-1) can be approximated by

$$x^*_{n,n} = \left[\frac{x^*_{n,n-1}}{\text{VAR}(x^*_{n,n-1})} + 0 \right] \frac{1}{1/\text{VAR}(x^*_{n,n-1}) + 0}$$

$$\doteq x^*_{n,n-1} \quad (2.3\text{-}3)$$

Thus the estimate $x^*_{n,n}$ is approximately equal to $x^*_{n,n-1}$, as it should be because the accuracy of $x^*_{n,n-1}$ is much better than that of y_n. For this case, in Figure 1.2-3 the combined estimate x^*_n is placed very close to the estimate $x^*_{n,n-1}$ (equal to it).

Equation (2.3-1) can be put in the form of one of the Kalman g–h tracking filters. Specifically, (2.3-1) can be rewritten as

$$x_{n,n}^* = x_{n,n-1}^* + \frac{\text{VAR}(x_{n,n}^*)}{\text{VAR}(y_n)}(y_n - x_{n,n-1}^*) \tag{2.3-4}$$

This in turn can be rewritten as

$$x_{n,n}^* = x_{n,n-1}^* + g_n(y_n - x_{n,n-1}^*) \tag{2.3-5}$$

This is the same form as (1.2-7) [and also (1.2-8b)] for the g–h tracking filter. Comparing (2.3-5) with (1.2-7) gives us the expression for the constant g_n. Specifically

$$g_n = \frac{\text{VAR}(x_{n,n}^*)}{\text{VAR}(y_n)} \tag{2.3-6}$$

Thus we have derived one of the Kalman tracking equations, the one for updating the target position. The equation for the tracking-filter parameter h_n is given by

$$h_n = \frac{\text{COV}(x_{n,n}^* \dot{x}_{n,n}^*)}{\text{VAR}(y_n)} \tag{2.3-7}$$

A derivation for (2.3-7) is given for the more general case in Section 2.6.

2.4 KALMAN FILTER IN MATRIX NOTATION

In this section we shall rework the Kalman filter in matrix notation. The Kalman filter in matrix notation looks more impressive. You can impress your friends when you give it in matrix form! Actually there are very good reasons for putting it in matrix form. First, it is often put in matrix notation in the literature, and hence it is essential to know it in this form in order to recognize it. Second, and more importantly, as shall be shown later, in the matrix notation form the Kalman filter applies to a more general case than the one-dimensional case given by (2.1-3) or (1.2-11).

First we will put the system dynamics model given by (1.1-1) into matrix notation. Then we will put the random system dynamics model of (2.1-1) into matrix notation. Equation (1.1-1) in matrix notation is

$$X_{n+1} = \Phi X_n \tag{2.4-1}$$

where

$$X_n = \begin{bmatrix} x_n \\ \dot{x}_n \end{bmatrix} = \text{state vector} \qquad (2.4\text{-}1a)$$

and

$$\Phi = \begin{bmatrix} 1 & T \\ 0 & 1 \end{bmatrix}$$

$\qquad$ = state transition matrix for constant-velocity trajectory [5, 43] $\qquad$ (2.4-1b)

To show that (2.4-1) is identical to (1.1-1), we just substitute (2.4-1a) and (2.4-1b) into (2.4-1) to obtain

$$\begin{bmatrix} x_{n+1} \\ \dot{x}_{n+1} \end{bmatrix} = \begin{bmatrix} 1 & T \\ 0 & 1 \end{bmatrix} \begin{bmatrix} x_n \\ \dot{x}_n \end{bmatrix} \qquad (2.4\text{-}1c)$$

which on carrying out the matrix multiplication yields

$$\begin{bmatrix} x_{n+1} \\ \dot{x}_{n+1} \end{bmatrix} = \begin{bmatrix} x_n + T\dot{x}_n \\ \dot{x}_n \end{bmatrix} \qquad (2.4\text{-}1d)$$

which we see is identical to (1.1-1).

As indicated in (2.4-1a), X_n is the target trajectory state vector. This state vector is represented by a column matrix. As pointed out in Section 1.4, it consists of the quantities being tracked. For the filter under consideration these quantities are the target position and velocity at time n. It is called a two-state vector because it consists of two target states: target position and target velocity. Here, Φ is the state transition matrix. This matrix transitions the state vector X_n at time n to the state vector X_{n+1} at time $n + 1$ a period T later.

It is now a simple matter to give the random system dynamics model represented by (2.1-1) in matrix form. Specifically, it becomes

$$X_{n+1} = \Phi X_n + U_n \qquad (2.4\text{-}2)$$

where

$$U_n = \begin{bmatrix} 0 \\ u_n \end{bmatrix}$$

$\qquad$ = dynamic model driving noise vector $\qquad$ (2.4-2a)

To show that (2.4-2) is identical to (2.1-1), we now substitute (2.4-1a), (2.4-1b), and (2.4-2a) into (2.4-2) to obtain directly from (2.4-1d)

$$\begin{bmatrix} x_{n+1} \\ \dot{x}_{n+1} \end{bmatrix} = \begin{bmatrix} x_n + T\dot{x}_n \\ \dot{x}_n \end{bmatrix} + \begin{bmatrix} 0 \\ u_n \end{bmatrix} \tag{2.4-2b}$$

which on adding the corresponding terms of the matrices on the right-hand side of (2.4-2b) yields

$$\begin{bmatrix} x_{n+1} \\ \dot{x}_{n+1} \end{bmatrix} = \begin{bmatrix} x_n + T\dot{x}_n \\ \dot{x}_n + u_n \end{bmatrix} \tag{2.4-2c}$$

which is identical to (2.1-1), as we desired to show.

We now put the trivial measurements equation given by (1.2-17) into matrix form. It is given by

$$Y_n = MX_n + N_n \tag{2.4-3}$$

where

$$M = \begin{bmatrix} 1 & 0 \end{bmatrix} = \text{observation matrix} \tag{2.4-3a}$$
$$N_n = \begin{bmatrix} \nu_n \end{bmatrix} = \text{observation error} \tag{2.4-3b}$$
$$Y_n = \begin{bmatrix} y_n \end{bmatrix} = \text{measurement matrix} \tag{2.4-3c}$$

Equation (2.4-3) is called the observation system equation. This is because it relates the quantities being estimated to the parameter being observed, which, as pointed out in Section 1.5, are not necessarily the same. In this example, the parameters x_n and $\dot{x}_n$ (target range and velocity) are being estimated (tracked) while only target range is observed. In the way of another example, one could track a target in rectangular coordinates (x, y, z) and make measurements on the target in spherical coordinates (R, θ, ϕ). In this case the observation matrix M would transform from the rectangular coordinates being used by the tracking filter to the spherical coordinates in which the radar makes its measurements.

To show that (2.4-3) is given by (1.2-17), we substitute (2.4-3a) to (2.4-3c), into (2.4-3) to obtain

$$\begin{bmatrix} y_n \end{bmatrix} = \begin{bmatrix} 1 & 0 \end{bmatrix} \begin{bmatrix} x_n \\ \dot{x}_n \end{bmatrix} + \begin{bmatrix} \nu_n \end{bmatrix} \tag{2.4-3d}$$

which on carrying out the multiplication becomes

$$\begin{bmatrix} y_n \end{bmatrix} = \begin{bmatrix} x_n \end{bmatrix} + \begin{bmatrix} \nu_n \end{bmatrix} \tag{2.4-3e}$$

Finally, carrying out the addition yields

$$\begin{bmatrix} y_n \end{bmatrix} = \begin{bmatrix} x_n + \nu_n \end{bmatrix} \tag{2.4-3f}$$

which is identical to (1.2-17).

Rather than put the g–h tracking equations as given by (1.2-11) in matrix form, we will put (1.2-8) and (1.2-10) into matrix form. These were the equations that were combined to obtain (1.2-11). Putting (1.2-10) into matrix form yields

$$X^*_{n+1,n} = \Phi X^*_{n,n} \tag{2.4-4a}$$

where

$$X^*_{n,n} = \begin{bmatrix} x^*_{n,n} \\ \dot{x}^*_{n,n} \end{bmatrix} \tag{2.4-4b}$$

$$X^*_{n+1,n} = \begin{bmatrix} x^*_{n+1,n} \\ \dot{x}^*_{n+1,n} \end{bmatrix} \tag{2.4-4c}$$

This is called the prediction equation because it predicts the position and velocity of the target at time $n + 1$ based on the position and velocity of the target at time n, the predicted position and velocity being given by the state vector of (2.4-4c). Putting (1.2-8) into matrix form yields

$$X^*_{n,n} = X^*_{n,n-1} + H_n(Y_n - MX^*_{n,n-1}) \tag{2.4-4d}$$

Equation (2.4-4d) is called the Kalman filtering equation because it provides the updated estimate of the present position and velocity of the target.

The matrix H_n is a matrix giving the tracking-filter constants g_n and h_n. It is given by

$$H_n = \begin{bmatrix} g_n \\ \dfrac{h_n}{T} \end{bmatrix} \tag{2.4-5}$$

for the two-state g–h or Kalman filter equations of (1.2-10). This form does not however tell us how to obtain g_n and h_n. The following form (which we shall derive shortly) does:

$$H_n = S^*_{n,n-1} M^T \left[R_n + M S^*_{n,n-1} M^T \right]^{-1} \tag{2.4-4e}$$

where

$$S^*_{n,n-1} = \Phi S^*_{n-1,n-1} \Phi^T + Q_n \quad \text{(predictor equation)} \tag{2.4-4f}$$

and

$$Q_n = \text{COV}[U_n] = E[U_n U_n^T] \quad \text{(dynamic model noise covariance)}$$
$$\tag{2.4-4g}$$

$$S_{n,n-1}^* = \text{COV}(X_{n,n-1}^*) = E[X_{n,n-1}^* X_{n,n-1}^{*T}] \tag{2.4-4h}$$

$$R_n = \text{COV}(N_n) = E[N_n N_n^T] \quad \text{(observation noise covariance)} \tag{2.4-4i}$$

$$S_{n-1,n-1}^* = \text{COV}(X_{n-1,n-1}^*)$$
$$= [I - H_{n-1}M]S_{n-1,n-2} \quad \text{(corrector equation)} \tag{2.4-4j}$$

As was the case for (1.4-1), covariances in (2.4-4g) and (2.4-4i) apply as long as the entries of the column matrices U_n and N_n have zero mean. Otherwise U_n and N_n have to be replaced by $U_n - E[U_n]$ and $N_n - E[N_n]$, respectively. These equations at first look formidable, but as we shall see, they are not that bad. We shall go through them step by step.

Physically, the matrix $S_{n,n-1}^*$ is an estimate of our accuracy in prediciting the target position and velocity at time n based on the measurements made at time $n-1$ and before. Here, $S_{n,n-1}^*$ is the covariance matrix of the state vector $X_{n,n-1}^*$. To get a better feel for $S_{n,n-1}^*$, let us write it out for our two-state $X_{n,n-1}^*$. From (1.4-1) and (2.4-4c) it follows that

$$\text{COV} X_{n,n-1}^* = \overline{X_{n,n-1}^* X_{n,n-1}^{*T}}$$

$$= \overline{\begin{bmatrix} x_{n,n-1}^* \\ \dot{x}_{n,n-1}^* \end{bmatrix} [x_{n,n-1}^* \quad \dot{x}_{n,n-1}^*]} = \begin{bmatrix} \overline{x_{n,n-1}^* x_{n,n-1}^*} & \overline{x_{n,n-1}^* \dot{x}_{n,n-1}^*} \\ \overline{\dot{x}_{n,n-1}^* x_{n,n-1}^*} & \overline{\dot{x}_{n,n-1}^* \dot{x}_{n,n-1}^*} \end{bmatrix}$$

$$= \begin{bmatrix} \overline{x_{n,n-1}^{*2}} & \overline{x_{n,n-1}^* \dot{x}_{n,n-1}^*} \\ \overline{\dot{x}_{n,n-1}^* x_{n,n-1}^*} & \overline{\dot{x}_{n,n-1}^{*2}} \end{bmatrix}$$

$$= \begin{bmatrix} s_{00}^* & s_{01}^* \\ s_{10}^* & s_{11}^* \end{bmatrix} = S_{n,n-1}^* \tag{2.4-4k}$$

where for convenience $E[Z]$ has been replaced by $\bar{Z}$, that is, $E[\cdot]$ is replaced by the overbar. Again, the assumption is made that mean of $X_{n,n-1}^*$ has been substracted out in the above.

The matrix R_n gives the accuracy of the radar measurements. It is the covariance matrix of the measurement error matrix N_n given by (2.4-4i). For our two-state filter with the measurement equation given by (2.4-3) to (2.4-3c),

$$R_n = \text{COV}[N_n] = \overline{[\nu_n][\nu_n]}^T = \overline{[\nu_n][\nu_n]}$$
$$= \overline{[\nu_n^2]} = [\overline{\nu_n^2}]$$
$$= [\sigma_\nu^2] = [\sigma_x^2] \tag{2.4-4l}$$

where it is assumed as in Section 1.2.4.4 that σ_ν and σ_x are the rms of ν_n independent of n. Thus σ_ν^2 and σ_x^2 are the variance of ν_n, the assumption being that the mean of ν_n is zero; see (1.2-18).

The matrix Q_n, which gives the magnitude of the target trajectory uncertainty or the equivalent maneuvering capability, is the covariance matrix of the dynamic model driving noise vector, that is, the random-velocity component of the target trajectory given by (2.4-2a); see also (2.1-1). To get a better feel for Q_n, let us evaluate it for our two-state Kalman filter, that is, for U_n given by (2.4-2a). Here

$$Q_n = \text{COV}U_n = \overline{U_n U_n^T} = \begin{bmatrix} 0 \\ u_n \end{bmatrix} \begin{bmatrix} 0 & u_n \end{bmatrix}$$

$$= \begin{bmatrix} 0 \cdot 0 & 0 \cdot u_n \\ u_n \cdot 0 & u_n \cdot u_n \end{bmatrix} = \begin{bmatrix} 0 & 0 \\ 0 & u_n^2 \end{bmatrix} \tag{2.4-4m}$$

Equation (2.4-4f) allows us to obtain the prediction covariance matrix $S_{n,n-1}^*$ from the covariance matrix of the filtered estimate of the target state vector at

TABLE 2.4-1. Kalman Equation

Predictor equation:

$$X_{n+1,n}^* = \Phi X_{n,n}^* \tag{2.4-4a}$$

Filtering equation:

$$X_{n,n}^* = X_{n,n-1}^* + H_n(Y_n - M X_{n,n-1}^*) \tag{2.4-4d}$$

Weight equation:

$$H_n = S_{n,n-1}^* M^T [R_n + M S_{n,n-1}^* M^T]^{-1} \tag{2.4-4e}$$

Predictor covariance matrix equation:

$$S_{n,n-1}^* = \text{COV}(X_{n,n-1}^*) \tag{2.4-4h}$$

$$S_{n,n-1}^* = \Phi S_{n-1,n-1}^* \Phi^T + Q_n \tag{2.4-4f}$$

Covariance of random system dynamics model noise vector U^a:

$$Q_n = \text{COV}(U_n) = E[U_n U_n^T] \tag{2.4-4g}$$

Covariance of measurement vector $Y_n = X_n + N_n^a$:

$$R_n = \text{COV}(Y_n) = \text{COV}(N_n) = E[N_n N_n^T] \tag{2.4-4i}$$

Corrector equation (covariance of smoothed estimate):

$$S_{n-1,n-1}^* = \text{COV}(X_{n-1,n-1}^*) = (I - H_{n-1}M)S_{n-1,n-2}^* \tag{2.4-4j}$$

a If $E[U] = E[N_n] = 0$.

time $n - 1$ given by $S^*_{n-1,n-1}$. The filtered estimate covariance matrix $S^*_{n-1,n-1}$ is in turn obtained from the previous prediction covariance matrix $S^*_{n-1,n-2}$ using (2.4-4j). Equations (2.4-4e), (2.4-4f), and (2.4-4j) allow us to obtain the filter weights H_n at successive observation intervals. For the two-state g–h filter discussed earlier, the observation matrix is given by (2.4-3a) and the filter coefficient matrix H_n is given by (2.4-5). The covariance matrix for the initial a priori estimates of the target position and velocity given by $S^*_{0,-1}$ allows initiation of the tracking equations given by (2.4-4d). First (2.4-4e) is used to calculate H_0 (assuming that $n = 0$ is the time for the first filter observation). For convenience the above Kalman filter equations are summarized in Table 2.4-1.

The beauty of the matrix form of the Kalman tracking-filter equations as given by (2.4-4) is, although presented here for our one-dimensional (range only), two-state (position and velocity) case, that the matrix form applies in general. That is, it applies for tracking in any number of dimensions for the measurement and state space and for general dynamics models. All that is necessary is the proper specification of the state vector, observation matrix, transition matrix, dynamics model, and measurement covariance matrix. For example, the equations apply when one is tracking a ballistic target in the atmosphere in three dimensions using rectangular coordinates (x, y, z) with a ten-state vector given by

$$X^*_{n,n-1} = \begin{bmatrix} x^*_{n,n-1} \\ \dot{x}^*_{n,n-1} \\ \ddot{x}^*_{n,n-1} \\ y^*_{n,n-1} \\ \dot{y}^*_{n,n-1} \\ \ddot{y}^*_{n,n-1} \\ z^*_{n,n-1} \\ \dot{z}^*_{n,n-1} \\ \ddot{z}^*_{n,n-1} \\ \beta^*_{n,n-1} \end{bmatrix} \qquad (2.4\text{-}6)$$

where β is the atmospheric drag on the target. One can assume that the sensor measures R, θ, ϕ, and the target Doppler $\dot{R}$ so that Y_n is given by

$$Y_n = \begin{bmatrix} R_n \\ \dot{R}_n \\ \theta_n \\ \phi_n \end{bmatrix} \qquad (2.4\text{-}7)$$

In general the vector Y_n would be given by

$$Y_n = \begin{bmatrix} y_{1n} \\ y_{2n} \\ \vdots \\ y_{mn} \end{bmatrix} \tag{2.4-8}$$

where y_{in} is the ith target parameter measured by the sensor at time n.

The atmosheric ballistic coefficient β is given by

$$\beta = \frac{m}{C_D A} \tag{2.4-9}$$

where m is the target mass, C_D is the atmospheric dimensionless drag coefficient dependent on the body shape, and A is the cross-sectional area of the target perpendicular to the direction of motion. [See (16.3-18), (16.3-19), (16.3-27) and (16.3-28) of Section 16.3 for the relation between drag constant and target atmospheric deceleration.]

For the g–h Kalman filter whose dynamics model is given by (2.1-1) or (2.4-2), the matrix Q is given by (2.4-4m), which becomes

$$Q = \begin{bmatrix} 0 & 0 \\ 0 & \sigma_u^2 \end{bmatrix} \tag{2.4-10}$$

if it is assumed that the mean of u_n is zero and its variance is σ_u^2 independent of n. For the equivalent g–h–k Kalman filter to our two-state g–h Kalman filter having the dynamic model of (2.4-2), the three-state dynamics model is given by (1.3-3) with (1.3-3a) replaced by

$$\dddot{x}^*_{n+1,n} = \dddot{x}^*_{n,n} + w_n \tag{2.4-11}$$

where w_n equals a random change in acceleration from time n to $n+1$. We assume w_n is independent from n to $n+1$ for all n and that it has a variance σ_w^2. Physically w_n represents a random-acceleration jump occurring just prior to the $n+1$ observation. For this case

$$Q = \begin{bmatrix} 0 & 0 & 0 \\ 0 & 0 & 0 \\ 0 & 0 & \sigma_w^2 \end{bmatrix} \tag{2.4-12}$$

The variance of the target acceleration dynamics σ_w^2 (also called σ_a^2) can be specified using an equation similar to that used for specifying the target velocity dynamics for the Kalman g–h filter. Specifically

$$\sigma_w = \frac{T \dddot{x}_{max}}{C} \tag{2.4-13}$$

where C is a constant and $\ddot{x}_{\max}$ is the maximum $\ddot{x}$. For the steady-state g–h–k Kalman filter for which Q is given by (2.4-12) g,h, and k are related by (1.3-10a) to (1.3-10c) [11, 14, 15] and σ_a^2, σ_x^2, and T are related to g and k by [14]

$$\frac{T^4 \sigma_a^2}{4 \sigma_x^2} = \frac{k^2}{1 - g} \tag{2.4-14}$$

For the general g–h–k Kalman filter (2.4-5) becomes [14]

$$H_n = \begin{bmatrix} g_n \\ \dfrac{h_n}{T} \\ \dfrac{2k_n}{T^2} \end{bmatrix} \tag{2.4-15}$$

This is a slightly underdamped filter, just as is the steady-state g–h Kalman filter that is the Benedict–Bordner filter. Its total error $E_{TN} = 3\sigma_{n+1,n} + b^*$ is less than that for the critically damped g–h–k filter, and its transient response is about as good as that of the critical damped filter [11]. In the literature, this steady-state Kalman filter has been called the optimum g–h–k filter [11].

If we set $\sigma_u^2 = 0$ in (2.4-10), that is, remove the random maneuvering part of the Kalman dynamics, then

$$Q = \begin{bmatrix} 0 & 0 \\ 0 & 0 \end{bmatrix} \tag{2.4-16}$$

and we get the growing-memory filter of Section 1.2.10, the filter used for track initiation of the constant g–h filters.

2.5 DERIVATION OF MINIMUM-VARIANCE EQUATION

In Section 2.3 we used the minimum-variance equation (2.3-1) to derive the two-state Kalman filter range-filtering equation. We will now give two derivations of the minimum-variance equation.

2.5.1 First Derivation

The first derivation parallels that of reference 7. For simplicity, designate the two independent estimates $x_{n,n-1}^*$ and y_n by respectively x_1^* and x_2^*. Designate $x_{n,n}^*$, the optimum combined estimate, by x_c^*. We desire to find an optimum linear estimate for x_c^*. We can designate this linear estimate as

$$x_c^* = k_1 x_1^* + k_2 x_2^* \tag{2.5-1}$$

We want this estimate x_c^* to be unbiased, it being assumed that x_1^* and x_2^* are unbiased. Designate as x the true value of x. Obtaining the mean of (2.5-1), it follows that for the estimate to be unbiased

$$x = k_1 x + k_2 x \tag{2.5-2}$$

which becomes

$$1 = k_1 + k_2 \tag{2.5-3}$$

Thus for the estimate to be unbiased we require

$$k_2 = 1 - k_1 \tag{2.5-4}$$

Substituting (2.5-4) into (2.5-1) yields

$$x_c^* = k_1 x_1^* + (1 - k_1) x_2^* \tag{2.5-5}$$

Let the variances of x_c^*, x_1^*, and x_2^* be designated as respectively σ_c^2, σ_1^2, and σ_2^2. Then (2.5-5) yields

$$\sigma_c^2 = k_1^2 \sigma_1^2 + (1 - k_1)^2 \sigma_2^2 \tag{2.5-6}$$

To find the k_1 that gives the minimum σ_c^2, we differentiate (2.5-6) with respect to k_1 and set the result to zero, obtaining

$$2 k_1 \sigma_1^2 - 2(1 - k_1) \sigma_2^2 = 0 \tag{2.5-7}$$

Hence

$$k_1 = \frac{\sigma_2^2}{\sigma_1^2 + \sigma_2^2} \tag{2.5-8}$$

Substituting (2.5-8) into (2.5-5) yields

$$x_c^* = \frac{\sigma_2^2}{\sigma_1^2 + \sigma_2^2} x_1^* + \frac{\sigma_1^2}{\sigma_1^2 + \sigma_2^2} x_2^* \tag{2.5-9}$$

Rewriting (2.5-9) yields

$$x_c^* = \left(\frac{x_1^*}{\sigma_1^2} + \frac{x_2^*}{\sigma_2^2} \right) \frac{1}{1/\sigma_1^2 + 1/\sigma_2^2} \tag{2.5-10}$$

which is identical to Eq. (2.3-1), as we desired to show. Note that substituting

(2.5-8) into (2.5-6) yields

$$\sigma_c^2 = \left(\frac{1}{\sigma_1^2} + \frac{1}{\sigma_2^2}\right)^{-1} \tag{2.5-11}$$

2.5.2 Second Derivation

The second derivation employs a weighted least-squares error estimate approach. In Figure 1.2-3 we have two estimates y_n and $x_{n,n-1}^*$ and desire here to replace these with a combined estimate $x_{n,n}^*$ that has a minimum weighted least-squares error. For an arbitrarily chosen $x_{n,n}^*$ shown in Figure 1.2-3 there are two errors. One is the distance of $x_{n,n}^*$ from y_n; the other is its distance of $x_{n,n}^*$ from $x_{n,n-1}^*$. For the minimum least-squares estimate in Section 1.2.6 we minimized the sum of the squares of the distances (errors) between the measurements and the best-fitting line (trajectory) to the measurements. We would like to similarly minimize the sum of the two errors here between $x_{n,n}^*$ and y_n and $x_{n,n-1}^*$ in some sense. One could minimize the sum of the squares of the errors as done in Section 1.2.6, but this is not the best tactic because the two errors are not always equally important. One of the estimates, either y_n or $x_{n,n-1}^*$, will typically be more accurate than the other. For convenience let us say $x_{n,n-1}^*$ is more accurate than y_n. In this case it is more important that $(x_{n,n-1}^* - x_{n,n}^*)^2$ be small, specifically smaller than $(y_n - x_{n,n}^*)^2$. This would be achieved if in finding the least sum of the squares of each of the two errors we weighted the former error by a larger constant than the latter error. We are thus obtaining a minimization of an appropriately weighted sum of the two errors wherein the former receives a larger weighting. A logical weighting is to weight each term by 1 over the accuracy of their respective estimates as the following equation does:

$$E = \frac{(y_n - x_{n,n}^*)^2}{\text{VAR}y_n} + \frac{(x_{n,n-1}^* - x_{n,n}^*)^2}{\text{VAR}(x_{n,n-1}^*)} \tag{2.5-12}$$

Here the error $(x_{n,n-1}^* - x_{n,n}^*)^2$ is weighted by 1 over the variance of $x_{n,n-1}^*$ and $(y_n - x_{n,n}^*)^2$ by 1 over the variance of y_n. Thus if $\text{VAR}(x_{n,n-1}^*) \ll \text{VAR}(y_n)$, then $1/\text{VAR}(x_{n,n-1}^*) \gg 1/\text{VAR}(y_n)$ and forces the error $(x_{n,n-1}^* - x_{n,n}^*)^2$ to be much smaller than the error $(y_n - x_{n,n}^*)^2$ when minimizing the weighted sum of errors E of (2.5-12). This thus forces $x_{n,n}^*$ to be close to $x_{n,n-1}^*$, as it should be. The more accurate $x_{n,n-1}^*$, the closer $x_{n,n}^*$ is to $x_{n,n-1}^*$. In (2.5-12) the two errors are automatically weighted according to their importance, the errors being divided by their respective variances. On finding the $x_{n,n}^*$ that minimizes E of (2.5-12), a weighted least-squares estimate instead of just a least-squares estimate is obtained. This is in contrast to the simple unweighted least-squares estimate obtained in Section 1.2.6.

It now remains to obtain the $x_{n,n}^*$ that minimizes (2.5-12). This is a straight-forward matter. Differentiating (2.5-12) with respect to $x_{n,n}^*$ and setting the result equal to zero yields

$$\frac{dE}{dx_{n,n}^*} = \frac{2(y_n - x_{n,n}^*)}{VAR(y_n)} + \frac{2(x_{nn-1}^* - x_{n,n}^*)}{VAR(x_{n,n-1}^*)} = 0 \qquad (2.5\text{-}13)$$

Solving for $x_{n,n}^*$ yields (2.3-1), the desired result.

2.6 EXACT DERIVATION OF r-DIMENSIONAL KALMAN FILTER

We will now extend the second derivation given in Section 2.5.2 to the case where a target is tracked in r-dimensions. An example of an r-dimensional state vector for which case $r = 10$ is given by (2.4-6). For this case the target is tracked in the three-dimensional rectangular (x, y, z) coordinates in position, velocity, and acceleration. In addition, the atmospheric drag parameter β is also kept track of to form the 10th parameter of the 10-dimensional state vector $X_{n,n-1}^*$.

The 10 states of the state vector are to be estimated. The measurements made on the target at time n are given by the measurement matrix Y_n. As indicated in Section 1.5, the measurements made on the target need not be made in the same coordinate system as the coordinate system used for the state vector $X_{n,n-1}^*$. For example the target measurements are often made in a spherical coordinate system consisting of slant range R, target azimuth angle θ, and target elevation angle ϕ, yet the target could be tracked in rectangular coordinates. If the target Doppler velocity $\dot{R}$ is measured, then Y_n becomes (2.4-7). The observation matrix M of (2.4-3) converts the predicted trajectory state vector $X_{n,n-1}^*$ from its coordinate system to the coordinate system used for making the radar measurements, that is, the coordinate system of Y_n.

For simplicity let us assume initially that the coordinates for the r-dimensional state vector $X_{n,n-1}^*$ is the same as for Y_n. Let $X_{n,n}^*$ be our desired combined estimate of the state vector after the measurement Y_n. The combined estimate $X_{n,n}^*$ will lie somewhere in between the predicted state vector $X_{n,n-1}^*$ and the measurement vector Y_n as was the case in the one-dimensional situation depicted in Figure 1.2-3. Figure 2.6-1 shows the situation for our present multidimensional case. As done in the second derivation for the one-dimensional case discussed in Section 2.5.2, we will choose for our best combined estimate the $X_{n,n}^*$ that minimizes the weighted sum of the error differences between Y_n and $X_{n,n}^*$ and between $X_{n,n-1}^*$ and $X_{n,n}^*$. Again the weighting of these errors will be made according to their importance. The more important an error is, the smaller it will be made. An error is deemed to be more important if it is based on a more accurate estimate of the position of the target.

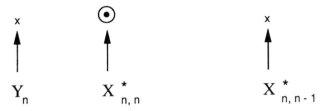

Figure 2.6-1 Filtering problem. Determination of $X_{n,n}^*$ based on measurement Y_n and prediction $X_{n,n-1}^*$.

Accordingly, as done for the one-dimensional case, we weight the square of the errors by 1 over the variance of the error. Thus the equivalent equation to (2.5-12) becomes

$$J = \frac{(Y_n - X_{n,n}^*)^2}{VAR\, Y_n} + \frac{(X_{n,n-1}^* - X_{n,n}^*)^2}{VAR\, X_{n,n-1}^*} \qquad (2.6\text{-}1)$$

This equation is only conceptually correct; the mathematically correct equivalent will be given shortly.

It remains now to use (2.6-1) to solve for the new combined estimate $X_{n,n}^*$ that minimizes the weighted sum of the squares of the errors. Conceptually, this is done just as it was done for the equivalent one-dimensional (2.5-12). Specifically, (2.6-1) is differentiated with respect to the combined estimate $X_{n,n}^*$ with the resulting equation set equal to zero in order to solve for the combined estimate $X_{n,n}^*$ that minimizes the weighted sum of the errors squared. There is one problem though: (2.6-1) is not correct when one is dealing with matrices. It is only conceptually correct as indicated.

When using matrix notation, the first term on the right of (2.6-1) must be written as

$$\frac{(Y_n - X_{n,n}^*)^2}{VAR\, Y_n} \equiv (Y_n - X_{n,n}^*)^T R_n^{-1} (Y_n - X_{n,n}^*) \qquad (2.6\text{-}2)$$

Where the matrix R_n is the covariance matrix Y_n, that is,

$$R_n = COV(Y_n) \qquad (2.6\text{-}3)$$

which is the same as defined by (2.4-4i). The inverse of the covariance matrix R_n, which is designed as R_n^{-1}, takes the place of dividing by the variance of Y_n when dealing with matrices. Note that if R_n is diagonal with all the diagonal terms equal to σ_x^2, then (2.6-2) becomes

$$\frac{(Y_n - X_{n,n}^*)^2}{\text{VAR } Y_n} \equiv \frac{(Y_n - X_{n,n}^*)^T (Y_n - X_{n,n}^*)}{\sigma_x^2}$$

$$= \frac{\sum_{i=1}^r \left((y_{in} - x_{i,nn}^*) \right)^2}{\sigma_x^2} \qquad (2.6\text{-}4)$$

If the coordinate system for Y_n and $X_{n,n}^*$ are not the same, then (2.6-2) becomes

$$\frac{(Y_n - X_{n,n}^*)^2}{\text{VAR } Y_n} \equiv \left(Y_n - MX_{n,n}^{*T} \right) R_n^{-1} (Y_n - MX_{n,n}^*) \qquad (2.6\text{-}5)$$

The corresponding correct form for the second term on the right of (2.6-1) is

$$\frac{(X_{n,n-1}^* - X_{n,n}^*)^2}{\text{VAR } X_{n,n-1}^*} \equiv (X_{n,n-1}^* - X_{n,n}^*)^T S_{n,n-1}^{*-1} (X_{n,n-1}^* - X_{n,n}^*) \qquad (2.6\text{-}6)$$

where $S_{n,n-1}^*$ is the covariance matrix of $X_{n,n-1}^*$; see (2.4-4h) and (2.4-4f). Substituting (2.6-5) and (2.6-6) into (2.6-1) yields

$$J = (Y_n - MX_{n,n}^*)^T R_n^{-1} (Y_n - MX_{n,n}^*) + (X_{n,n-1}^* - X_{n,n}^*)^T S_{n,n-1}^{*-1} (X_{n,n-1}^* - X_{n,n}^*) \qquad (2.6\text{-}7)$$

Now we are in a position to solve for $X_{n,n}^*$. As discussed before, this is done by differentiating (2.6-7) with respect to $X_{n,n}^*$, setting the resulting equation equal to zero in solving for $X_{n,n}^*$. The details of this are carried out in the remaining paragraphs of this section. The results are the full-blown Kalman filter equations given by (2.4-4) and summarized in Table 2.4-1. The reader may forgo the detailed mathematical derivation that follows in the next few paragraphs. However, the derivation is relatively simple and straight forward. At some point it is recommended that the reader at least glance at it, the Kalman filter having had such a major impact on filtering theory and the derivation given here being of the simplest of the full-blown Kalman filter that this author has seen.[†] The derivation makes use of matrix differentiation. The reader not familiar with matrix differentiation will be able to learn it by following the steps of the derivation given. Matrix differentiation is really very simple, paralleling standard algebraic differentiation in which

$$\frac{d(x^2)}{dx} = 2x \qquad (2.6\text{-}8)$$

[†] This derivation was pointed out to the author by Fred Daum of the Raytheon Company.

Differentiation of a matrix equation, such as that of (2.6-7), is achieved by obtaining the gradient of the matrix equation as given by

$$\text{Gradient of } J \triangleq \frac{\partial J}{\partial X^*_{n,n}} = \left[\frac{\partial J}{\partial x_1}, \frac{\partial J}{\partial x_2}, \cdots, \frac{\partial J}{\partial x_n} \right] \qquad (2.6\text{-}9)$$

Applying (2.6-9) to (2.6-7) yields

$$\frac{\partial J}{\partial X} = 2(X^*_{n,n} - X^*_{n,n-1})^T S^{*-1}_{n,n-1} + 2(Y_n - MX^*_{n,n})^T R_n^{-1}(-M) = 0 \qquad (2.6\text{-}10)$$

This can be rewritten as

$$X^{*T}_{n,n}\left(S^{*-1}_{n,n} + M^T R_n^{-1} M\right) = X^{*T}_{n,n-1} S^{*-1}_{n,n-1} + Y_n^T R_n^{-1} M \qquad (2.6\text{-}11)$$

which on taking the transpose of both sides and using $(AB)^T = B^T A^T$ yields

$$\left(S^{*-1}_{n,n} + M^T R_n^{-1} M\right) X^*_{n,n} = S^{*-1}_{n,n-1} X^*_{n,n-1} + M^T R_n^{-1} Y_n \qquad (2.6\text{-}12)$$

or

$$X^*_{n,n} = \left(S^{*-1}_{n,n-1} + M^T R_n^{-1} M\right)^{-1}\left(S^{*-1}_{n,n-1} X^*_{n,n-1} + M^T R_n^{-1} Y_n\right) \qquad (2.6\text{-}13)$$

The well-known matrix inversion lemma [5] states

$$(S^*_{n,n} + M^T R_n^{-1} M)^{-1} = S^*_{n,n} - S^*_{n,n} M^T (R_n + M S^*_{n,n} M^T)^{-1} M S^*_{n,n} \qquad (2.6\text{-}14)$$

This can be rewritten as

$$\left(S^{*-1}_{n,n} + M^T R_n^{-1} M\right)^{-1} = S^*_{n,n-1} - H_n M S^*_{n,n-1} \qquad (2.6\text{-}15)$$

where, as given by (2.4-4e),

$$H_n = S^*_{n,n-1} M^T (R_n + M S^*_{n,n-1} M^T)^{-1} \qquad (2.6\text{-}15a)$$

Substituting (2.6-15) into (2.6-13) yields

$$\begin{aligned} X^*_{n,n} &= \left[S^*_{n,n-1} - H_n M S^*_{n,n-1}\right]\left[S^{*-1}_{n,n-1} X^*_{n,n-1} + M^T R^{-1} Y_n\right] \\ &= X^*_{n,n-1} - H_n M X^*_{n,n-1} + (S^*_{n,n-1} - H_n M S^*_{n,n-1}) M^T R^{-1} Y_n \qquad (2.6\text{-}16) \end{aligned}$$

But as shall be shown shortly,

$$H_n = (S^*_{n,n-1} - H_n M S^*_{n,n-1}) M^T R^{-1} \qquad (2.6\text{-}17)$$

Hence (2.6-16) can be written as

$$X^*_{n,n} = X^*_{n,n-1} + H_n Y_n - H_n M X^*_{n,n-1} \qquad (2.6\text{-}18)$$

or

$$X^*_{n,n} = X^*_{n,n-1} + H_n(Y_n - M X^*_{n,n-1}) \qquad (2.6\text{-}19)$$

But (2.6-19) is identical to the Kalman filter equation given by (2.4-4d), which is what we set out to prove.

We will now prove (2.6-17). From (2.6-15a) it follows that

$$S^*_{n,n-1} M^T = H_n(R_n + M S^*_{n,n-1} M^T) \qquad (2.6\text{-}20)$$

This equation can be rewritten as

$$S^*_{n,n-1} M^T R_n^{-1} - H_n M S^*_{n,n-1} M^T R_n^{-1} = H_n \qquad (2.6\text{-}21)$$

which in turn becomes (2.6-17), as we set out to derive.

The corrector equation (2.4-4j) follows from (2.6-19) and (2.6-21). The predictor equation (2.4-4f) follows from (2.4-2) and (2.4-4a). This completes out derivation of the Kalman equations given by (2.4-4a) through (2.4-4j).

2.7   TABLE LOOKUP APPROXIMATION TO THE KALMAN FILTER

An approximation to the Kalman filter can be used that involves a table lookup instead of the use of (2.4-4e) to calculate the coefficients of the matrix H_n in (2.4-4d). One such approximate lookup table is given in Table 2.7-1. As indicated in the table the coefficients g and h are determined by the sequence of detection hits and misses observed for the target in track. Also given in this table is the size of the search window to be used for a given track update. The approximate lookup procedure given in Table 2.7-1 is similar to that used for the ARTS III filter [21].

2.8   ASQUITH–FRIEDLAND STEADY-STATE g–h KALMAN FILTER

It was indicated earlier that the steady-state Kalman filter for the target dynamics model given by (2.1-1) yields the Benedict–Bordner g–h filter for

TABLE 2.7-1. Table Lookup Approximation to Kalman Filter

Firmness, F_n	Tracking Parameter Lookup		
	Position Smooth, g_n	Velocity Smooth, h_n	Search Bin, Δ_n
0	1.000	.000	—
1	1.000	1.000	$\Delta t \cdot v_{max}$
2	.833	.700	21.0σ
3	.700	.409	11.7σ
4	.600	.270	8.8σ
$\vdots$	$\vdots$	$\vdots$	$\vdots$
15	.228	.030	4.5σ

Note: If "hit" at n: $F_{n+1} = F_n + 1$ $(F_{n,max} = 15)$. If "miss": $F_{n+1} = F_n - 2$ $(F_{n,min} = 1)$.
Initial hit: $F_0 = 0 \rightarrow F_1 = 1$. These are similar but not identical to ARTS III filter.
Source: After Sittler [21].

which g and h are related by (1.2-27). An alternate target dynamics model for the two-state Kalman filter is given by [22, 23]

$$x_{n+1} = x_n + \dot{x}_n T + \tfrac{1}{2} a_n T^2 \qquad (2.8\text{-}1a)$$

$$\dot{x}_{n+1} = \dot{x}_n + a_n T \qquad (2.8\text{-}1b)$$

where a_n is a random acceleration occurring between time n and $n+1$. The random acceleration a_n has the autocorrelation function given by

$$\overline{a_n a_m} = \begin{cases} \sigma_a^2 & \text{for } n = m \\ 0 & \text{for } n \neq m \end{cases} \qquad (2.8\text{-}2)$$

Hence a_n is characterized as white noise. This model differs from that of (2.1-1) in that here a_n is a random acceleration that is constant between the time n and $n + 1$ measurements whereas in (2.1-1b) a random-velocity jump occurs just before the $(n + 1)$st observation.

Figure 2.8-1 compares the possible g, h pairings for the Asquith–Friedland filter [(2.8-4)] obtained using the dynamics model of (2.8-1) and (2.8-2) with those of the Benedict–Bordner filter [(2.1-4)] having the dynamic model given by (2.1-1) or (2.4-2) with COV(U_n) given by (2.4-10). Also shown for comparison are the g and h weight pairings for the critically damped filter [1.2-36)] and the expanding-memory polynomial filter described in Sections 1.2.6 and 1.2.10, respectively. Recall that the expanding-memory polynomial filter [(1.2-38)] is a Kalman filter for which there is no noise term in the target dynamics model, that is, $u_n = 0$ in (2.1-1b); see (2.4-16) for the resulting Q for this target model. Note that the steady-state Asquith–Friedland filter (referred to as the discrete Kalman filter in reference 22) has viable designs for $h > 1$;

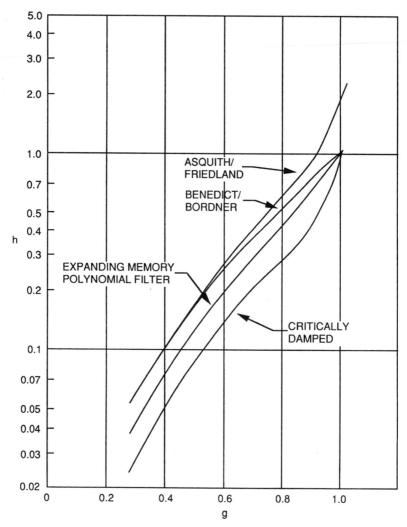

Figure 2.8-1   Comparison of *g* and *h* parameters for several *g–h* filters. (After Asquith [22].)

specifically it can have $h = 1.8755$ and $g = 0.9990$. This pair of values still leads to a stable *g–h* filter; that is, it falls in the triangular region of Figure 1.2-17 where *g–h* filters are stable. In fact, all the curves of Figure 2.8-1 fall within the triangular region. The figure indicates that the Asquith–Friedland and Benedict–Bordner filter are approximately the same for $g < 0.5$. Of these four constant *g–h* filters, the Asquith–Friedland filter is the most underdamped followed by the Benedict–Bordner filter, the expanding-memory filter, and the critically damped filter. Figure 2.8-2 plots a normalized dynamic error $b^* = b^*_{n+1,n}$ for three of these *g–h* filters versus the rms predicted position error.

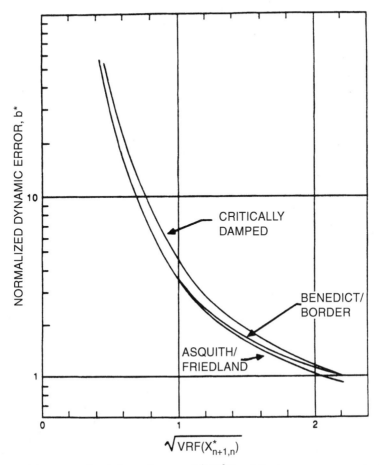

Figure 2.8-2   Normalized dynamic error $b^*/T^2\ddot{x} = 1/h$ versus square root of one-step prediction variance reduction factor for several g–h filters. (Asquith [22].)

It is easy to show that the Q matrix of (2.4-4g) for a target having the dynamic model given by (2.8-1a) and (2.8-1b) is given by [22]

$$Q = T^2 \begin{bmatrix} \sigma_a^2 \dfrac{T^2}{4} & \sigma_a^2 \dfrac{T}{2} \\[2ex] \sigma_a^2 \dfrac{T}{2} & \sigma_a^2 \end{bmatrix} \tag{2.8-3}$$

In steady state [22]

$$h = 4 - 2g - 4\sqrt{1 - g} \tag{2.8-4}$$

and

$$T^2 \frac{\sigma_a^2}{\sigma_x^2} = \frac{h^2}{1 - g} \qquad (2.8\text{-}5)$$

which note is similar to (2.1-5) obtained for Q given by (2.4-10).

2.9   SINGER g–h–k KALMAN FILTER

In this section and the next we will give some feel for the Singer g–h–k Kalman filter [6, 8, 24], indicating the type of maneuvering target for which it is designed and then give some performance results.

For this filter Singer specified a target dynamics model for which the acceleration is a random function of time whose autocorrelation function is given by

$$E\left[\ddot{x}(t)\ddot{x}(t + t')\right] = \sigma_a^2 \exp\left(-\frac{|t'|}{\tau}\right) \qquad (2.9\text{-}1)$$

where τ is the correlation time of the acceleration that could be due to a target maneuver, a target turn, or atmospheric turbulence. For a lazy turn τ is typically up to 60 sec, for an evasive maneuver τ is typically between 10 and 30 sec, while atmospheric turbulence results in a correlation time of the order of 1 or 2 sec. It is further assumed by Singer that the target acceleration has the probability density function given by Figure 2.9-1. This figure indicates that the target can have a maximum acceleration of $\pm A_{\max}$ with probability $P_{\max}$, no acceleration with probability P_0, and an acceleration between $\pm A_{\max}$ with a probability given by the uniform density function amplitude given in Figure 2.9-1.

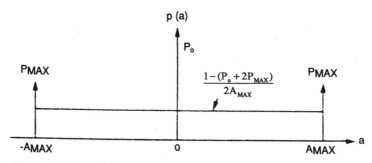

Figure 2.9-1   Model used by Singer [24] for target acceleration probability density function. (After Singer, R. A. "Estimating Optimal Tracking Filter Performance for Manned Maneuvering Targets," IEEE Trans. on Aerospace and Electronic Systems, Vol. AES-6(4), 1970. © 1970, IEEE.)

The total variance of the acceleration is given by

$$\sigma_a^2 = \frac{A_{\max}^2}{3}(1 + 4P_{\max} - P_0) \tag{2.9-2}$$

To apply the Kalman filter as given by (2.4-4), we need a white noise for the target model velocity jump term; that is, the velocity jump u_n of (2.4-2a) must be independent from time n to time $n + 1$. For the target dynamics as given by (2.9-1) we find that the velocity jump u_n of (2.4-2a) is correlated. Even though the actual acceleration term is correlated, a white-noise acceleration forcing term can be generated in its place. This is done by finding the circuit that when driven by white noise $n_a(\tau)$ gives the correlated acceleration $\ddot{x}(\tau)$ as its output with the autocorrelation function for the output given by (2.9-1).

The transfer function for the filter that achieves this is given by

$$H(\omega) = \frac{\tau}{1 + j\omega\tau} \tag{2.9-3}$$

where $\omega = 2\pi f$. The inverse of the above filter $H_a(\omega)$ is the Wiener–Kolmogorov whitening filter. The differential equation for this filter is given by

$$\frac{d}{dt}(\ddot{x}) = -\frac{1}{\tau}\ddot{x} + n_a(t) \tag{2.9-4}$$

To use the driving term with the white-noise acceleration $n_a(\tau)$ in place of the autocorrelated acceleration $\ddot{x}_n$ requires the augmentation of the number of states in the tracking filter from 2 to 3. Thus

$$X_n = \begin{bmatrix} x_n \\ \dot{x}_n \\ \ddot{x}_n \end{bmatrix} \tag{2.9-5}$$

instead of $X_n = [x_n \quad \dot{x}_n]^T$. Consequently, a g–h–k filter results instead of a g–h filter.

We shall show shortly that for the Singer dynamics model the three-state dynamics equation equivalent to (2.1-1) then takes the form

$$X_{n+1} = \Phi X_n + U_n \tag{2.9-6}$$

when the whitening filter of (2.9-4) is used, which allows us to replace the correlated acceleration with white-noise acceleration. The three-state dynamics driving noise U_n is now

$$U_n = \begin{bmatrix} u_{1,n} \\ u_{2,n} \\ u_{3,n} \end{bmatrix} \tag{2.9-7}$$

where $u_{i,n}$ is independent of $u_{i,n+1}$; that is U_n is now a white-noise vector as required in order to apply the Kalman filter of (2.4-4). The terms $u_{i,n}$ and $u_{j,n}$ are, however, correlated for given n, as we shall see shortly. Applying the Kalman filter to a target having the target dynamics given by (2.9-7) provides the best performance in terms of minimizing the mean-square estimation error given by (2.1-2).

The transition matrix is now given by

$$\Phi(T,\tau) = \begin{bmatrix} 1 & T & \tau^2\left[-1+\dfrac{T}{\tau}+\exp\left(-\dfrac{T}{\tau}\right)\right] \\[2mm] 0 & 1 & \tau\left[1-\exp\left(-\dfrac{T}{\tau}\right)\right] \\[2mm] 0 & 0 & \exp\left(-\dfrac{T}{\tau}\right) \end{bmatrix} \qquad (2.9\text{-}8)$$

When T/τ is small so the target can be considered to have a constant acceleration between sample update periods, (2.9-8) reduces to

$$\Phi(T,\tau) = \begin{bmatrix} 1 & T & \frac{1}{2}T^2 \\ 0 & 1 & T \\ 0 & 0 & 1 \end{bmatrix} \qquad (2.9\text{-}9)$$

This, as expected, is identical to the transition matrix for the constant-accelerating target obtained from (1.3-1), the acceleration being constant from time n to $n+1$. The above matrix is called a Newtonian matrix [24].

The covariance of the white-noise maneuver excitation vector U_n is given by [24]

$$Q_n = E[U(n)U^T(n)] = 2\alpha\sigma_a^2 \begin{bmatrix} q_{11} & q_{12} & q_{13} \\ q_{12} & q_{22} & q_{23} \\ q_{13} & q_{23} & q_{33} \end{bmatrix} \qquad (2.9\text{-}10)$$

where

$$\alpha = \frac{1}{\tau} \qquad (2.9\text{-}10a)$$

and

$$q_{11} = \frac{1}{2\alpha^5}\left(1 - e^{-2\alpha T} + 2\alpha T + \frac{2\alpha^3 T^3}{3} - 2\alpha^2 T^2 - 4\alpha Te^{-\alpha T}\right) \qquad (2.9\text{-}10b)$$

$$q_{12} = \frac{1}{2\alpha^4}\left(e^{-2\alpha T} + 1 - 2e^{-\alpha T} + 2\alpha Te^{-\alpha T} - 2\alpha T + \alpha^2 T^2\right) \qquad (2.9\text{-}10c)$$

$$q_{13} = \frac{1}{2\alpha^3}(1 - e^{-2\alpha T} - 2\alpha T e^{-\alpha T}) \tag{2.9-10d}$$

$$q_{22} = \frac{1}{2\alpha^3}(4e^{-\alpha T} - 3 - e^{-2\alpha T} + 2\alpha T) \tag{2.9-10e}$$

$$q_{23} = \frac{1}{2\alpha^2}(e^{-2\alpha T} + 1 - 2e^{-\alpha T}) \tag{2.9-10f}$$

$$q_{33} = \frac{1}{2\alpha}(1 - e^{-2\alpha T}) \tag{2.9-10g}$$

For τ and T fixed the maneuver excitation covariance matrix Q_n is independent of n and designated as Q. Assuming again, that T/τ is small, specifically $T/\tau \ll \frac{1}{2}$, as done for (2.9-9), then [24]

$$Q = 2\alpha\sigma_a^2 \begin{bmatrix} \frac{1}{20}T^5 & \frac{1}{8}T^4 & \frac{1}{6}T^3 \\ \frac{1}{8}T^4 & \frac{1}{3}T^3 & \frac{1}{2}T^2 \\ \frac{1}{6}T^3 & \frac{1}{2}T^2 & T \end{bmatrix} \tag{2.9-11}$$

When T/τ is large, that is $T/\tau \gg 1$, the acceleration is independent from n, $n+1$, and

$$Q = \begin{bmatrix} 0 & 0 & 0 \\ 0 & 0 & 0 \\ 0 & 0 & \sigma_a^2 \end{bmatrix} \tag{2.9-12}$$

which is identical to (2.4-12), σ_a^2 being used in place of σ_w^2 for the variance of the acceleration jump from time n to $n+1$.

The Kalman filter of (2.4-4) for this target model has an observation matrix given by

$$M = [1 \quad 0 \quad 0] \tag{2.9-13}$$

It is initialized using

$$x_{1,1}^* = y_1 \tag{2.9-14a}$$

$$\dot{x}_{1,1}^* = \frac{y_1 - y_0}{T} \tag{2.9-14b}$$

$$\ddot{x}_{1,1}^* = 0 \tag{2.9-14c}$$

where y_0 and y_1 are the first two range measurements. The covariance matrix for $x^*_{n,n}$, that is, $S^*_{n,n}$, is initialized using for $n = 1$[24]

$$[S^*_{1,1}]_{00} = \sigma^2_x \qquad (2.9\text{-}15a)$$

$$[S^*_{1,1}]_{01} = [S^*_{1,1}]_{10} = \frac{\sigma^2_x}{T} \qquad (2.9\text{-}15b)$$

$$[S^*_{1,1}]_{02} = [S^*_{1,1}]_{20} = 0 \qquad (2.9\text{-}15c)$$

$$[S^*_{1,1}]_{11} = \frac{2\sigma^2_x}{T^2} + \frac{\sigma^2_a}{\alpha^4 T^2}\left(2 - \alpha^2 T^2 + \frac{2\alpha^3 T^3}{3} - 2e^{-\alpha T} - 2\alpha T e^{-\alpha T}\right) \qquad (2.9\text{-}15d)$$

$$[S^*_{1,1}]_{12} = [S^*_{1,1}]_{21} = \frac{\sigma^2_a}{\alpha^2 T}(e^{-\alpha T} + \alpha T - 1) \qquad (2.9\text{-}15e)$$

$$[S^*_{1,1}]_{22} = \sigma^2_a \qquad (2.9\text{-}15f)$$

where $[S^*_{1,1}]_{i,j}$ is the i, j element of the covariance matrix $S^*_{1,1}$ and we index the rows and columns starting with the first being 0, the second 1, and so on, to be consistent with reference 5 and Chapters 5 and 7.

If the filter acquisition occurs before the target maneuvers, that is, during a time when the target has a constant velocity, as is usually the case, then the above covariance initialization equations become simply [24]

$$[S^*_{1,1}]_{00} = \sigma^2_x \qquad (2.9\text{-}16a)$$

$$[S^*_{1,1}]_{01} = [S^*_{1,1}]_{10} = \frac{\sigma^2_x}{T} \qquad (2.9\text{-}16b)$$

$$[S^*_{1,1}]_{11} = \frac{2\sigma^2_x}{T^2} \qquad (2.9\text{-}16c)$$

$$[S^*_{1,1}]_{02} = [S^*_{1,1}]_{20} = [S^*_{1,1}]_{12}$$
$$= [S^*_{1,1}]_{21} = [S^*_{1,1}]_{22} = 0 \qquad (2.9\text{-}16d)$$

The next section gives convenient normalized design curves for the steady-state Singer g–h–k Kalman filters. Section 3.5.2 shows how the use of a chirp waveform affects the performance of a Singer g–h–k Kalman filter. Before giving the design curves we will indicate how (2.9-6) is obtained.

We start by rewriting the target dynamics equation given by (2.9-4) in state matrix form. Doing this yields

$$\dot{X}(t) = AX(t) + BN_a(t) \qquad (2.9\text{-}17)$$

where

$$X(t) = \begin{bmatrix} x(t) \\ \dot{x}(t) \\ \ddot{x}(t) \end{bmatrix} \tag{2.9-17a}$$

$$A = \begin{bmatrix} 0 & 1 & 0 \\ 0 & 0 & 1 \\ 0 & 0 & -\alpha \end{bmatrix} \tag{2.9-17b}$$

$$B = [0 \ 0 \ 1] \tag{2.9-17c}$$

$$N_a(t) = \begin{bmatrix} 0 \\ 0 \\ n_a(t) \end{bmatrix} \tag{2.9-17d}$$

For $BN_a(t) = 0$ in (2.9-17) and $\alpha = 0$ in A of (2.9-17b) we obtain the target dynamics for the case of a constant-accelerating target with no random target dynamics. We already know the solution of (2.9-17) for this case. It is given by (2.9-6) with $U_n = 0$ and the transition matrix given by (2.9-9). To develop the solution to (2.9-17) when $BN_a(t) \neq 0$, we will make use of the solution to (2.9-17) for $BN_a(t) = 0$ obtained in Section 8.1. In Section 8.1 it is shown that the solution to (2.9-17) for $BN_a(t) = 0$ is given by

$$X(t + T) = [\exp(TA)]X(t) \tag{2.9-18}$$

see (8.1-19), where here $\varsigma = T$. Alternatively, we can write

$$X(t) = [\exp(tA)]X(0) \tag{2.9-19}$$

Although A is a matrix, it is shown in Section 8.1 that $\exp(tA)$ holds with "exp = e" to a matrix power being defined by (8.1-15). We can rewrite (2.9-18) as

$$X(t_n + T) = \Phi(T)X(t_n) \tag{2.9-20}$$

or

$$X_{n+1} = \Phi X_n \tag{2.9-20a}$$

for $t = t_n$ and where

$$\Phi = \Phi(t) = \exp TA \tag{2.9-20b}$$

is the transition matrix for the target dynamics with $n_a(t) = 0$, which from (2.9-17) implies a constant-accelerating target dynamics model. For this case, as already indicated, Φ is given by (2.9-9). It is also verified in Section 8.1 that for A given by (2.9-17b) with $\alpha = 0$, (2.9-20b) becomes (2.9-9); see (8.1-10a) and problem 8.1-3. We can easily verify that for $BN_a(t) = 0$, (2.9-20) is the solution to (2.9-17) for a general matrix A by substituting (2.9-20) into (2.9-17). We carry the differentiation by treating $X(t)$ and A as if they were not matrices, in which case we get, by substituting (2.9-20) into (2.9-17),

$$d\{[\exp(tA)]X(0)\} = AX(t) \tag{2.9-21a}$$

$$[\exp(tA)]X(0)\frac{d(tA)}{dt} = AX(t) \tag{2.9-21b}$$

$$A[\exp(tA)]X(0) = AX(t) \tag{2.9-21c}$$

$$AX(t) = AX(t) \tag{2.9-21d}$$

as we desired to show.

The solution of (2.9-17) for $BN_a(t) \neq 0$ is given by the sum of the solution (2.9-20) obtained above for $BN_a(t) = 0$, which is called the homogeneous solution and which we will designate as $X(t)_H$, plus an inhomogeneous solution given by

$$W(t) = \int_0^t [\exp A(t - \varepsilon)]BN_a(\varepsilon)d\varepsilon \tag{2.9-22}$$

Thus the total solution to (2.9-17) is

$$X(t) = [\exp(At)]X(0) + \int_0^t [\exp A(t - \varepsilon)]BN_a(\varepsilon)d\varepsilon \tag{2.9-23}$$

Substituting (2.9-23) into (2.9-17) and carrying out the differentiation as done for the homogenous solution above, we verify that (2.9-23) is the solution to (2.9-17).

Comparing (2.9-23) with (2.9-6), it follows that

$$U_n = \int_{nT}^{(n+1)T} [\exp A(\tau - \varepsilon)]BN_a(\varepsilon)d\varepsilon \tag{2.9-24}$$

and $\Phi(T, \tau)$ for $\alpha \neq 0$ is given by

$$\Phi(T, \tau) = \exp AT \tag{2.9-25}$$

Substituting (2.9-17b) into (2.9-25) and using the definition for exp AT given by (8.1-15) yields (2.9-8). From (2.9-24) the independence of U_{n+1} from U_n immediately follows and in turn (2.9-10) to (2.9-10g) follow; see [24] for details.

2.10 CONVENIENT STEADY-STATE *g–h–k* FILTER DESIGN CURVES

The constant (steady-state) *g–h–k* filter was introduced in Section 1.3. The Singer–Kalman filter described in Section 2.9 is a *g–h–k* filter. The steady-state Singer filter designs can be used as the basis for constant *g–h–k* filters. Toward this end, Fitzgerald has developed useful normalized curves that provide the steady-state weights *g*, *h*, and *k* for the steady-state Singer *g–h–k* Kalman filters [25]. For these curves the filter weights are given in terms of two dimensionless Fitzgerald parameters p_1 and p_2:

$$p_1 = \frac{\tau}{T} \tag{2.10-1}$$

$$p_2 = T^2 \frac{\sigma_a}{\sigma_x} \tag{2.10-2}$$

These curves are given in Figures 2.10-1 through 2.10-3. Fitzgerald also showed that the filter steady-state rms predicted and filtered position, velocity, and

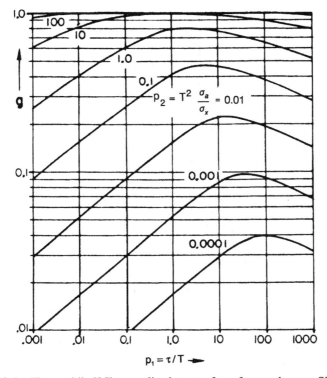

Figure 2.10-1 Fitzgerald's [25] normalized curves for *g* for steady-state Singer *g–h–k* filter. (After Fitzgerald, R. J., "Simple Tracking Filters: Steady-State Filtering and Smoothing Performance," IEEE Transactions on Aerospace and Electronic Systems, Vol. AES-16, No. 6, November 1980. © 1980 IEEE.)

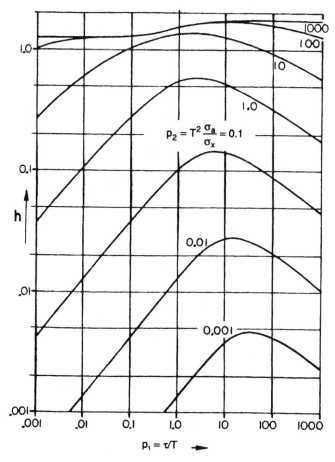

Figure 2.10-2 Fitzgerald's [25] normalized curves for h for Singer g–h–k Kalman filter. (After Fitzgerald, R. J., "Simple Tracking Filters: Steady-State Filtering and Smoothing Performance," IEEE Transactions on Aerospace and Electronic Systems, Vol. AES-16, No. 6, November 1980. © 1980 IEEE.)

acceleration errors for these g–h–k filters can be expressed in terms of the two above dimensionless parameters; see Figures 2.10-4 through 2.10-7. In these figures the filtered position error is referred to by Fitzgerald as the position error after measurement update while the predicted position error is referred to as the error before the measurement update. These normalized error curves are useful for preliminary performance prediction of constant g–h–k filters.

Fitzgerald also developed normalized curves showing the accuracy the Singer filter gives for the target smoothed position near the middle of its

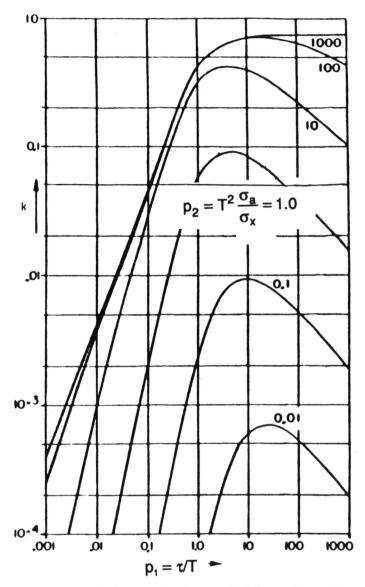

Figure 2.10-3   Fitzgerald's [25] normalized curves for k for steady-state Singer g–h–k Kalman filter. (After Fitzgerald, R. J, "Simple Tracking Filters: Steady-State Filtering and Smoothing Performance," IEEE Transactions on Aerospace and Electronic Systems, Vol. AES-16, No. 6, November 1980. © 1980 IEEE.)

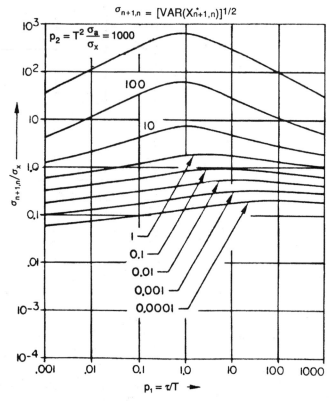

Figure 2.10-4 Fitzgerald's [25] normalized curves for one-step prediction rms error for steady-state Singer g–h–k Kalman filter. (After Fitzgerald, R. J., "Simple Tracking Filters: Steady-State Filtering and Smoothing Performance," IEEE Transactions on Aerospace and Electronic Systems, Vol. AES-16, No. 6, November 1980. © 1980 IEEE.)

observed trajectory. Figure 2.10-8 shows these normalized curves obtained by Fitzgerald. These curves provide an indication of the improvement in the target position estimate that can be obtained post-flight after all the data have been collected. The curves apply for any point far from the endpoints of the filter-smoothing interval. Specifically, assume that the data are smoothed over the interval $n = 0, \ldots J$. Then Figure 2.10-8 applies for time n as long as $0 \ll n \ll J$. The smoothed position accuracy given in Figure 2.10-8 is normalized to the filtered target position accuracy at the trajectory endpoint $n = J$ based on the use of all the collected data.

Fitzgerald also obtained normalized curves similar to those of Figure 2.10-8 for the smoothed velocity and acceleration of the target somewhere toward the middle of the target trajectory. Figures 2.10-9 and 2.10-10 give these curves.

Physically p_1 is the random-acceleration time constant τ normalized to the track update time T. The parameter p_2 is physically 2 times the motion expected

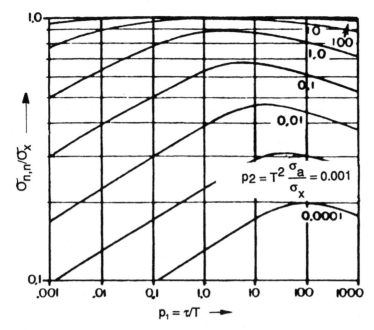

Figure 2.10-5 Fitzgerald's [25] normalized curve of rms filtered position error (i.e., position error just after update measurement has been made and incorporated into target position estimation) for steady-state Singer *g–h–k* Kalman filter. (After Fitzgerald, R. J., "Simple Tracking Filters: Steady-State Filtering and Smoothing Performance," IEEE Transactions on Aerospace and Electronic Systems, Vol. AES-16, No. 6, November 1980. © 1980 IEEE.)

from the random acceleration $2(\frac{1}{2})\sigma_a T^2$ normalized by the measurement uncertainty σ_x. Sometimes p_2 is called the tracking index.

Example   The use of the above normalized curves will now be illustrated for a $g–h–k$ angle tracker. Assume that the target is at a slant range $R = 50$ km; that the target is characterized by having $\tau = 3$ sec and $\sigma_a = 30$ m/sec^2, and that the radar 3-dB beamwidth $\theta_3 = 20$ mrad and the angle measurement error $\sigma_\theta = 1$ mrad, θ replacing x here in Figures 2.10-4 to 2.10-10. Assume that the 3σ predicted tracking position in angle is to be less than $\frac{1}{2}\theta_3$ for good track maintenance. The problem is to determine the tracker sampling rate required, the tracker rms error in predicting position, the ability of the tracker to predict position after track update (the target's filtered position), and the ability of the tracker to predict the target rms position by postflight smoothing of the data (to some point far from the trajectory endpoints, as discussed above).

Solution: We desire the predicted angle normalized position error $\sigma_{n+1,n}(\theta)$, which we here disignate as $\sigma_{n+1,n}$ for convenience [noting that this term serves

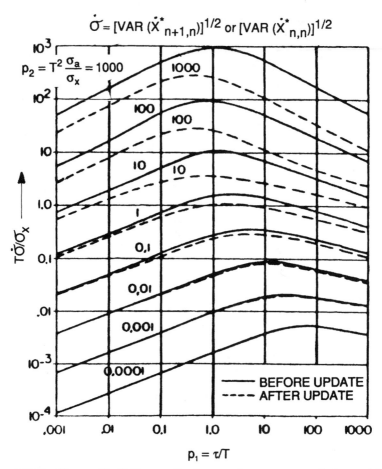

Figure 2.10-6 Fitzgerald's [25] normalized curves for one-step prediction and filtered velocity rms errors (i.e., before and after update respectively) for steady-state Singer g–h–k Kalman filter. (After Fitzgerald, R. J., "Simple Tracking Filters: Steady-State Filtering and Smoothing Performance," IEEE Transactions on Aerospace and Electronic Systems, Vol. AES-16, No. 6, November 1980. © 1980 IEEE.)

double duty; see (1.2-23)], to be given by

$$3\sigma_{n+1,n} \leq \left(\tfrac{1}{2}\right)\theta_3 = 20\,\text{mrad}/2 = 10\,\text{mrad} \qquad (2.10\text{-}3)$$

Normalizing the above by dividing both sides by $3\sigma_\theta$ yields

$$\frac{\sigma_{n+1,n}}{\sigma_\theta} \leq \frac{\theta_3}{2}\frac{1}{3(\sigma_\theta)} = \frac{10\,\text{mrad}}{3(1\,\text{mrad})} = 3.33 \qquad (2.10\text{-}4)$$

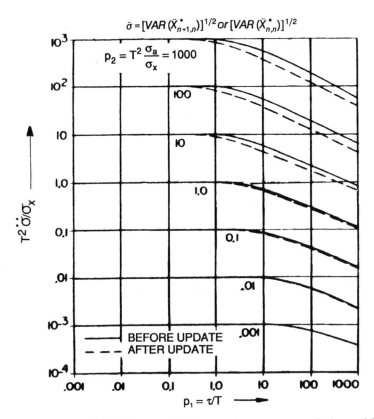

Figure 2.10-7 Fitzgerald's [25] normalized curves for one-step prediction and filtered acceleration rms errors (i.e., before and after update respectively) for steady-state Singer *g–h–k* Kalman filter. (After Fitzgerald, R. J., "Simple Tracking Filters: Steady-State Filtering and Smoothing Performance," IEEE Transactions on Aerospace and Electronic Systems, Vol. AES-16, No. 6, November 1980. © 1980 IEEE.)

In Figure 2.10-4, p_1 and p_2 are not known. Also, because T is in both p_1 and p_2, an iteration trial-and-error process is needed generally to solve for T. A good starting point is to assume that the solution is at the peak of a $p_2 = $ constant curve. As a result it follows from Figure 2.10-4 that if (2.10-4) is to be true it is necessary that $p_2 \leq 2.5$. As a result

$$p_2 = T^2 \frac{\sigma_a}{\sigma_{cx}} = T^2 \frac{30\,\text{m/sec}^2}{(1\,\text{mrad})\,(50\,\text{km})} \leq 2.5 \qquad (2.10\text{-}5)$$

where σ_{cx} is the cross-range position rms measurement accuracy given by

$$\sigma_{cx} = \sigma_\theta R = (1\,\text{mrad})\,(50\,\text{km}) = 50\,\text{m} \qquad (2.10\text{-}6)$$

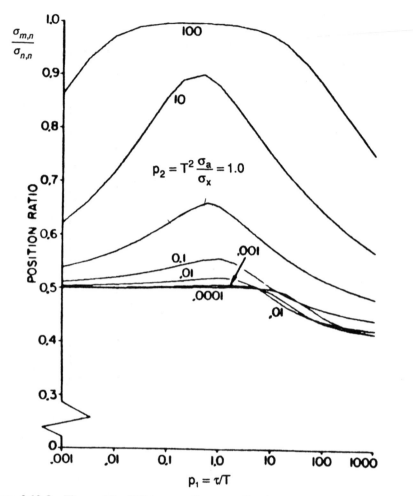

Figure 2.10-8 Fitzgerald's [25] curves for normalized accuracy to which target's position can be obtained for observed point somewhere in middle of target track history, the smoothed position accuracy. Normalization is with respect to steady-state filtered position error $\sigma_{n,n}$ given in Figure 2.10-5. (After Fitzgerald, R. J., "Simple Tracking Filters: Steady-State Filtering and Smoothing Performance," IEEE Transactions on Aerospace and Electronic Systems, Vol. AES-16, No. 6, November 1980. © 1980 IEEE.)

From the above (2.10-5) it follows that $T \leq 2.04$ sec. We will choose $T = 2$ sec as our first iteration estimate for T.

For the above value of T it follows that $p_1 = \tau/T = (3\,\mathrm{sec})/(2\,\mathrm{sec}) = 1.5$. We now check if this value for p_1 is consistent with the value for p_1 obtained from Figure 2.10-4 for $p_2 = 2.5$. If it is not, then another iteration is required in

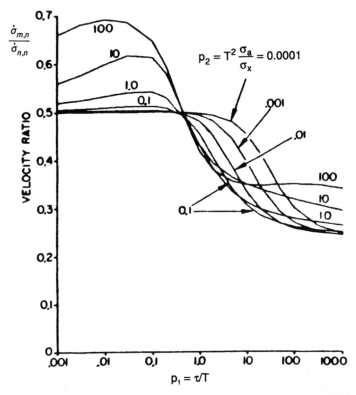

Figure 2.10-9 Fitzgerald's [25] normalized curves for accuracy to which target's velocity can be obtained at point somewhere in middle of target track history. Normalization is with respect to filtered velocity estimate given in Figure 2.10-6. (After Fitzgerald, R. J., "Simple Tracking Filters: Steady-State Filtering and Smoothing Performance," IEEE Transaction on Aerospace and Electronic Systems, Vol. AES-16, No. 6, November 1980. © 1980 IEEE.)

the selection of T. Figure 2.10-4 yields a value of $p_1 = 1.5$. Hence the value for p_1 obtained from Figure 2.10-4 is consistent with the value obtained for $T = 2\,\text{sec}$. As a result a second iteration is not needed for our example. In general, the requirement for an iteration being carried out in order to solve for T could be eliminated if the curve of Figure 2.10-4 is replaced with a different normalized curve, specifically, if p_2 is replaced using curves of constant $\tau^2 \sigma_a / \sigma_x$ [25].

Figure 2.10-5 yields the normalized after-measurement (filtered) estimate of the target position as 0.93. Hence the unnormalized estimate of the filtered target position is given by 0.93 (50 m) = 46.5 m. From Figure 2.10-8 it follows that the postflight rms estimate of position is given by 0.73(46.5) = 34 m.

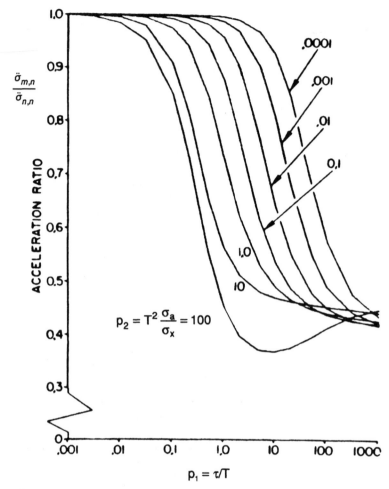

Figure 2.10-10 Fitzgerald's normalized curves providing accuracy to which target's acceleration can be obtained for point somewhere in middle of target track history. Normalization is with respect to filtered acceleration error given in Figure 2.10-7. (After Fitzgerald, R. J., "Simple Tracking Filters: Steady-State Filtering and Smoothing Performance," IEEE Transaction on Aerospace and Electronic Systems, Vol. AES-16, No. 6, November 1980. © 1980 IEEE.)

2.11   SELECTION OF TRACKING FILTER

Generally one would like to use the Kalman filter because it gives the best performance. However, the Kalman filter also imposes the greatest computer complexity. Table 2.11-1 gives a comparison of the accuracy of five tracking filters as obtained for various missions in reference 26. The Kalman filter is seen to provide the best performance. The two-point extrapolator uses the last

TABLE 2.11-1. Synopsis of the Accuracy Comparison of the Five Tracking Filters (After Singer and Behnke [26])

Target Type		Air		Surface or Subsurface
Filter type	Sensor Type	Air Search Radar	Surface and Air Search Radar	Surface and Air Search Radar	Radar, Sonar
Two-point extrapolator		3	3	3	3
Wiener filter		0	0	0	0
g–h Filter		2	0	2	1
Simplified Kalman filter		0	0	0	0
Kalman filter		0	0	0	0

Note: $0 =$ within 20% of Kalman filter; $1 = 20$–40% worse than Kalman filter; $2 = 40$–70% worse than Kalman filter; $3 =$ more than 70% worse than Kalman filter.

TABLE 2.11-2. Comparison of the Computer Requirements for the Five Filters

	Initialization		Main Loop	
Filter Type	Time[a]	Memory Locations[b]	Time[a]	Memory Location[b]
Two-point extrapolator	7	15	7	15
Wiener filter	8	29	21	33
g–h Filter	40	46	44	58
Simplified Kalman filter	51	54	81	71
Kalman Filter	54	67	100	100

[a] Percentage of the computer time required by the Kalman filter in the main loop.
[b] Percentage of the memory locations required by the Kalman filter in the main loop.
Source: From Singer and Behnke [26].

received data point to determine the target range and bearing and the last two data points to determine the target range-rate and bearing rate. This is the simplest filter for estimating target kinematics, Table 2.11-2 gives a comparison of the computer time and memory requirements for the five filters. All the filters of Table 2.11-1 except the two-point extrapolator met the system requirements (of reference 26) relative to prediction accuracy. The constant g–h filter meets the accuracy requirements and at the same time requires very little computer time. Hence this filter might be selected. However, if the radar is in addition required to calculate weapon kill probabilities, then the constant g–h filter would probably not be used. This is because this filter does not by itself provide inputs for the calculation of weapon kill probabilities (like accuracies of the

TABLE 2.11-3. Filter and System Parameters for the Second Example

	Prediction Accuracy (1σ)				Percent of Computer Time Devoted to Tracking	Auxiliary Functions
	Range (yd)	Bearing (deg)	Speed (knot)	Course (deg)		
System requirements	150	0.80	2.5	7.0	10	Kill probability calculations
Two-point extrapolator	210	1.04	3.7	9.7	0.8	—
Wiener Filter	142	0.71	2.4	6.8	2.4	—
α–β Filter	173	0.80	2.8	8.2	5.2	—
Simplified Kalman filter	138	0.65	2.4	6.5	9.5	a
Kalman filter	130	0.63	2.1	6.1	12	a

[a] Tracking accuracy statistics suitable for the auxiliary functions are calculated automatically.
Source: From Singer and Behnke [26].

target predicted location). On the other hand, the Kalman filter provides the accuracy of the predicted position of the target as a matter of course in computing the filter weights. For the constant g–h filter an additional computation has to be added to the standard filter computations. With these additional computations using the constant g–h filter requires as many computations as does the simplified Kalman filter. As a result, the simplified Kalman filter would be preferred for the application where weapon kill probabilities have to be calculated, this filter providing tracking accuracies within 20% of the Kalman filter. The Kalman filter was not selected because its computer time computations were greater than specified by the system requirements for the application of reference 26 and for the time at which the work for reference 26 was performed (prior to 1971). With the great strides made in computer throughput and memory over the last two decades the choice of filter today may be different. Table 2.11-3 gives the comparison of the performance of the various filters for a second example of reference 26.

Another problem with using the constant g–h filter is that it does not provide good performance when the target is maneuvering. However, aircraft targets generally go in straight lines, rarely doing a maneuver. Hence, what one would like to do is to use a Kalman filter when the target maneuvers, which is rarely, and to use a simple constant g–h filter when the target is not maneuvering. This can be done if a means is provided for detecting when a target is maneuvering. In the literature this has been done by noting the tracking-filter residual error, that is, the difference between the target predicted position and the measured position on the nth observation. The detection of the presence of a maneuver

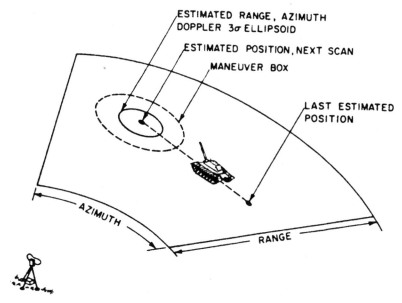

Figure 2.11-1 Nonmaneuvering and maneuvering detection windows. (From Mirkin et al., "Automated Tracking with Wetted Ground Surveillance Radars," Proceedings of 1980 IEEE International Radar Conference. © 1980 IEEE.)

could be based either on the last residual error or some function of the last m residual errors.

An alternate approach is to switch when a maneuver is detected from a steady-state $g–h$ filter with modest or low g and h values to a $g–h$ filter with high g and h values, similar to those used for track initiation. This type of approach was employed by Lincoln Laboratory for its netted ground surveillance radar system [27]. They used two prediction windows to detect a target maneuver, see Figure 2.11-1. If the target was detected in the smaller window, then it was assumed that the target had not maneuvered and the values of g and h used were kept at those of the steady-state $g–h$ filter. Specifically g and h were selected to be 0.5 and 0.2, respectively, in reference 27. If the target fell outside of this smaller 3σ window but inside the larger window called the maneuver window, the target was assumed to have maneuvered. In this case the weights g and h of the filter were set equal to 1.0, thus yielding a filter with a shorter memory and providing a better response to a maneuvering target.

Even when using a Kalman filter, there is still the question as to what value to use for the variance of the maneuver σ_a^2 of (2.9-1). Lincoln Laboratory evaluated a Kalman filter that used $\sigma_a = 0$ when the target was not maneuvering and a nonzero value when a maneuver was detected [28]. A weighted sum of the residuals from the previous observations was used for detecting the presence of a maneuver. When a maneuver was detected, the appropriate change on the filter driving noise variance was made and the filter

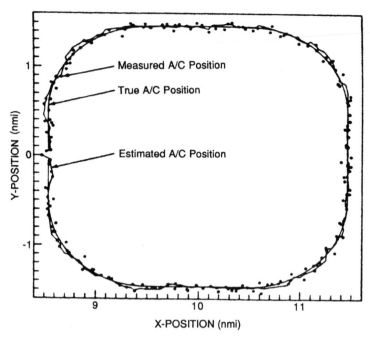

Figure 2.11-2 Performance of Kalman filter in tracking aircraft in holding pattern. Target periodically goes into turn maneuver followed by straight-line trajectory. (From McAulay and Denlinger, "A Decision Directed Adaptive Tracker," IEEE Transactions on Aerospace and Electronic Systems, March 1973. © IEEE.)

reinitialized based on the most recent measurements. Figure 2.11-2 shows how well the filter worked when tracking an aircraft in a holding pattern and doing turns as indicated. The radar was assumed to be 10 miles from the center of the holding pattern shown in Figure 2.11-2 and assumed to have an accuracy of 150 ft in range and 0.2° in angle. The target was assumed to be flying at 180 knots when flying in a straight line. This filter was intended for an air traffic control (ATC) application. Its performance was evaluated by seeing how well it could maintain track on an aircraft in a modified holding pattern involving 90° turns at 3° per second, as shown in Figure 2.11-2. It is seen that the filter worked very well.

Another approach to handling changing target dynamics is to continually change the Kalman filter dynamics based on the weighted average of the most recent residual errors. This approach was used by the Applied Physics Laboratory (APL) [17]. Still other approaches are discussed elsewhere [29].

A final comparison obtained from reference 8 between the performance of a g–h Kalman filter and constant g–h filters will now be given. The g–h Kalman filter design was obtained for the dynamics model given by (2.1-1) or (2.4-2) and (2.4-2a) for which the maneuver covariance excitation matrix is given by (2.4-10). The Kalman filter design was obtained assuming $\sigma_u = 1\,\text{m/sec}$ for

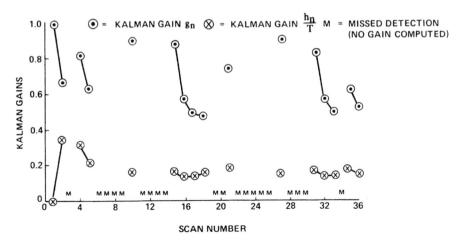

Figure 2.11-3 Two-state Kalman filter simulation; $T = 1$ sec, hence Kalman gain $h_n/T = h_n$. (Reprinted with permission from *Multiple-Target Tracking with Radar Applications*, by S. S. Blackman. Artech House, Inc., Norwood, MA, USA, 1986.)

(2.4-10), $\sigma_x = 5$ m and $T = 1$ sec. The filter was initiated assuming the rms of $x_{0,0}^*$ and $\dot{x}_{0,0}^*$ are 10 m and 5 m/sec, respectively. It was also assumed that a track detection was achieved on each revisit with probability 0.5. Figure 2.11-3 shows the results of a simulation for this Kalman filter. The position gain g_n starts high (0.997) and then oscillates between about 0.5 and 0.9. The velocity gain h_n starts at zero, goes to 0.3 for the next two observations, and then remains at about 0.15. The effects of missed detections on g_n and h_n is clearly indicated in Figure 2.11-3, g_n in increasing significantly after missed detections.

Based on the results of Figure 2.11-3 a constant g–h filter was designed as an approximate replacement for the Kalman filter [8]. For this approximate filter $g = 0.7$ and $h = 0.15$ were chosen. It was found that the approximate constant-gain filter had a $\sigma_{n+1,n}$ that was only about 5% larger than obtained for the Kalman filter. For a larger σ_u, $\sigma_{n+1,n}$ increased at least 10%. It was found that using a Benedict–Bordner filter design or a g–h filter design based on reference 15 gave severely degraded performance. The difference is that the Kalman filter design, and as a result its approximate replacement, picked g and h to compensate for missed data. Hence if a constant g–h filter or g–h–k filter is to be used for a tracking application where there can be missed updates with high probability, the weights should be determined taking into account the possibility of missed updates as done above.

With the enormous advances made in computer throughput, some have argued that g–h and g–h–k filters should never be used, that we should now use just Kalman filters [30,31]. This point of view may be a bit extreme. Reference 8 argues for using simplifications and approximations to the Kalman filter in an attempt to reduce computational requirements significantly without degrading tracking performance. One technique is the use of a table lookup for the weights, as described in Section 2.7. Another is the approximation of the Kalman filter weights using fixed weights that account for the possibility of missed updates, as done above for the example of Figure 2.11-3. With this approach the full-blown Kalman filter is simulated with missed updates to determine the approximate steady-state weights to use. Still another approach is the use of a lower order filter, such as a g–h instead of a g–h–k [8]. The reader is referred to reference 8 for further discussions on obtaining approximations to the Kalman filter. Certainly studying and designing g–h and g–h–k filters for a given application will give a physical feel for what to expect from a Kalman filter.

2.12 Conditions for Which Kalman Filter is Optimal

The minimization of the quadratic form of J in (2.6-7) is optimal, and hence the resulting Kalman filter is optimal, in the sense that it provides an unbiased minimum variance estimate of $X_{n,n}^*$ from the class of all nonlinear as well as linear estimates as long as $Y_0, Y_2, \ldots, Y_{n+1}, Y_n$ have a Gaussian distribution and are independent of each other and the initial condition X_n. In this case, the optimum estimate is a linear function of the measurements, i.e., of $Y_0, \ldots, Y_n$. In the literature, this filter is sometimes referred to as a Minimum Mean Square Estimator (MMSE) [141]. (The plant noise U_n also needs to have a Gaussian distribution and U_n has to be independent of U_m for all $n \neq m$.) For this case, the exponent of the a posteriori probability distribution is given by J so that minimizing J is tantamount to maximizing the a posteriori distribution and therefore also yields a MMSE [141]. If $Y_0, \ldots, Y_n$ (and the U_n) do not have a Gaussian distribution, then the Kalman filter is merely the optimal MMSE among the class of all possible linear estimates only (excluding possible benefits of a nonlinear filter or nonlinear estimator). It is often called the linear MMSE (LMMSE) [141].

2.13 Other Forms of Kalman Filter

To reduce the sensitivity of the Kalman filter to computer errors other forms of the Kalman filter than those of Table 2.4-1 exist. One is the "Joseph form" which uses the following form for the covariance update, instead of (2.4-4j), [119]

$$S_{n-1,n-1}^* = (1 - H_{n-1}M)S_{n-1,n-2}^{*-1}(1 - H_{n-1}M)^T + H_{n-1}R_{n-1}H_{n-1}^T \qquad (2.13-1)$$

This form requires a greater number of computations (multiplies and adds) then does (2.4-4j). For the Kalman filter the covariance matrix update calculation is done usually in double precision to maintain numerical accuracy independent of the Kalman filter form. Typically the rest of the computations are done using single precision [119]. The square-root Kalman filter algorithm discussed in Section 14.5 will provide the same numerical accuracy as does the Joseph form with a reduced computer word length while using an equivalent number of computations. Finally, the U-D covariance factorization form of Section 14.5 provides the same numerical accuracy with a smaller computation load. Reference [158] found the Joseph form to be important for the extended Kalman filter discussed in Chapter 16.

3

PRACTICAL ISSUES FOR RADAR TRACKING

3.1 TRACK INITIATION AND CLUTTER REJECTION

The problems of clutter rejection and track initiation are very much interwined. It is possible to use track initiation to help in the rejection of clutter. On the other hand it is possible to use appropriate clutter rejection techniques to reduce the track initiation load. Examples of these are discussed in the following sections.

3.1.1 Use of Track Initiation to Eliminate Clutter

The radar track initiator can be used to eliminate clutter by passing the clutter returns as well as target returns, which are indistinguishable initially, into the track initiation filter. However, only those returns that behave like a moving target would be passed into track at the output of the track initiation filter, thus finally eliminating the clutter returns. For example, the stationary clutter returns would be dropped by the track initiation filter (or classified as clutter returns for association with future such returns).

The use of this approach can in some cases potentially provide about an order of magnitude or more increase in system sensitivity. This is the case when dealing with spiky clutter. In order to achieve a low false-alarm probability due to spiky clutter returns at the input to the track initiation filter, it is necessary that the detector threshold be increased by 10 dB or more above what would be required if the clutter were not spiky, that is, if the clutter were Rayleigh distributed at the output of the receiver envelope detector (which implies that it has a Gaussian distribution at the input to the envelope detector). This is shown to be the case in Figure 3.1-1 for spiky sea clutter, rain clutter, lognormal

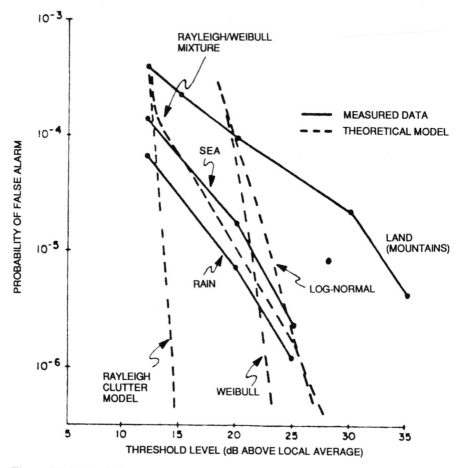

Figure 3.1-1 Typical measured and theoretical clutter probability distributions. (After Prengaman, et al., "A Retrospective Detection Algorithm for Extraction of Weak Targets in Clutter and Interference Environments," IEE 1982 International Radar Conference, London, 1982.)

clutter, and Weibull clutter for low probabilities of false alarm of 10^{-6}. For mountain land clutter the situation is even worse. Here the threshold has to be increased 20 dB for a 20-dB worse sensitivity than for Rayleigh clutter when a low probability of false alarm of 10^{-5} is desired. Extrapolating the curve of Figure 3.1-1 indicates that if a higher false-alarm probability of 3×10^{-3} is used, then there is essentially no loss of sensitivity for spiky sea and rain clutter and a low loss for lognormal clutter and Weibull clutter. The idea is to use a higher false-alarm probability for detection and to then use the track initiator to eliminate the false alarms resulting from the spiky clutter.

The APL retrospective detection is an example of such a system. It uses the track initiator to eliminate spiky sea clutter for an ocean surveillance radar without penalizing system sensitivity [32, 33]. In fact system sensitivity is increased because scan-to-scan video integration is used. For the retrospective detector the returns from eight successive scans are examined to determine if they form a moving-target track. If they do, a target is declared present. If they do not, then it is assumed that only clutter was observed. (As we shall see shortly, the retrospective detector even permits the detection of stationary targets, such as buoys, and zero Doppler targets.) The details of the procedure now follows.

The radar coverage is broken down into small range and bearing regions. For each of these regions the returns from eight scans are stored. By way of illustration, Figure 3.1-2 plots an example set of such returns seen as range versus bearing and time. These are the returns from one of the range–bearing regions mentioned. It is next determined if any set of these returns correspond to a target having a constant velocity. Return numbers 1, 4, 6, 10, 12, and 14 form the returns from such a constant-velocity target, the target having a velocity in the band between 28 and 35 knots. The rule used for declaring a target present in a Doppler band is that M out of N returns be detected in the band. Here $N = 8$ and typically M would be of the order of 5.

The echo from a clutter spike generally will not generate a false-alarm return at the output of the retrospective detector. This is because a spiky echo range resolution of 50 ft and azimuth beamwidth of $0.9°$. Its scan period is 4 sec with no coherent Doppler processing being used. Figure 3.1-3 shows the raw data observed in the lower southwest quadrant after constant false-alarm rate (CFAR) processing. The returns from 100 scans are displayed in the figure. The coverage is out to a range of 9 nmi. A digital range-averaging logarithmic CFAR was used by the system for the results shown in Figure 3.1-3. The results shown in Figure 3.1-3 were obtained without the use of the retrospective detector. A high false-alarm rate, about 10^{-3}, was used in obtaining the results of Figure 3.1-3. This resulted in about 2000 false alarms per scan. In this figure the sea clutter false alarms are indistinguishable from small target returns resulting from a single scan. Figure 3.1-4 shows the results obtained after retrospective detector processing; again 100 scans of data are displayed. The retrospective detector has reduced the false-alarm rate by at least four orders of magnitude. The ships and boats in the channel are clearly visible. The reduction in the false-alarm rate provided by the retrospective detector is further demonstrated in Figure 3.1-5, which displays 1000 scans of data (about 1 h of data). Very few false alarms are displayed.

The echo from a clutter spike generally will not generated a false-alarm return at the output of the retrospective detector. this is because a spiky echo return typically does not have a duration of more than about 10 sec [34–36]. If it did, a retrospective detector with a longer integration time and higher threshold M would be used. As mentioned, the detector even allows the detection of targets having a zero Doppler velocity. Such returns could arise from ships moving perpendicularly to the radar line of sight and from buoys. The detection of buoys is essential for shipboard radars navigating shallow waters.

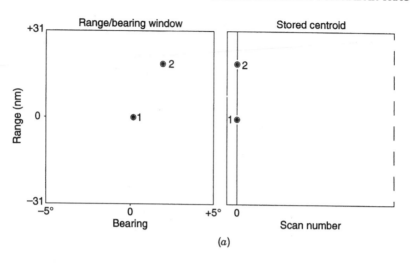

(a)

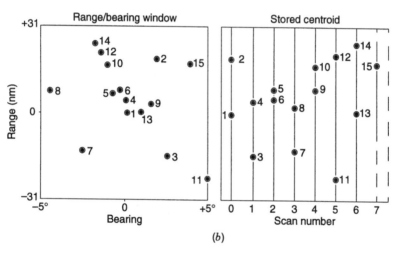

(b)

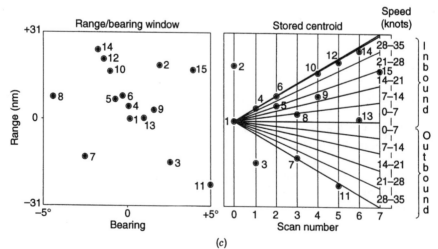

(c)

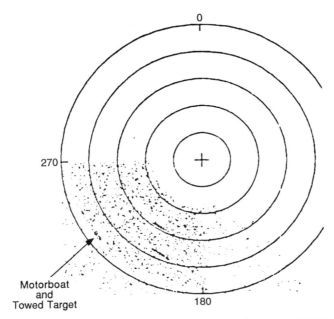

Figure 3.1-3 Raw data observed in lower southwest quadrant after CFAR processing (100 scans of data; 9 nmi total range). (After Prengaman, et al., "A Retrospective Detection Algorithm for Extraction of Weak Targets in Clutter and Interference Environments," IEE 1982 International Radar Conference, London, 1982.)

The above data showed the effectiveness of the retrospective detector in detection of targets in an ocean clutter. This type of detector would also be very effective in detecting targets over spiky land clutter, potentially reducing or eliminating the large detection sensitivity loss otherwise needed to eliminate spiky clutter.

The hardware implementation of the above retrospective detector for the AN/FPS-114 radar consisted of 6 × 6-in. wire wrap cards containing 250 small and medium-scale integrated circuits (late 1970s/early 1980s technology). The total power consumption is 30 W. This implementation is limited to 2000 contacts per scan by the memory size available. With the use of modern very large scale integrated (VLSI) circuitry, the size of the signal processor will decrease and its capability increase. Table 3.1-1 summarizes the characteristics of the retrospective detector and its performance.

Figure 3.1-2 Operation of retrospective processor: (*a*) returns from single scan; (*b*) returns from Eight Scans; (*c*) eight scans of data with trajectory filters applied. (After Prengaman, et al., "A Retrospective Detection Algornithm for Extraction of Weak Targets in Clutter and Interference Environments," IEE 1982 International Radar Conference, London, 1982.)

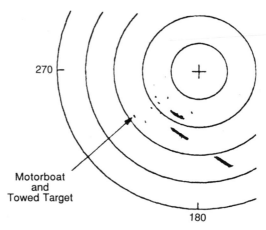

Figure 3.1-4   Results obtained after retrospective detector processing using 100 scans of data. (After Prengaman, et al., "A Retrospective Detection Algorithm for Extraction of Weak Targets in Clutter and Interference Environments," IEE 1982 International Radar Conference, London, 1982.)

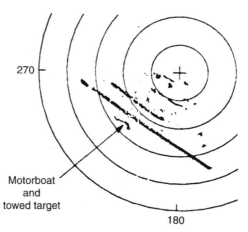

Figure 3.1-5   Retrospective detector output after 1000 scans of data (about 1 hr of data). (After Prengaman, et al., "A Retrospective Detection Algorithm for Extraction of Weak Targets in Clutter and Interference Environments," IEE 1982 International Radar Conference, London, 1982.)

3.1.2   Clutter Rejection and Observation-Merging Algorithms for Reducing Track Initiation Load

In this section we shall describe how clutter rejection and observation-merging algorithms are used to reduce the track initiation load for a coherent ground two-dimensional surveillance radar. A two-dimensional radar typically is a radar that has a vertically oriented narrow fan beam (see Figure 1.1-1) that is

TABLE 3.1-1. Retrospective Detector

Radar demonstrated on:
  S-band AN/FPS-114 at Laguna Peak, CA
  Resolution: 0.1 μsec ×0.9°
  4-sec scan-to-scan period
  1500 ft altitude

Retrospective processor: special purpose, consisting of six 6 × 6-in. wire wrap cards containing 250 small and medium integrated circuits; total power: 30 W (Late–1970s/early–1980s technology.)

Performance results: with single-scan false-alarm rate set at 2000 per scan, after 100 scans false-alarm rate reduced by at least four orders of magnitude, after 1000 scans ($\sim$ 1 hr) only a few alarms visible

---

scanned 360° mechanically in azimuth about the local vertical axis [1]. Such a radar provides two-dimensional information: slant range and the bearing angle to the target. (These algorithms are also applicable to three-dimensional radars, that is, radars that measure target range, azimuth, and elevation simultaneously as done with the GE AN/FPS-117 [1, 37], the Marconi Martello stacked beam radars [1, 38–40], and the Westinghouse ARSR-4 [1].) The next four paragraphs will describe how the Lincoln Laboratory moving-target detector (MTD) Doppler suppression technique works [41, 42].

3.1.2.1 Moving-Target Detector

For a radar using the Lincoln Laboratory MTD clutter rejection technique, the radar pulse repetition rate (PRF), that is, the rate at which the radar pulses are transmitted measured in pulses per second, and antenna scan rate are chosen so that the target is illuminated by more than $2N$ pulses on one scan across the target. For the first N pulses PRF $=$ PRF$_1$ is used. For the second set of N pulses PRF $=$ PRF$_2$ is used. The purpose of the two PRFs is to remove pulse-Doppler ambiguities and blind velocities that result if only are PRF is used as shall be explained shortly. Briefly the pulse Doppler velocity ambiguity and blind velocities arise because typically the PRFs used are too low to permit unambiguous pulse Doppler velocity measurements, that is, the PRF is lower than the Nyquist sampling rate needed to measure the maximum Doppler shift from the target unambiguously.

An MTD system is coherent, meaning the transmitted train of N pulses can be thought of as generated by gating a stable radio frequency (RF) oscillator. The system can be thought of as a sample data system where the sampling is done on transmit. One could just as well have transmitted a continuous wave (CW) and done the sampling on receive. However, range information would then not be available. In order not to have range ambiguities, the pulse-to-pulse period is made larger than the round-trip distance to the longest range target from which echoes are expected. For example, if the maximum range for the radar is 100 nmi, then the pulse-to-pulse period would be (100 nmi) (12.355 nmi/µsec) = 1235 µsec or greater. The system PRF would then be 1/1235 µsec = 810 Hz. For a 1.3-GHz carrier frequency L-band radar an approaching target having a Doppler velocity of 182 knots would give rise to a Doppler-shifted echo having a Doppler shift equal to the PRF of 810 Hz. Because we have in effect a sampled data system with a sample data rate of 810 Hz, any target having a target velocity producing a Doppler shift higher than the sampling rate would be ambiguous with a target having a Doppler shift lower than the sampling rate. For example, a target having a Doppler velocity of 202 knots would appear as a Doppler-shifted signal produced by a target having a Doppler velocity of 202 knots modulo the sampling rate of 182 knots, or equivalent 20 knots. Thus we would not know if the target actually was going at 202 knots or 20 knots, hence the ambiguity. The use of the second PRF for the second set of N pulses removes this ambiguity problem. This is done by the application of the Chinese remainder theorem.

A blind velocity problem also arises from the Doppler ambiguity problem. If the target had a Doppler velocity of 182 knots, that is, equal to the sampling rate, then it would be ambiguous with the zero Doppler echo returns. But the zero Doppler returns are primarily from strong ground clutter. Hence, the echo from a 182-knot target is ambiguous with the strong ground clutter. As a result, often if would be most likely masked by the ground clutter return. This results in the second reason for use of the second set of N pulses at a second PRF. A target ambiguous in Doppler with the zero Doppler clutter on the first PRF will not be ambiguous on the second PRF. This same problem occurs for near zero Doppler rain clutter and a Doppler ambiguous aircraft echo. The rain could mask the ambiguous aircraft echo. This masking also is eliminated by the second set of N pulses having the second PRF.

A typical value for N is 8. To measure the Doppler velocity of the target in a given range cell, the $N = 8$ echoes from this range cell are passed into a bank of $N = 8$ narrow Doppler filters covering the band from 0 to 182 knots, with each filter having a bandwidth of (182 knots)/8 = 22.8 knots. The nonzero Doppler filters would have a frequency transfer characteristic with a notch at zero Doppler frequency so as to better reject the zero Doppler ground clutter while at the same time passing the signal in the Doppler band of that filter. One would think that the output of the zero Doppler filter is ignored, having the strong ground clutter return. Actually it is not ignored. Instead the filter centered at zero Doppler is used to detect aircraft targets moving perpendicularly to the radar line of sight, which as a result have a zero Doppler velocity. Such targets can often be detected in the clutter because when the target has a zero Doppler velocity it is being viewed broadside. For this aspect angle the target generally has a very large cross section. Hence it is possible that its return echo will be stronger than that of the ground clutter. When this occurs, the target is detected. To prevent the detection of ground clutter echoes in the zero Doppler filter, the threshold in this filter is set higher than the ground clutter. The setting of this threshold varies with range and azimuth since the clutter strength varies with range and azimuth. To determine what value the threshold should be set at for a given range–azimuth cell, a clutter map is generated. For this clutter map the strength of the clutter for each range–azimuth cell of the radar is stored in memory. Typically the power of the echo in a particular range–azimuth cell from the last H scans (where H might be of the order of 7 to 10) are averaged to generate the clutter strength for this cell. An exponentially decaying average is typically used for ease of implementation, it then being possible to implement the filter with a simple feedback infinite-impulse response filter rather than the more complicated finite-impulse response filter. The above described processor is the Lincoln Laboratory MTD [41, 42].

Lincoln Laboratory first implemented an MTD for the FAA experimental AN/FPS-18 air traffic control radar at the National Aviation Facilities Engineering Center (NAFEC). It was installed in 1975. The FPS-18 is an S-band (2.7 to 2.9 GHz) radar having a PRF of 1000 to 1200 Hz [41, 42, 44]. For this radar, in effect, $N = 8$, with eight Doppler filters used to process eight echo pulses. The coherent processing of a set of $N = 8$ echoes having a specified PRF is called a coherent procesing interval (CPI). There are thus eight Doppler outputs per range cell per CPI. Figure 3.1-6 shows a typical system

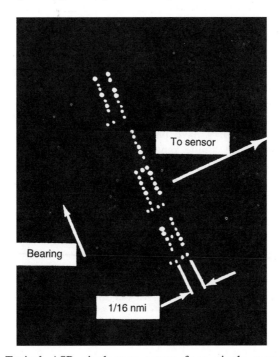

Figure 3.1-6 Typical ASR single-scan return from single target. Radar range resolution was $\frac{1}{16}$ nmi while Doppler resolution was about 16 knots. (From Castella, F. R. and J. T. Miller, Jr., "Moving Target Detector Data Utilization Study," IEE Radar—77, London, 1977.)

return for a single target resulting from a single scan across the target with a two-dimensional TWS radar. Each dot represents a Doppler filter crossing. For the first range cell return from the target there are four CPIs observed for the target. For the next range cell three CPI detections are observed for the same target. For each CPI the target is also detected in more than one Doppler cell. In some cases all eight Doppler cells show detections. Apparently a low detection threshold was used for the illustrated scan given.

3.1.2.2 Observation Merging (Redundancy-Elimination, Clustering)

Castella and Miller [44] noticed that in general, for a single scan across a target, detection occurred for more than one range–Doppler–azimuth cell. It is imperative that all these detections from the same targets not be reported as separate target returns as this would lead to an overload of the tracker. The association of such detections, called redundant detections, with a single target and reporting them as a single range–Doppler–bearing report for one target is called observation merging [8], redundancy elimination [8], and clustering [16]. Techniques to be used for observation merging are discussed in detail in reference 8.

Redundant detections, while being a problem, can be used to effectively eliminate false clutter returns. Very effective algorithms developed by Castella and Miller (of the Johns Hopkins University APL) for doing this are described in reference 44. These are now summarized.

Castella and Miller found that aircraft targets on which there is a firm track typically give rise to responses in two or more of either range cells, Doppler cells, or azimuth cells during one pass across the target. Clutter returns on the other hand typically give rise to only one range, Doppler, or azimuth cell response. As a result they suggested the use of this characteristic to eliminate clutter returns before they have passed onto the track initiator. Table 3.1-2 shows the statistics for single range, Doppler, and bearing CPI returns for aircraft targets for which there is a firm track and for cell echoes which are primarily comprised of clutter returns. About 79% of the clutter echoes consist of only one range cell response, one Doppler cell response, or one bearing cell response. In contrast, for aircraft targets for which there is a firm track, only 15

TABLE 3.1-2. Sample Centroid Statistics

Characteristics	Percent of All Centroids	Percent of Firm Track Centroids Only
Number of CPIs $= 1$	78.7	15.2
Maximum number of Doppler cell detections per CPI per range cell $= 1$	78.7	13.7
Maximum range extent $= 1$	79.3	21.0
Total centroids considered	34,445.0	3485.0

Source: After Castella and Miller [44].

to 21% of the echoes will consist of only one range cell response, one Doppler cell response, or one bearing response. Figure 3.1-7 shows how effectively clutter returns can be eliminated by using an algorithm that requires two or more bearing returns (that is, CPI returns) for the target to be declared detected. Figure 3.1-7a shows all the target detections obtained in a 50-sec interval if a target detection is based on the observation of one CPI return. Figure 3.1-7b shows the radar target detections if it is required that two or more CPIs be observed. More than 75% of the original centroids have been eliminated using this simple algorithm. Yet the number of target tracks is seen to be virtually identical for both displays.

Figure 3.1-8 shows the benefits accrued from using the following two constraints for declaring a target detection: (a) detection on two or more CPIs and (b) detection on two or more Doppler cells per range–bearing cell. Figure 3.1-8a gives the resulting detections of 335 scans when no constraints are used for declaring a detection; that is, detection is based on observing one or more range–Doppler CPI detections. For this case 54 tracks were observed. Figure 3.1-8b gives the results when the above two constraints were used. The number of tracks observed was reduced from 54 to 44. However, the mean track duration increased from 39.1 to 47.6 sec. Reference 44 concludes from results like these that only insignificant short-term clutter tracks are being eliminated. More significantly, the number of tentative tracks were reduced from 14,000 to a little over 800, a 17:1 reduction in the track load. Table 3.1-3 summarizes the results obtained with the above two constraints.

3.1.3 Editing for Inconsistencies

Sometimes the clutter returns will overlap the signal returns and corrupt the range, Doppler, and bearing estimations for the target. It is desirable to know when the data are corrupted. Corrupted data could then be appropriately treated in the track update filter, either by not using it or by giving it a low weighting for determining the next update position. The presence of such corruption can be determined by looking for inconsistencies in the data [44]. For example, the presence of inconsistencies in the Doppler between adjacent range cells would be indicative of possible corruption from clutter. Another possible reason for this inconsistency could be that two targets are overlapping. In either case noticing such inconsistencies alerts one to the fact that the Doppler estimate is being corrupted. Similarly, inconsistencies in the Doppler for successive CPIs would be indicative of possible clutter corruption. Finally inconsistencies between the amplitudes of the return between successive CPIs in one scan would also be indicative of possible clutter corruption.

3.1.4 Combined Clutter Suppression and Track Initiation

It is apparent from Sections 3.1.1 and 3.1.2 that clutter suppression and track initiation go hand in hand. Often the engineer generating the clutter suppression

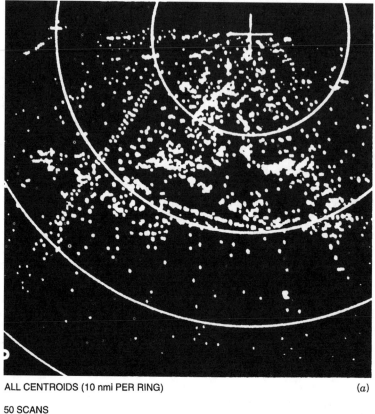

ALL CENTROIDS (10 nmi PER RING)   (a)

50 SCANS

2880 CENTROIDS

Figure 3.1-7 (*a*) Target detections obtained in 50-sec interval when target detections are based on observation of only one CPI return. (*b*) Target detections when two or more CPIs are required for declaring its presence. (From Castella, F. R. and J. T. Miller, Jr., "Moving Target Detector Data Utilization Study," IEE Radar—77, London, 1977.)

algorithms works independently of the engineer generating the track initiation algorithms. The two must work hand in hand to achieve best results. The literature is rich in clutter rejection techniques, and it is the designers task to pick those most applicable to a situation; for example, see References 16, 27, 29, 32, 33, 41, 42, 44 to 61, and 135.

The clutter suppression and track initiation algorithms act as filters for reducing the number of false tracks. Figure 3.1-9 illustrates this for a system designed by APL [56]. This system applies various algorithms for suppressing real-world clutter for a two-dimensional shipboard surveillance radar. For this system there are 6×10^6 samples per scan. These samples are filtered down to just 392 initial contacts. After further screening these in turn are filtered down

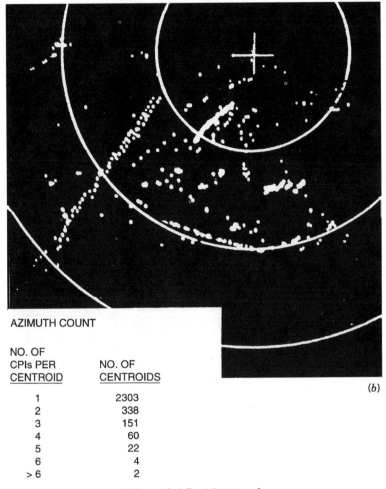

AZIMUTH COUNT

NO. OF CPIs PER CENTROID	NO. OF CENTROIDS
1	2303
2	338
3	151
4	60
5	22
6	4
> 6	2

(b)

Figure 3.1-7   (*Continued*)

to 35 possible track updates and 6 tentative tracks. On the average, for the system under consideration only 1 out of 15 of the tentative tracks becomes a firm track after further processing. The 35 possible track updates are filtered down to just 21 firm track updates.

Feedback can be used at various stages of a well-designed system in order to reduce the number of false tracks and enhanced the target detectability. The enhancement of target detectability is achieved by lowering the threshold in prediction windows where the track updates are expected to appear. This procedure is referred to as "coached" detection; see reference 8.

Feedback for controlling the threshold over large areas of coverage have been used in order to lower the false-alarm rate per scan [16]. This is called area

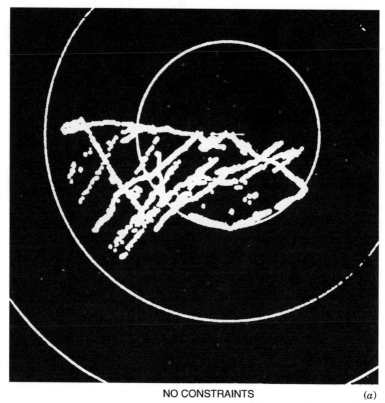

NO CONSTRAINTS   (*a*)

Figure 3.1-8 (*a*) Detections observed after 335 scans when target detection is based on observation of one or more range–Doppler CPIs. (*b*) Detections observed when two constraints are imposed: That two or more CPIs be detected and that detections on two or more Doppler cells per range–bearing cell be observed. Number of target tracks reduced from 54 to 44. Mean target track duration increased from 39.1 to 47.6 sec. Number of tentative tracks reduced from 14,000 to 800—17:1 track load reduction. (From Castella, F. R. and J. T. Miller, Jr., "Moving Target Detector Data Utilization Study," IEE Radar—77, London, 1977.)

CFAR control. It was used in the Lincoln Laboratory ATC experimental ASR-7 MTD processor to eliminate heavy angel clutter (birds, insects, etc.); see Figure 3.2-10. Figure 3.1-10*a* shows the radar display without the use of the feedback adaptive thresholding, that is, without area CFAR. Illustrated is a situation of heavy angel clutter. Figure 3.1-10*b* shows the same display with the area CFAR being used.

Two types of feedback were used in above Lincoln Laboratory system. The first, fast feedback having a time constant of about 5 sec, was applied to control the threshold in an area consisting of all ranges less than 20 nmi. The data from three Doppler bins were used. Its purpose was to respond quickly to the onset of

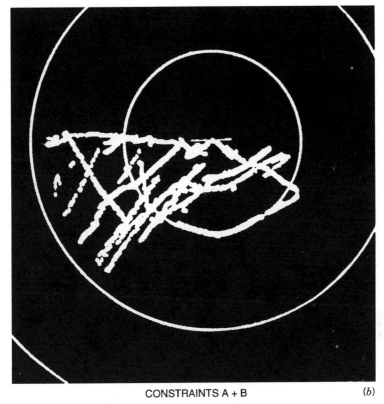

CONSTRAINTS A + B   (*b*)

Figure 3.1-8   (*Continued*)

TABLE 3.1-3. FPS-18 MTD System Results

Clutter removal algorithm
    Require detection on ≥ 2 CPIs
    Require detection on ≥ 2 Doppler cells per range–azimuth cell

Results
    Number of firm tracks decreased from 54 to 44 in 335-scan run
    Number of tentative tracks during entire run reduced from 14,000 to 800 — a
        17:1 track load reduction

Source: From Castella and Miller [44].

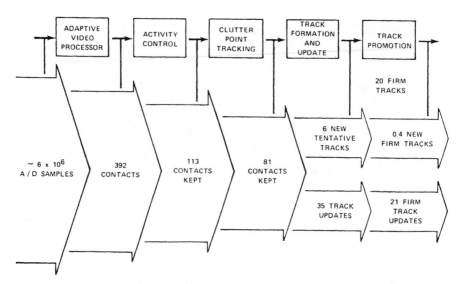

Figure 3.1-9 Joint action of clutter suppression and track initiation as filters for reducing number of false alarms due to severe clutter and radio frequency interference (RFI). (From Bath et al., "False Alarm Control in Automated Radar Surveillance Systems," IEE 1982 International Radar Conference, Radar—82, London, 1982.)

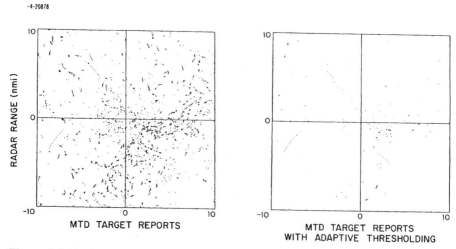

Figure 3.1-10 Improvement provided by feedback area CFAR processing: (*a*) radar display without area CFAR; (*b*) display with area CFAR. (From Anderson, J. R. and D. Karp, "Evaluation of the MTD in a High-Clutter Environment," Proceedings of IEE International Radar Conference, © 1980 IEEE.)

false alarms due to angel clutter or other clutter. Nominally the system was designed so as to yield no more than about 60 false alarms per scan.

The second feedback threshold control was used for controlling the false-alarm rate over smaller regions, regions $16\,\text{nmi}^2$ by three Doppler cells. The time constant for these regions is much longer, being about 200 sec. Its purpose was to eliminate localized long-duration false returns. The stationary discrete clutter returns that exceeded these two thresholds were passed along to the tracker, where they were eliminated by using a logic that dropped all returns that did not move more than $\frac{1}{4}\,\text{nmi}$ from the position of track initiation.

Techniques for tracking through clutter are discussed in detail in references 6, 8, and 9.

3.2 TRACK-START AND TRACK-DROP RULES

The question arises as to when does one declare that a target is finally in track after track initiation has started. The question also arises as to when a target that has been in track and is no longer in the radar coverage region is declared not to be in track. A typical simple rule for declaring that a target is in track is the detection of M out of N observations, where, for example, $M = 3$ and $N = 5$ [8]. When Doppler information is available, confirmation based on one additional observation is sufficient often [8]. This is because the Doppler information allows us to use a very small prediction window for the confirmation observation. Also, with the additional dimension of velocity, the likelihood of two false alarms creating a track is unlikely. Reference 8 gives details of a sophisticated track detection rule that is dependent on the details of the track initiation history. A sequential probability ratio test (SPRT) and a Bayes rule detection criterion are used for establishing a track.

Track deletion can be based on a simple rule consisting of deciding that the track has ended if there are M consecutive misses. Typically M is of the order of 3 or 4, the value of 3 being used in the Lincoln Laboratory MTD processor of reference 16. Alternately a sequential test could be used similar to the one used for establishing a track [8]. Finally, a measure of the quality of the track could be used. If the quality falls below some value, then the track would be dropped [8]. The quality of the track can be based on some function of the tracking-filter residue over the last few observations.

3.3 DATA ASSOCIATION

3.3.1 Nearest-Neighbor Approach

As indicated previously, a prediction window is used to look for the target return on the next track update. The window is made large enough so that the probability is high enough (e.g., $> 99\%$) that the target will be in this window

[8, 62]. If a return is detected in this prediction window, then it is associated with the track that established this prediction window. A problem arises, however, when more than one return falls in the prediction window either due to clutter or the presence of a nearby target. The question then arises of with which return to associate the track.

One solution is to use the nearest-neighbor approach. First we address the question as to how to measure distance. One measure is the statistical distance given by [62, 63]

$$D^2 = \frac{(r_m - r^*)^2}{\sigma_r^2} + \frac{(\theta_m - \theta^*)^2}{\sigma_\theta^2} \tag{3.3-1}$$

where r_m and θ_m are respectively the range and bearing measurement at the time of the track update, r^* and q^* are respectively the predicted range and bearing angles for the track update period in question, and finally σ_r^2 and σ_θ^2 are respectively the variances of $r_m - r^*$ and $\theta_m - \theta^*$. The statistical distance is used in place of the Euclidean distance [which (3.3-1) gives when the variance σ_r^2 and $\sigma_\theta^2 = 1$] because of the large differences in the accuracies of the different coordinates, the range accuracy usually being much better than the bearing accuracy.

The example of reference 62 [also 63 and 64] will be used to further illustrate the nearest-neighbor approach. Figure 3.3-1 shows the example. For this example there are two detections in window 1, three in window 2, and one in window 3. Table 3.3-1 gives the statistical distances for each track relative to the detections in their prediction window, these detections being given in order of increasing distance. Initially the closest detections are associated with each track. Next the conflict resulting from associating a detection with more than

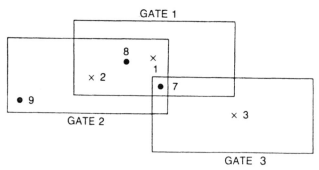

Figure 3.3-1 Difficulties caused by occurrence of multiple detections and tracks in close proximity. (From Cantrell, Trunk, and Wilson [69]; sec also Trunk [62–64].)

TABLE 3.3-1. Association Table for Example Shown in Figure 3.3-1

Track Number	Closest Association Detection Number	D^2	Second Association Detection Number	D^2	Third Association Detection Number	D^2
1	8	1.2	7	4.2		
2	8	3.1	7	5.4	9	7.2
3	7	6.3				

Source: From Trunk [64].

one track is resolved. For example, detection 8 associated with tracks 1 and 2. It is decided to associate detection 8 with track 1, it having the smaller distance. The association of detection 8 with all of the other tracks is then eliminated, in this case track 2. Next consider detection 7. It is in the windows of tracks 1, 2, and 3. However, detection 7 is not associated with track 1 because track 1 has already been associated with detection 8. This leaves detection 7 with tracks 2 and 3. This conflict is eliminated by associating detection 7 with track 2 because it is closer to track 2. This leaves track 3 with no detections for an update. Hence, track 3 is not updated at this time.

An alternate logic exists that assigns a track to each detection. With this logic track 3 is updated with detection 7, it being the only detection in the track 3 prediction window. As before detection 8 is associated with track 1. This leaves detection 9 for updating track 2. Reference 2 elaborates more fully on the nearest-neighbor association procedure.

3.3.2 Other Association Approaches

With the nearest-neighbor approach described above a decision is made relative to which track a detection is to be associated with at the time of the update, a decision that is irrevocable. A more optimum approach is to defer a decision on one or more scans. This could be done for all track updates or at least for the cases where association is difficult. One simple version of this approach is called branching or track splitting [8]. A more sophisticated approach uses multiple probability hypothesis testing. Other versions of this test are the all-neighbors data association approach, probabilistic data association [PDA], and joint probabilistic data association [JPDA] [8]. These techniques provide better tracking through heavy clutter [8]. Reference 8 presents a simplified recursive approximation to the probabilistic hypothesis testing approach. Future tracking can be expected to go in the direction of the more optimum but more complex hypothesis testing association techniques as very high speed integrated circuit (VHSIC) and VLSI technology [1, 65, 66] increase the radar computer capability.

In the future target signatures such as target length, target cross section, target shape, and target-induced modulation, such as through airplane vibration and jet engine modulation (JEM) [8, 47, 67], will be used to aid in keeping track of targets. This is especially true when tracking through outages, such as due to ground shadowing, or crossing targets, or targets in a dense environment. The use of target shape for aiding in the tracking of ocean surface ships is presented in reference 68.

3.4   TRACK-WHILE-SCAN SYSTEM

A typical TWS system is the fan-beam surveillance radar system described in Section 1.1 and shown in Figure 1.1-1. For this type of system the fan beam periodically scans the 360° search region (such as once every 10 sec) looking for targets. When a target is detected, it is put into track in the receiver processor. A new track update is made (if the target is detected) every 10 sec each time the search volume is searched, hence the name track while scan.

A methodology in which the clutter, tentative, and firm tracks for a TWS system can be efficiently handled is described in this section [62–64]. Assume a clockwise rotation of the antenna beam with it presently searching out sector 12, as illustrated in Figure 3.4-1. While gathering in the data from section 12 it is processing the data from previous sectors; see Figure 3.4-1. The sequencing of this processing of past sectors is selected so as to best eliminate clutter detections from the signal detection file and to prevent tentative tracks from stealing detections belonging to firm tracks.

For the situation shown in Figure 3.4-1, the detections made in sectors 9, 10, and 11 are correlated (associated) if possible with clutter points (stationary

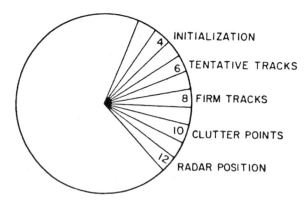

Figure 3.4-1 Parallel operations in TWS system of detection, establishment of clutter scatterers, firm tracks, and tentative tracks, and initiation of new tracks in different sectors of system's coverage. (From Cantrell, Trunk, and Wilson [69]; see also Trunk [62, 63].)

tracks) in the clutter file. Those detections that correlate are declared to be clutter point detections and deleted from the detection file and used instead for updating the clutter points. While this is going on a determination is made if detections in sectors 7, 8, and 9 (for which clutter point detections have already been deleted) correlate with firm tracks in sector 8. For those detections for which such an association is made, they are deleted from the detection file and used to update the corresponding firm tracks. At the same time tentative tracks are being established for sector 6 using detections not associated with firm tracks. In turn the detections not associated with tentative tracks are used to initiate new possible tracks, this processing going on simultaneously in sector 4; see Figure 3.4-1. A track initiation formed on one scan will on a later scan become tentative track and still later possibly a firm track if enough correlating detections are made and if the target has an appropriate nonzero velocity. If the target's velocity is zero, it will be declared a clutter point and tracked as such. On the other hand, if insufficient correlated returns are received on ensuing scans, the track initiation and tentative tracks are dropped.

For some systems an initial detection is used to simultaneously establish a tentative target and clutter track. Future detections will cause one of these to be dropped depending on the target's velocity. Alternately both will be dropped if insufficient correlating detections are observed on later scans. When most of the detections are a result of clutter residues, the approach wherein simultaneously a target and clutter track are started on the detection of a new observation often requires less computer computation.

When a track is established, it is assigned a track number. A set of track parameters are associated with each track or equivalently a track number. For example, assigned with a track may be the parameters of filtered and predicted target position and velocity, time of last update, track quality, covariance matrix if a Kalman filter is used, and track history, that is, the number of track detections. A track number is assigned to a sector for ease of correlation with new data. Clutter returns are treated in a similar way to target tracks, having track numbers, parameters, and sectors assigned to them.

TABLE 3.4-1. Track Number Parameters

Parameter	Description
NT	Track number
DROPT	Obtain (1) or drop (0) a track number NT
FULLT	Number of available track numbers
NEXTT	Next track number available
LASTT	Last track number not being used
LISTT(M)	File whose M locations correspond to track numbers
M	Maximum number of tracks

Source: From Trunk [63, 64].

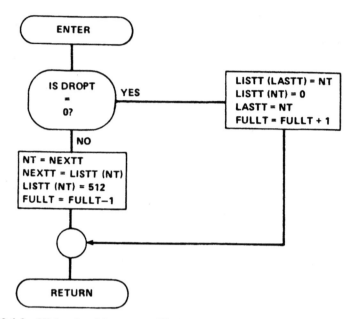

Figure 3.4-2   Higher level languages like Pascal provide pointer systems that permit efficient implementation of flow charts such at shown. (From Cantrell, Trunk, and Wilson [69].)

Table 3.4-1 shows a typical track number file written for a FORTRAN program [63, 64]. Figure 3.4-2 shows the flow chart for the track number file. A similar file and flow chart is developed for the clutter points tracked. Higher level languages like Pascal provide pointer systems that permit the efficient implementation of flow charts such as shown in Figure 3.4-2 [63, 64, 69].

3.5   TRACKING WITH A CHIRP WAVEFORM

The chirp waveform is one of the most extensively used waveforms for rader systems [57, 58, 70]. We shall first describe chirp waveform, pulse compression, and match filtering and then discuss the effects of using such a waveform on tracking accuracy.

3.5.1   Chirp Waveform, Pulse Compression, Match Filtering of Chirp Waveform, and Pulse Compression

Figure 3.5-1 shows the typical chirp pulse waveform. It consists of a pulse having a width T_U that has a linear frequency modulation (LFM) of bandwidth B_s. The LFM chirp waveform is what is power amplified and passed through the transmitting antenna of chirp radar systems. The chirp waveform has the

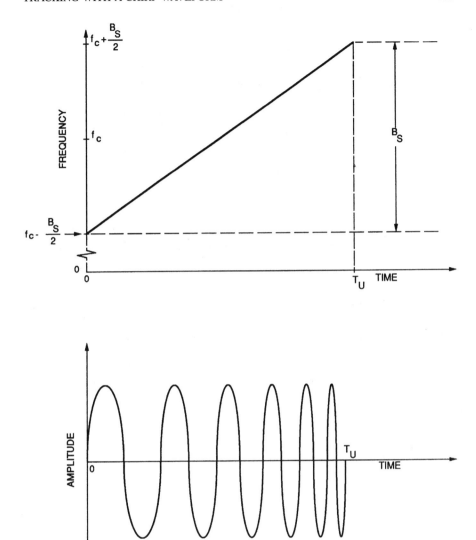

Figure 3.5-1 Chirp waveform.

property that if it is passed through a time dispersive network in the receiver, as shown in Figure 3.5-2, the output of this dispersive network will be a narrow pulse, as indicated. The width of this pulse is equal to 1 over the signal bandwidth B_s. Thus if $B_s = 1 \, \text{MHz}$ and the uncompressed pulse width $T_U = 100 \, \mu\text{sec}$, then the compressed-pulse 4-dB width will be $1 \, \mu\text{sec}$. Thus the output pulse is one-hundredth the width of the input pulse for a pulse

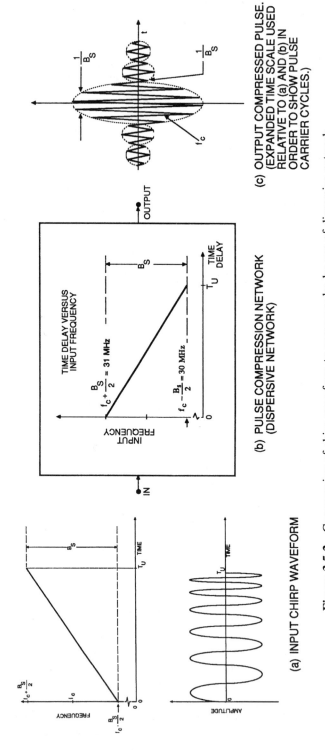

Figure 3.5-2 Compression of chirp waveform to narrow pulse by use of dispersive network. Note that time scale at output of dispersive network is not the same as for its input, timed scale for output expanded for clarity.

(a) INPUT CHIRP WAVEFORM

(b) PULSE COMPRESSION NETWORK (DISPERSIVE NETWORK)

TIME DELAY VERSUS INPUT FREQUENCY

$f_c + \dfrac{B_S}{2} = 31$ MHz

$f_c - \dfrac{B_S}{2} = 30$ MHz

(c) OUTPUT COMPRESSED PULSE. (EXPANDED TIME SCALE USED RELATIVE TO (a) AND (b) IN ORDER TO SHOW PULSE CARRIER CYCLES.)

134

compression of 100:1. This output pulse is called the compressed pulse because it has a compressed width. The envelope of the compressed pulse has a sin x/x shape where $x = \pi t/\tau_c$ and where τ_c is the compressed 4-dB width.

Normally the range resolution of a radar is equal to the width of the transmitted pulse. For the chirp waveform the width $T_U = 100\,\mu\,\text{sec}$, which normally would imply a time resolution of $100\,\mu\,\text{sec}$ or equivalently range resolution of about 8 nmi! Because of the pulse width compression that the chirp waveform echo sees in the receiver, its resolution after pulse compression will be $1\,\mu\,\text{sec}$ or equivalently 0.08 nmi (about 490 ft), a substantial improvement over the uncompressed-pulse width resolution.

One major advantage of the chirp waveform is that it permits one to use a radar transmitter having a low peak power to detect a weak target at a long range. This is because the ability to detect a target is dependent on the transmitted pulse total energy and not on its instantaneous peak power or shape. Thus a long pulse having a low peak power will provide the same ability to detect a small target at long range as a short pulse having a high peak power as long as the two pulses have the same energy. For example, assume that the peak power for the transmitted chirp pulse of Figure 3.5-1 is 1 MW. The total transmitted energy is thus 100 J if $T_U = 100\,\mu\text{sec}$. A standard normal unmodulated pulse having a 1-μsec width (rectangular) would have to have a peak power of 100 MW in order to have the same total transmitted energy as that of the transmitted chirp pulse.

What we have described above is the basis for pulse compression. Generally, pulse compression involves the coding of a long pulse (in the case of the chirp waveform with a LFM coding) that has a bandwidth equal to that of the short pulse that normally would be used to give the desired range resolution for the radar. For a short-pulse uncoded radar the bandwidth is equal to 1 over the pulse width, that is, $B_s = 1/\tau_c$, where τ_c is the pulse width for the uncoded short-pulse radar. Thus the bandwidth for the coded pulse radar has to equal B_s. If the long pulse has a coding that provides a bandwidth of B_s, then after receiver pulse compression its width will be equal to $\tau_c = 1/B_s$, the same as for the narrow-pulse uncoded waveform system, thus having the same resolution as the latter system. The ability to achieve a high range resolution with a long pulse (which allows the use of a low peak power) is the second major advantage of the chirp waveform.

As indicated before, using a long pulse permits the use of a low-peak-power transmitter, the peak power being lower by the pulse compression ratio of the coded waveform, that is, by T_U/τ_c. This makes it possible to use a less expensive transmitter, the transmitter cost typically increasing with the peak power required. It also makes it possible to use solid-state transmitters that prefer to have a low peak power [1, 3]. Also in some instances it makes an unfeasible transmitter feasible. This is the case of the L-band 95-ft phased-array Cobra Dane radar of Figure 1.1-12 that employs a 15-MW peak-power transmitter. It transmits a 1-msec, 200-MHz LFM chirp waveform that enables it to see a grapefruit-size target (silver coated) having a cross section of $0.01\,\text{m}^2$

at a range of 1000 nmi. This chirp waveform permits a range resolution of about 3 ft; that is, it can resolve two grapefruits separated by about 3 ft [1]. If instead of using the long chirp waveform a simple uncoded 5-nsec rectangular pulse were used, the peak power of the transmitter would have to be 3 trillion watts in order that the total transmitted energy be the same. If such a peak power were transmitted through the 35,000-element array, one would get one big bang and probably find that all the radiating elements are welded together. It would not be possible to use the antenna again. To transmit another pulse, a second Cobra Dane radar would have to be built. To transmit 10 pulses, 10 Cobra Dane radars would have to be built. The Cobra Dane cost about $50 million in mid-1970s dollars. Ten such radars would have thus cost about $500 million in mid-1970s dollars. With inflation the cost would almost have doubled for the late 1980s dollars. The government would not go for this expendable 3-trillion-watt version of the system. Pulse compression allowed the use of a low-power 15-MW peak-power transmitter that did not destroy the antenna. This is the version built.

It was pointed out that the pulse compression (PC) is given by the ratio of the uncompressed to the compressed pulse width:

$$PC = \frac{T_U}{\tau_c} \tag{3.5-1}$$

But $\tau_c = 1/B_s$. Substituting this in Eq. (3.5-1) yields

$$PC = T_U B_s \tag{3.5-2}$$

that is, the pulse compression ratio is equal to the waveform time-bandwidth product, the product of the uncompressed pulse width and the waveform bandwidth

If two targets are present, then two chirp echoes arrive at the receiver. These will be compressed, resulting in two compressed pulses of width τ_c that are separated in time, or equivalently in range, by the separation between the targets. Also the peak power for these two echoes will be proportional to the cross sections of their respective targets. The above properties result because the pulse compression process is a linear one so that superposition holds.

Let us now explain why the dispersive network of Figure 3.5-2 compresses the echo pulse. Assume the echo chirp signal has been mixed down to as carrier intermediate-frequency (IF) of 30.5 MHz with the LFM 1-MHz echo going up linearly from 30 MHz to 31 MHz in 100 µsec; see Figures 3.5-1 and 3.5-2. The dispersive network delays the different parts of the chirp waveform so that all components come out of the dispersive network at the same time. Specifically, assume the component at the IF of 30 MHz comes into the network first at $t = 0$ and is delayed by $T_U = 100$ µsec so that it arrives at the output of the dispersive network at the same time that the highest IF component (31 MHz) arrives at the input to the network. The highest IF component of the input signal arrives at time $t = 100\,\mu\text{sec}$ and is not delayed at all so it arrives at the output of the dispersive network at the same time that the lowest frequency arrives at the output of the dispersive network, that is, at time $t = 100\,\mu\text{sec}$. The signal component at the center IF of 30.5 MHz arrives at the input to the dispersive network at time $t = 50\,\mu\text{sec}$ and is delayed by the dispersive network by 50 μsec so that it also arrives at the output of the dispersive network at time $t = 100\,\mu\text{sec}$. The component at frequency 30.25 MHz that arrives at time $t = 25\,\mu\text{sec}$ gets delayed by an amount 75 μsec by the dispersive network so as to arrive at the output at the same time $t = 100\,\mu\text{sec}$. In the same manner all other input components of the chirp waveform are delayed so that they arrive at the output at time $t = 100\,\mu\text{sec}$. As a consequence, the long 100-μsec input chirp signal appears at the output of the dispersive network as a narrow pulse, as shown in Figure 3.5-2. Specifically, it appears as a narrow IF pulse centered at 30.5 MHz having a $\sin x/x$ envelope as shown. (The reader is referred to references 57 and 59 for a detailed mathematical proof.)

3.5.1.1   Matched Filter

The pulse compression dispersive network of Figure 3.5-2 has the important property that its output SNR is maximum. That is, no network will result in a higher SNR than that of the network shown in the figure. This network is referred to as the matched filter for the chirp waveform. [Implied in the statement that the filter provides the maximum SNR is that the input IF noise is white (flat) over the signal bandwidth.] Any other filter results in a lower SNR and is called a mismatched filter. Often a mismatched filter is used to realize lower sidelobes for the compressed pulse than obtained with the $\sin x/x$ waveform indicated. This mismatch is obtained by not having the dispersive network have a flat gain versus frequency over the signal bandwidth but instead letting it have a bell-shaped gain over the signal bandwidth. Specifically, if it has the well-known Hamming weighting [3, 48, 88], the sidelobes of the compressed pulse will be 40 dB down. This improved sidelobe level is achieved at a loss in the SNR at the output of the network of about 1.3 dB, called the mismatched loss.

3.5.1.2   Alternate Way to View Pulse Compression and Pulse Coding

The chirp waveform can be generated in the transmitter by passing a waveform having the shape of the compressed pulse at the output of the dispersive network in Figure 3.5-2 into another dispersive network, that shown in Figure 3.5-3. For this dispersive network the dispersion increases linearly with increasing frequency instead of decreasing with increasing frequency, as it does for the network of Figure 3.5-2. This input waveform has an approximately rectangular spectrum over the signal bandwidth B_s (it would be exactly rectangular if not for the fact that the input $\sin x/x$ waveform is not infinite in duration, being truncated). The dispersive network of Figure 3.5-3 will delay the different frequency components of the input signal by different

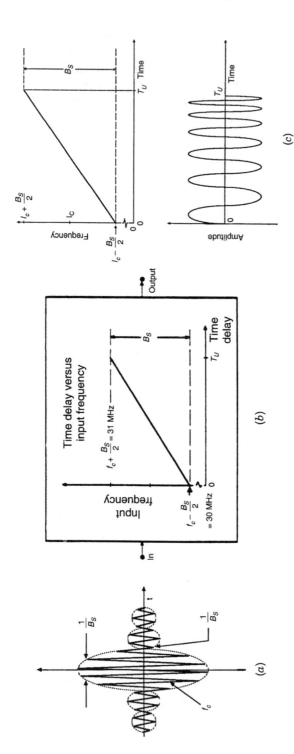

Figure 3.5-3 Generation of chirp waveform by use of dispersive network in transmitter. Time scales at input and output are not the same. (*a*) Input pulse on carrier $\mathbf{f}_C$; expanded time scale used relative to (*b*), and (*c*) in order to show pulse carrier cycles. (*b*) Pulse coding network; dispersive network. (*c*) Output coded waveform; chirp waveform.

amounts. Specifically, it will delay the lowest frequency component (that at the 30 MHz IF) by a zero amount, that at the highest carrier frequency (of 31 MHz) by 100 µsec, that at the middle frequency (30.5 MHz) by 50 µsec, and so on. As a result, the output consists of a pulse of 100 µsec in length whose carrier frequency increases linearly with time, as shown in Figure 3.5-3. Thus, we have generated our chirp waveform having a LFM that increases with time and has 100 µsec duration. It is this waveform that is power amplified and then transmitted out of the antenna for detecting targets.

The dispersive network of Figure 3.5-3 can be viewed as a distortion network that stretches the pulse having a width $1/B_s = \tau_c = 1$ µsec to a pulse $T_U = 100$ µsec for a pulse expansion of 1 to $100 = T_U/\tau_c = \mathrm{PC} = T_U B_s =$ time–bandwidth product of the pulse. The pulse compression dispersive network of Figure 3.5-2 can correspondingly be viewed as a network that removes the distortion introduced in the transmitter dispersive network. In order for a network not to introduce any distortion, it is necessary that its delay be constant with frequency over the signal bandwidth. The transmitter pulse dispersion network has a varying delay with frequency. Thus this network distorts a signal. If we cascade the transmitter pulse expansion network with the receiver pulse compression network as shown in Figure 3.5-4, we have a combined network that provides a constant delay with frequency over the signal bandwidth with the total delay being equal to $T_U = 100$ µsec at all frequencies. Such a network is just a delay line that introduces no distortion. At the output of the first dispersive network (the pulse expansion network) of Figure 3.5-4, the signal is distorted as shown, being stretched out to 100 sec. This is the chirp signal that is power amplified and sent out of the radar antenna for detecting targets. Its echo is sent through the second dispersive network in the receiver whereupon it is compressed to a narrow pulse providing the narrow-pulse resolution capability desired. The receiver pulse compression network can thus be viewed as a network equalizing the transmitter network distortion just as a communication network equalizer removes channel distortion. Because the system is linear, if one has two targets present, the output of the pulse compression network will consist of two narrow pulses whose amplitudes are proportional to the strengths of the targets and whose spacings will be equal to the range separations of the targets.

The pulse compression technique described above is dependent on two factors. First, the ability of the radar to detect a target is only dependent on the total waveform energy, it being independent of the waveform shape. Second, the resolution of the waveform is dependent on the signal bandwidth, not on the shape of the waveform envelope. It is possible to use other modulations than the LFM shown in Figure 3.5-1. Of course, it is possible to have a LFM that is decreasing in frequency with time instead of increasing. Other possible modulations are a nonlinear FM [3, 59, 61, 70], a binary phased modulation [3, 59, 61, 70], a frequency hopped waveform, and finally a noise phase modulation, to name just a few.

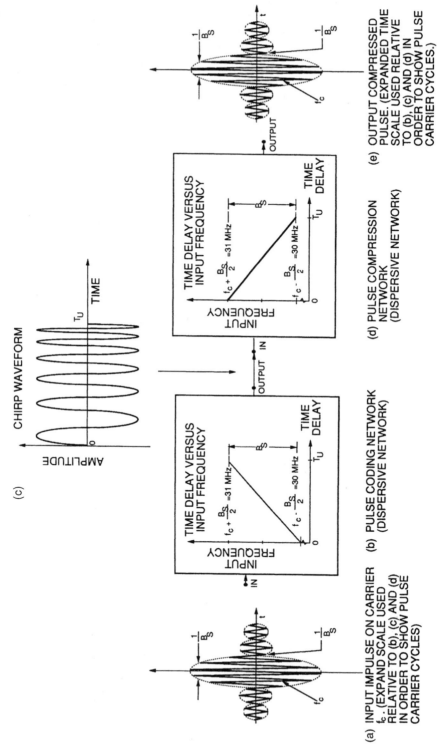

Figure 3.5-4 Cascade of pulse expansion and pulse compression networks.

(a) INPUT IMPULSE ON CARRIER f_c. (EXPAND SCALE USED RELATIVE TO (b), (c) AND (d) IN ORDER TO SHOW PULSE CARRIER CYCLES)

(b) PULSE CODING NETWORK (DISPERSIVE NETWORK)

(c) CHIRP WAVEFORM

(d) PULSE COMPRESSION NETWORK (DISPERSIVE NETWORK)

(e) OUTPUT COMPRESSED PULSE. (EXPANDED TIME SCALE USED RELATIVE TO (b), (c) AND (d) IN ORDER TO SHOW PULSE CARRIER CYCLES.)

140

3.5.1.3 Effect of Tracking with a Chirp Waveform

Using a chirp waveform results in the well-known range–Doppler coupling error that shall be described shortly. However, it has been shown by Fitzgerald [71] that if the chirp waveform is chosen to have an upchirp instead of a downchirp, then the tracking accuracy obtained when using this waveform can be better in steady state than the accuracy obtained with a simple uncoded pulse having the same power accuracy. This is indeed an amazing result! Before describing this result, we will first describe the range–Doppler coupling problem.

3.5.1.4 Range–Doppler Ambiguity Problem of Chirp Waveform

Consider that an upchirp waveform is used. Figure 3.5-5 shows the echoes received from a variety of possible targets. The solid curve is the echo from a

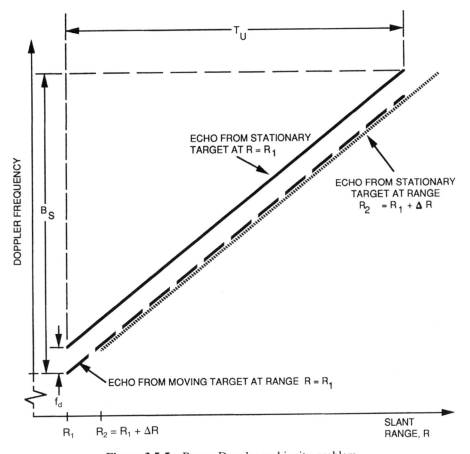

Figure 3.5-5 Range-Doppler ambiguity problem.

stationary target. The long-dash curve is the echo from a moving target at the same range as that of the target that resulted in the solid curve.

The moving target is assumed to be going away from the radar so as to produce a negative Doppler velocity shift v_d with the result that the return echo is at a lower frequency for the Doppler-shifted target than for the stationary target. The Doppler shift is given by

$$f_d = \frac{2v_d}{\lambda}$$
(3.5-3)

where λ is the wavelength of the signal carrier frequency. Specifically

$$\lambda = \frac{c}{f_c}$$
(3.5-4)

where c is the velocity of light and f_c is the signal carrier frequency.

The dotted curve represents the echo from another zero Doppler target, a target further in range away from the radar, specifically, a range ΔR further away with its echo arriving $\Delta \tau$ later in time from the other stationary target, where

$$\Delta \tau = \frac{2\Delta R}{c}$$
(3.5-5)

On examining Figure 3.5-5 we see the Doppler-shifted waveform of the target at the closer range is almost identical to the zero Doppler-shifted echo from the target at a range ΔR further away. Only a small difference exists between these two echoes, a difference occurring at the lowest and highest frequencies, that is, at the ends of the waveform. As a result the outputs of the pulse compression network for these two chirp waveforms will be almost identical. (The compressed pulse for the Doppler-shifted chirp waveform could be slightly wider than for the stationary chirp waveform echo in Figure 3.5-5. This would result if the pulse compression network had a passband that only extended from the lowest to the maximum frequency of the echoes for a stationary target, which was from 30 to 31 MHz in Figure 3.5-2. As a result the frequencies for the Doppler-shifted echo below 30 MHz are not processed, since these are outside the pulse compression network passband. If the pulse compression network were assumed to operate as a dispersive network for these lower frequencies as well, then the envelope of the compressed pulse for the Doppler-shifted echo would actually be identical to that for the stationary echo.)

Because the echoes from the stationary target and the moving target are essentially identical, one has an ambiguity problem. It is not known if the echo is coming from a target moving away from the radar at a range let us say R_1 or else from a stationary target at a range $R_2 = R_1 + \Delta R$. It is a simple matter to show that for a moving target having a Doppler v_d, the displacement ΔR of target position is given by

$$\Delta R = \frac{c}{-\lambda} v_d \frac{T_U}{B_s} \tag{3.5-6}$$

where the convention used here is that a positive v_d indicates a target moving toward the radar. Instead of indicating the target to be at the true range R at the time of the measurement, the radar is indicating it to be at the range R_c, given by

$$R_c = R + \Delta R \tag{3.5-7}$$

It was pointed out by Klauder et al. [57] in their now classical paper on the chirp waveform that the range–Doppler ambiguity can be made into a nonproblem in those situations where it is satisfactory to know the target range at a time other than the time at which the measurement is made. Specifically, the range R_c indicated by the radar measurement actually is the range the target will have at a time Δt later for a target moving away from the radar. To see this, note that ΔR of (3.5-6) can be written as

$$\Delta R = -\Delta t \cdot v_d \tag{3.5-8}$$

where, from (3.5-6), Δt is defined as

$$\Delta t = \frac{c}{\lambda} \frac{T_U}{B_s} = f_c \frac{T_U}{B_s} \tag{3.5-9}$$

Combining (3.5-7) and (3.5-8) yields

$$R_c = R - v_d \Delta t \tag{3.5-10}$$

For a target moving away v_d is negative, so that

$$R_c = R + |v_d| \Delta t \tag{3.5-11}$$

If the target is moving toward the radar, one obtains the same result, that is, the range R_c measured is that which the target would have at time Δt after the time of the measurement. In this case v_d is positive and (3.5-10) becomes

$$R_c = R - v_d \Delta t \tag{3.5-12}$$

The range R_c is called the extrapolated range [57, 61]. Note that Δt is independent of the target velocity, depending only on the waveform parameters; see (3.5-9). Thus for any upchirp, all range measurements R_c are the measurements of the range that the target would have at a time Δt after the measurement was actually made.

If a downchirp waveform is used instead of an upchirp waveform, then R_c of (3.9-10) becomes

$$R_c = R + v_d \Delta t \qquad (3.5\text{-}13)$$

and R_c now physically represents the range that the target would have at a time Δt earlier than the time of the measurement. Which is better, an upchirp or a downchirp? Or does it matter? For tracking target an upchirp waveform should be better, since for tracking we prefer a measurement of the range at a later time than at an earlier time, thereby reducing the prediction effort for the track update position. Furthermore, even if we only obtain an estimate of the present position, an upchirp waveform is better since it is better to do smoothing to obtain the present target position than to do prediction, as would be the case if a downchirp waveform were used. These are the conclusions first pointed out by Fitzgerald [71] and that will shortly be summarized.

An unchirp is also desirable for separating a fast target from a slow target, which is in the same resolution cell or close behind it. An example is an approaching reentry vehicle (RV) with its associated ionized wake at its base as illustrated in Figure 3.5-6. The RV travels at a velocity of about 20,000 ft/sec while the ionized wake at its base is going at nearly zero velocity. If a nonchirped waveform is used, such as a simple short pulse, then the echo from the RV and the wake will nearly overlap, as shown in Figure 3.5-6. However, if an upchirp waveform is used, the echo from the RV is shifted ahead ΔR (as shown in Figure 3.5-6) while the echo from the essentially stationary wake remains put. On the other hand, a downchirp may be useful for separating two approaching targets just before closest approach in a command-guidance intercept application or for collision avoidance.

It is useful to substitute some typical numbers into (3.5-8) and (3.5-9) to get a feel as to how big ΔR and Δt are. For a first example assume $T_U = 200\,\mu\text{sec}$, $B_S = 1\,\text{MHz}$, and $f_c = 5\,\text{GHz}$. Then (3.5-9) yields $\Delta t = 1\,\text{sec}$. Hence if the target Doppler velocity was 300 knots $= 506\,\text{ft/sec}$ away from the radar, then, from (3.5-8), $\Delta R = 506\,\text{ft}$, which is about equal to the slant range resolution cell width $\Delta r = c/2B_S = 492\,\text{ft}$ for the 1-MHz signal bandwidth. If the signal bandwidth B_S were increased but T_U kept constant, then ΔR would still be about equal to the range resolution, because both Δr and ΔR vary inversely with B_S. On the other hand, if T_U were increased while B_S was kept constant, then Δt and ΔR would increase in proportion while the range resolution would remain unchanged. If T_U and B_S were kept at their original values but the target velocity was increased by a factor of 4, then ΔR would become 2024 ft or about 4 times the range resolution of 492 ft.

For a second example assume that $T_U = 150\,\mu\text{sec}$, $B_S = 4\,\text{MHz}$, and $f_c = 1.3\,\text{GHz}$ (the parameters of the Martello S723 stacked-beam solid-state three-dimensional radar [1]). Now $\Delta t = 0.0488\,\text{sec}$ and $\Delta R = 24.7\,\text{ft}$ for the same 300 knots target Doppler velocity. Here ΔR is about one-fifth of the range resolution of 123 ft. The decrease in $\Delta R/\Delta r$ relative to the first example is

—————— ECHO WITH UNCODED SHORT PULSE

⸝⸝⸝⸝⸝⸝⸝⸝ ECHO WITH CHIRP WAVEFORM

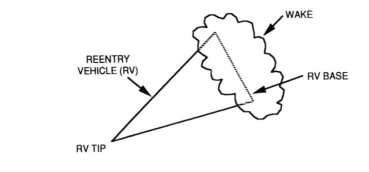

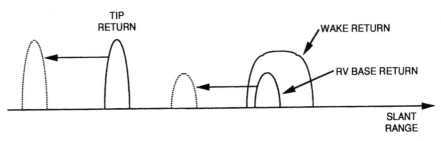

Figure 3.5-6 Comparison of echoes seen by simple uncoded narrow pulse and by upchirp waveform having same range resolution. Target is an approaching RV with its ionized based wake. Solid drawing shows echos for uncoded narrow pulse. Wake echo masks base RV echo. Dashed drawings show shift to left of echo from RV tip and base when upchirp is used. Wake return remains put. As a result, echo from RV base is visible.

primarily due to the large decrease in f_c with the small decrease in T_c contributing also.

Note that ΔR as given by (3.5-8) is only true if the target acceleration were zero, that is, if the target were truly a constant-velocity target. If in actuality the target has an acceleration a_d in the direction of the line of sight, then the target actual change in range at time Δt later would be

$$\Delta R = -v_d \, \Delta t - \tfrac{1}{2} a_d (\Delta t)^2 \qquad (3.5\text{-}14)$$

The component of ΔR due to the acceleration is given by

$$\Delta R_a = -\tfrac{1}{2} a_d (\Delta t)^2 \qquad (3.5\text{-}15)$$

Assume that $a_d = 1g$. Then for our first example given above $\Delta R_a = 16$ ft so that $\Delta R = 522$ ft instead of 506 ft, a small difference. For our second example $\Delta R_a = 0.04$ ft so that $\Delta R = 24.7$ ft, is essentially unchanged.

We shall now give the results obtained by Fitzgerald [71] as to the effects of tracking with a chirp waveform on the Kalman filter tracking accuracy.

3.5.2   Effect of Using Chirp Waveform on Tracker Filtering Accuracy

Fitzgerald analyzed the filtering accuracy achieved with a g–h–k Kalman filter when a chirp waveform is used for tracking [71]. He used the Singer target dynamics model described in Section 2.9. Fitzgerald modeled the Kalman filter to account for the shift Δt in the effective time for which the range measurement applies. He did this by modifying the observation matrix M [see e.g., (2.4-3a)] from $M = [1, 0, 0]$ to $M = [1, \Delta t, 0]$. Figure 3.5-7 gives the steady-state filtering accuracy obtained by Fitzgerald for the Kalman filter as a function of the normalized parameter p_2 of (2.10-2) and a new Fitzgerald normalized parameter

$$p_3 = \frac{\Delta t}{T} \tag{3.5-16}$$

Specifically, Figure 3.5-7 gives the steady-state rms error in true position (range) estimate, R, immediately after an update. The $\Delta t/T = 0$ curve applies for the case of a nonchirped pulse, that is, a simple unmodulated short pulse. Two points are worth noting from Figure 3.5-7. One is that an upchirp waveform (positive Δt) gives better accuracy than a downchirp waveform (Δt negative), especially for large $\Delta t/T$ (like 5 to 10) when p_2 is small. This is equivalent to having the acceleration small because $p_2 = T^2\sigma_a/\sigma_x = 2\Delta R_a/\sigma_x$. The other point worth noting is that tracking with an upchirp waveform gives better tracking accuracy than tracking with a nonchirped waveform when p_2 is small enough, less than 10 for $p_2 = 10$. Assume the range measurement accuracy σ_x is equal to about one-tenth of the radar range resolution Δr. It then follows that for this case p_2 is given by

$$p_2 = \frac{20\Delta R_a}{\Delta r} \tag{3.5-17}$$

Thus for p_2 to be less than about 10 the ΔR_a must be less than about one-half the radar range resolution Δr.

Intuitively it might be expected that alternately using an upchirp waveform followed by a downchirped waveform might provide a better filtering and prediction performance than obtainable by the use of an upchirp or downchirp waveform alone for tracking. This is because the use of an upchirp followed by a downchirp waveform removes the range–Doppler ambiguity problem, providing an unambiguous measurement of both range and Doppler velocity.

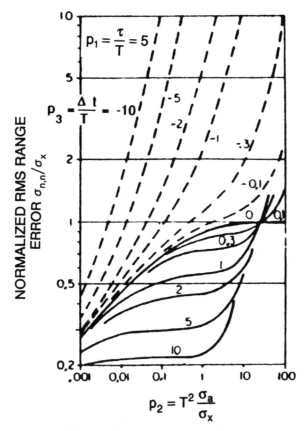

Figure 3.5-7 Effect of different chirp waveform range-Doppler couplings (as given by $\Delta t/T$) on rms of Singer g–h–k Kalman tracker steady-state filtered range estimate (normalized). Normalization is with respect to measurement accuracy, that is, rms error in measured range. (From Fitzgerald, R. J., "Effect of Range-Doppler Coupling on Chirp Radar Tracking Accuracy," IEEE Transactions on Aerospace and Electronic Systems, July, 1974. © 1974 IEEE.)

Fitzgerald [71], however, shows that for the optimum modified Singer Kalman filter this was not the case. Using only the upchirp $\Delta t/T = +1$ waveform gives the best performance, as indicated in Figure 3.5-8, which will now be explained. This figure gives the normalized rms filtering and prediction errors as a function of time. The $\Delta t/T = -1$ sawtooth curve gives the results obtained when only a downchirp waveform is used. The peak of the sawtooth obtained at the time just before $t = nT$ gives the normalized rms prediction error, designated as $\sigma_{n+1,n}$ in Figure 3.5-8, just before the update measurement is incorporated into the filter. The minimum occurring at the time just after $t = nT$ represents the filtered normalized rms error just after the update measurement has been included. Actually the sawtooth jump at time $t = nT$ will occur almost instantaneously

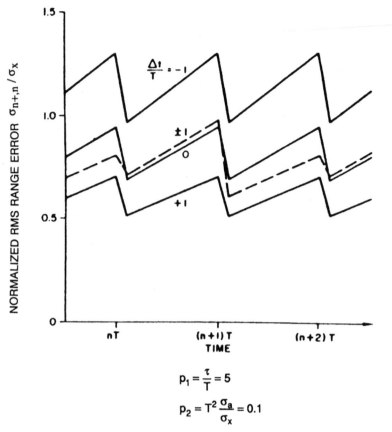

$$P_1 = \frac{\tau}{T} = 5$$

$$P_2 = T^2 \frac{\sigma_a}{\sigma_x} = 0.1$$

Figure 3.5-8 Comparison of use of alternately an upchirp and downchirp waveform, only upchirp waveform, only downchirp waveform, or simple uncoded waveform to track target. (From Fitzgerald, R. J., "Effect of Range-Doppler Coupling on Chirp Radar Tracking Accuracy," IEEE Transactions on Aerospace and Electronic Systems, July, 1974. © 1974 IEEE.)

rather than being spread over time, as shown by Fitzgerald in Figure 3.5-8 for clarity. The normalized rms prediction error, designated as $\sigma_{n+,n}$ in Figure 3.5-8, increases from just after $t = nT$ until just before $t = (n + 1)T$, at which time another measurement update is to be included and the sawtooth cycle repeated. In steady state, which is the condition for which Figure 3.5-8 applies, the sawtooth cycling shown applies. This sawtooth curve is the classical rms filtering and prediction error plot obtained for a linear tracking filter.

As indicated above, the $\Delta t/T = +1$ represents the corresponding curve obtained if an upchirp waveform is used solely for tracking the target. As expected from Figure 3.5-7, tracking with the upchirp waveform provides a much better performance. The $\Delta t/T = \pm 1$ curve represents the performance obtained if the upchirp/downchirp waveform was used. As seen, this waveform gives a performance in between that obtained using the upchirp or downchirp waveform alone. Moreover, its performance is about the same as that obtained when using a nonchirped waveform having the same range resolution as given by the $\Delta t/T = 0$ dashed curve.

Good filtering and prediction are important for collision avoidance systems, weapon delivery systems, instrumentation systems, mortar and artillery locations systems, impact prediction systems, and docking systems, among

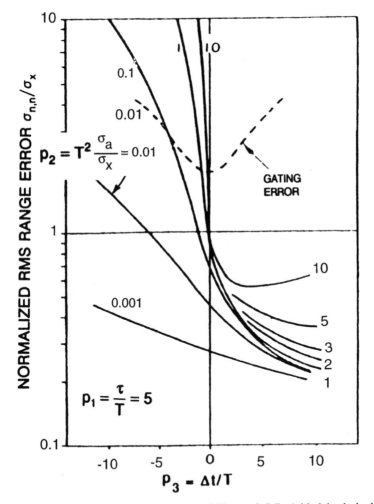

Figure 3.5-9 Replot of normalized curves of Figure 3.5-7. Added is dashed curve giving accuracy of prediction of measured range R_c. (From Fitzgerald, R. J., "Effect of Range-Doppler Coupling on Chirp Radar Tracking Accuracy," IEEE Transactions on Aerospace and Electronic Systems, July, 1974. © 1974 IEEE.)

others. If the problem is just tracking of the target, then one is only interested in the accuracy of the prediction of R_c for the next update time so that an appropriate range window might be set up for locating the next observation of the target. To keep track of the target, one needs to predict where the radar will see the target echo on the next scan. This is the apparent range $R_c = R - v_d \Delta t$. The accuracy of the prediction of R_c, as might be expected, is independent of whether a downchirp or upchirp waveform is used for tracking. To show this, the curves of Figure 3.5-7 are replotted in Figure 3.5-9 versus $\Delta t / T$. Plotted on top of these curves is a dashed curve giving the accuracy (normalized relative to σ_x) of the prediction of R_c one sample period T ahead. As the figure indicates, the accuracy of the prediction of R_c is independent of the sign of Δt.

Fitzgerald [71] also simulated the use of the upchirp and downchirp waveforms for tracking an RV. For this case the target motion is represented by a nonlinear dynamics model resulting from the nonlinear variation of

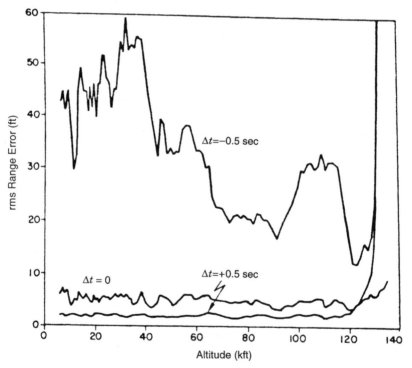

Figure 3.5-10 Simulation giving comparison of RV filtered range tracking accuracies obtained when using alternately upchirp waveform, downchirp waveform, and nonchirped waveform, all having same range resolution and measurement accuracy. (From Fitzgerald, R. J., "Effect of Range-Doppler Coupling on Chirp Radar Tracking Accuracy," IEEE Transactions on Aerospace and Electronic Systems, July, 1974. © 1974 IEEE.)

atmospheric drag as a function of altitude. A fully coupled seven-state range–direction–cosine filter was used. This seven-state filter is comprised of three position coordinates (range and the two direction cosines, which are essentially the azimuth and elevation angles of the target), the derivatives of the position coordinates, and finally a function of the RV ballistic coefficient β, or equivalently, the atmospheric drag. Figure 3.5-10 shows the results obtained. Again the upchirp waveform provides much better filtering performance than the downchirp waveform, about an order of magnitude better over part of the trajectory. The $\Delta t = 0$ curve represents the results obtained with a nonchirped waveform that has the same range resolution as the chirped waveforms. (Fitzgerald cautions that comparison with the nonchirped waveform is not as meaningful because other waveform parameters changes are implied.)

For the simulation the trajectory was started at an altitude of 135,000 ft with a velocity of 24,000 ft/sec and reentry angle of 35°. The ballistic coefficient β was assumed to be 1500 lb/ft^2 nominally with two random components: a random-bias component that is different from one simulation flight to another and an exponentially correlated random component. Also, $T = 0.1$ sec, the rms of R_c was 8 ft, and the rms of the direction cosine angles was 1 millisine.

Modifying M from [1, 0, 0] to [1, Δt, 0] as described at the beginning of this section, causes the Kalman filter to track the true target range R rather than the measured (apparent) range R_c of (3.5-7). This is important when the target is to be handed off from one radar to another, when doing target interception, or when changing radar waveforms.

PART II

LEAST-SQUARES FILTERING, VOLTAGE PROCESSING, ADAPTIVE ARRAY PROCESSING, AND EXTENDED KALMAN FILTER

4

LEAST-SQUARES AND MINIMUM–VARIANCE ESTIMATES FOR LINEAR TIME-INVARIANT SYSTEMS

4.1 GENERAL LEAST-SQUARES ESTIMATION RESULTS

In Section 2.4 we developed (2.4-3), relating the 1×1 measurement matrix Y_n to the 2×1 state vector X_n through the 1×2 observation matrix M as given by

$$Y_n = MX_n + N_n \qquad (4.1\text{-}1)$$

It was also pointed out in Sections 2.4 and 2.10 that this linear time-invariant equation (i.e., M is independent of time or equivalently n) applies to more general cases that we generalize further here. Specifically we assume Y_n is a $1 \times (r + 1)$ measurement matrix, X_n a $m \times 1$ state matrix, and M an $(r + 1) \times m$ observation matrix [see (2.4-3a)], that is,

$$Y_n = \begin{bmatrix} y_0 \\ y_1 \\ \vdots \\ y_r \end{bmatrix}_n \qquad (4.1\text{-}1a)$$

$$X_n = \begin{bmatrix} x_0(t) \\ x_1(t) \\ \vdots \\ x_{m-1}(t) \end{bmatrix} \qquad (4.1\text{-}1b)$$

and in turn

$$N_n = \begin{bmatrix} \nu_0 \\ \nu_1 \\ \vdots \\ \nu_r \end{bmatrix}_n \tag{4.1-1c}$$

As in Section 2.4, $x_0(t_n), \ldots, x_{m-1}(t_n)$ are the m different states of the target being tracked. By way of example, the states could be the x, y, z coordinates and their derivatives as given by (2.4-6). Alternately, if we were tracking only a one-dimensional coordinate, then the states could be the coordinate x itself followed by its m derivatives, that is,

$$X_n = X(t_n) = \begin{bmatrix} x \\ Dx \\ \vdots \\ D^m x \end{bmatrix}_n \tag{4.1-2}$$

where

$$D^j x_n = \frac{d^j}{dt^j} x(t) \Big|_{t=t_n} \tag{4.1-2a}$$

The example of (2.4-1a) is such a case with $m = 1$. Let m' always designate the number of states of $X(t_n)$ or X_n; then, for $X(t_n)$ of (4.1-2), $m' = m + 1$. Another example for $m = 2$ is that of (1.3-1a) to (1.3-1c), which gives the equations of motion for a target having a constant acceleration. Here (1.3-1a) to (1.33-1c) can be put into the form of (2.4-1) with

$$X_n = \begin{bmatrix} x_n \\ \dot{x}_n \\ \ddot{x}_n \end{bmatrix} \tag{4.1-3}$$

and

$$\Phi = \begin{bmatrix} 1 & T & T^2/2 \\ 0 & 1 & T \\ 0 & 0 & 1 \end{bmatrix} \tag{4.1-4}$$

Assume that measurements such as given by (4.1-1a) were also made at the L preceding times at $n-1, \ldots, n-L$. Then the totality of $L+1$ measurements can be written as

$$
\begin{bmatrix} Y_n \\ \hline Y_{n-1} \\ \hline \vdots \\ Y_{n-L} \end{bmatrix} = \begin{bmatrix} MX_n \\ \hline MX_{n-1} \\ \hline \vdots \\ MX_{n-L} \end{bmatrix} + \begin{bmatrix} N_n \\ \hline N_{n-1} \\ \hline \vdots \\ N_{n-L} \end{bmatrix}
\qquad (4.1\text{-}5)
$$

Assume that the transition matrix for transitioning from the state vector X_{n-1} at time $n-1$ to the state vector X_n at time n is given by Φ [see (2.4-1) of Section 2.4, which gives Φ for a constant-velocity trajectory; see also Section 5.4]. Then the equation for transitioning from X_{n-i} to X_n is given by

$$
X_n = \Phi_i X_{n-i} = \Phi^i X_{n-i} \qquad (4.1\text{-}6)
$$

where Φ_i is the transition matrix for transitioning from X_{n-i} to X_n. It is given by

$$
\Phi_i = \Phi^i \qquad (4.1\text{-}7)
$$

It thus follows that

$$
X_{n-1} = \Phi^{-i} X_n \qquad (4.1\text{-}8)
$$

where $\Phi^{-i} = (\Phi^{-1})^i$. Thus (4.1-5) can be written as

$$
\begin{bmatrix} Y_n \\ \hline Y_{n-1} \\ \hline \vdots \\ Y_{n-L} \end{bmatrix} = \begin{bmatrix} MX_n \\ \hline M\Phi^{-1}X_n \\ \hline \vdots \\ M\Phi^{-L}X_n \end{bmatrix} + \begin{bmatrix} N_n \\ \hline N_{n-1} \\ \hline \vdots \\ N_{n-L} \end{bmatrix}
\qquad (4.1\text{-}9)
$$

or

$$
\begin{bmatrix} Y_n \\ \hline Y_{n-1} \\ \hline \vdots \\ Y_{n-L} \end{bmatrix} = \underbrace{\begin{bmatrix} M \\ \hline M\Phi^{-1} \\ \hline \vdots \\ M\Phi^{-L} \end{bmatrix}}_{m'} X_n + \underbrace{\left.\begin{bmatrix} N_n \\ \hline N_{n-1} \\ \hline \vdots \\ N_{n-L} \end{bmatrix}\right\}}_{1} (L+1)(r+1) = s \qquad (4.1\text{-}10)
$$

$$
\underbrace{\phantom{\begin{bmatrix} Y_n \end{bmatrix}}}_{1}
$$

which we rewrite as

$$Y_{(n)} = TX_n + N_{(n)} \tag{4.1-11}$$

where

$$Y_{(n)} = \begin{bmatrix} Y_n \\ \hline Y_{n-1} \\ \hline \vdots \\ \hline Y_{n-L} \end{bmatrix} \qquad N_{(n)} = \begin{bmatrix} N_n \\ \hline N_{n-1} \\ \hline \vdots \\ \hline N_{n-L} \end{bmatrix} \tag{4.1-11a}$$

$$T = \left.\begin{bmatrix} M \\ \hline M\Phi^{-1} \\ \hline \vdots \\ \hline M\Phi^{-L} \end{bmatrix}\right\} s \tag{4.1-11b}$$

$$\underbrace{\phantom{M\Phi^{-L}}}_{m'}$$

    Equation (4.1-1) is the measurement equation when the measurement is only made at a single time. Equation (4.1-11) represents the corresponding measurement equation when measurements are available from more than one time. Correspondingly M is the observation matrix [see (2.4-3a)] when a measurement is available at only one time whereas T is the observation matrix when measurements are available from $L + 1$ times. Both observation matrices transform the state vector X_n into the observation space. Specifically X_n is transformed to a noise-free Y_n in (4.1-1) when measurements are available at one time or to $Y_{(n)}$ in (4.1-11) when measurements are available at $L + 1$ time instances. We see that the observation equation (4.1-11) is identical to that of (4.1-1) except for T replacing M.

    [In Part 1 and (4.1-4), T was used to represent the time between measurements. Here it is used to represent the observation matrix given by (4.1-11b). Unfortunately T will be used in Part II of this text to represent these two things. Moreover, as was done in Sections 1.4 and 2.4 and as shall be done later in Part II, it is also used as an exponent to indicate the transpose of a matrix. Although this multiple use for T is unfortunate, which meaning T has should be clear from the context in which it is used.]

By way of example of T, assume $L = 1$ in (4.1-11a) and (4.1-11b); then

$$Y_{(n)} = \begin{bmatrix} Y_n \\ Y_{n-1} \end{bmatrix} \tag{4.1-12}$$

$$T = \begin{bmatrix} M \\ M\Phi^{-1} \end{bmatrix} \tag{4.1-13}$$

Assume the target motion is being modeled by a constant-velocity trajectory. That is, $m = 1$ in (4.1-2) so that X_n is given by (2.4-1a) and Φ is given by (2.4-1b). From (1.1-1a) and (1.1-1b), it follows that

$$x_{n-1} = x_n - T\dot{x}_n \tag{4.1-14a}$$
$$\dot{x}_{n-1} = \dot{x}_n \tag{4.1-14b}$$

On comparing (4.1-14a) and (4.1-14b) with (4.1-8) we see that we can rewrite (4.1-14a) and (4.1-14b) as (4.1-8) with X_n given by (2.4-1a) and

$$\Phi^{-1} = \begin{bmatrix} 1 & -T \\ 0 & 1 \end{bmatrix} \tag{4.1-15}$$

We can check that Φ^{-1} is given by (4.1-15) by verifying that

$$\Phi\Phi^{-1} = I \tag{4.1-16}$$

where I is the identify matrix and Φ is given by (2.4-1b).

As done in Section 2.4, assume a radar sensor with only the target range being observed, with x_n representing the target range. Then M is given by (2.4-3a) and Y_n and N_n are given by respectively (2.4-3c) and (2.4-3b). Substituting (4.1-15) and (2.4-3a) into (4.1-13) yields

$$T = \begin{bmatrix} 1 & 0 \\ 1 & -T \end{bmatrix} \tag{4.1-17}$$

Equation (4.1-17) applies for $L = 1$ in (4.1-11b). It is easily extended to the case where $L = n$ to yield

$$T = \begin{bmatrix} 1 & 0 \\ 1 & -T \\ 1 & -2T \\ \vdots & \vdots \\ 1 & -nT \end{bmatrix} \tag{4.1-18}$$

It is instructive to write out (4.1-11) for this example. In this case (4.1-11) becomes

$$Y_{(n)} = \begin{bmatrix} y_n \\ y_{n-1} \\ y_{n-2} \\ \vdots \\ y_0 \end{bmatrix} = \begin{bmatrix} 1 & 0 \\ 1 & -T \\ 1 & -2T \\ & \vdots \\ 1 & -nT \end{bmatrix} \begin{bmatrix} x_n \\ \dot{x}_n \end{bmatrix} + \begin{bmatrix} \nu_n \\ \nu_{n-1} \\ \nu_{n-2} \\ \vdots \\ \nu_0 \end{bmatrix} \qquad (4.1\text{-}19)$$

where use was made of (2.4-3b) and (2.4-3c), which hold for arbitrary n; specifically,

$$Y_{n-i} = [y_{n-i}] \qquad (4.1\text{-}20)$$
$$N_{n-i} = [\nu_{n-i}] \qquad (4.1\text{-}21)$$

Evaulating y_{n-i} in (4.1-19) yields

$$y_{n-i} = x_n - iT\dot{x}_n + \nu_{n-i} \qquad (4.1\text{-}22)$$

The above physically makes sense. For a constant-velocity target it relates the measurement y_{n-i} at time $n - i$ to the true target position and velocity x_n and $\dot{x}_n$ at time n and the measurement error ν_{n-i}. The above example thus gives us a physical feel for the observation matrix T. For the above example, the $(i + 1)$st row of T physically in effect first transforms X_n back in time to time $n - i$ through the inverse of the transition matrix Φ to the ith power, that is, through Φ^{-i} by premultiplying X_n to yield X_{n-i}, that is,

$$X_{n-i} = \Phi^{-i}X_n \qquad (4.1\text{-}23)$$

Next X_{n-i} is effectively transformed to the noise-free Y_{n-i} measurement at time $n - i$ by means of premultiplying by the observation matrix M to yield the noise-free Y_{n-i}, designated as Y'_{n-i} and given by

$$Y'_{n-i} = M\Phi^{-i}X_n \qquad (4.1\text{-}24)$$

Thus T is really more than an observation matrix. It also incorporates the target dynamics through Φ. We shall thus refer to it as the transition–observation matrix.

By way of a second example, assume that the target motion is modeled by a constant-accelerating trajectory. Then $m = 2$ in (4.1-2), $m' = 3$, and X_n is given by (4.1-3) with Φ given by (4.1-4). From (1.3-1) it follows that

$$x_{n-1} = x_n - \dot{x}_nT + \ddot{x}_n(\tfrac{1}{2}T^2) \qquad (4.1\text{-}25a)$$
$$\dot{x}_{n-1} = \dot{x}_n - \ddot{x}_nT \qquad (4.1\text{-}25b)$$
$$\ddot{x}_{n-1} = \ddot{x}_n \qquad (4.1\text{-}25c)$$

We can now rewrite (4.1-25a) to (4.1-25c) as (4.1-8) with X_n given by (4.1-3) and

$$\Phi^{-1} = \begin{bmatrix} 1 & -T & \frac{1}{2}T^2 \\ 0 & 1 & -T \\ 0 & 0 & 1 \end{bmatrix} \tag{4.1-26}$$

Again we can check that Φ^{-1} is given by (4.1-26) by verifying that (4.1-16) is satisfied.

As done for the constant-velocity target example above, assume a radar sensor with only target range being observed, with x_n again representing target range. Then M is given by

$$M = \begin{bmatrix} 1 & 0 & 0 \end{bmatrix} \tag{4.1-27}$$

and Y_n and N_n are given by respectively (2.4-3c) and (2.4-3b). Substituting (4.1-26) and (4.1-27) into (4.1-11b) yields finally, for $L = n$,

$$T = \begin{bmatrix} 1 & 0 & 0 \\ 1 & -T & \frac{1}{2}T^2 \\ 1 & -2T & \frac{1}{2}(2T)^2 \\ \vdots & \vdots & \vdots \\ 1 & -nT & \frac{1}{2}(nT)^2 \end{bmatrix} \tag{4.1-28}$$

For this second example (4.1-11) becomes

$$\begin{bmatrix} y_n \\ y_{n-1} \\ y_{n-2} \\ \vdots \\ y_0 \end{bmatrix} = \begin{bmatrix} 1 & 0 & 0 \\ 1 & -T & \frac{1}{2}T^2 \\ 1 & -2T & \frac{1}{2}(2T)^2 \\ \vdots & \vdots & \vdots \\ 1 & -nT & \frac{1}{2}(nT)^2 \end{bmatrix} \begin{bmatrix} x_n \\ \dot{x}_n \\ \ddot{x}_n \end{bmatrix} + \begin{bmatrix} \nu_n \\ \nu_{n-1} \\ \nu_{n-2} \\ \vdots \\ \nu_0 \end{bmatrix} \tag{4.1-29}$$

Again, we see from the above equation that the transition–observation matrix makes physical sense. Its $(i + 1)$st row transforms the state vector at time X_n back in time to X_{n-i} at time $n - i$ for the case of the constant-accelerating target. Next it transforms the resulting X_{n-i} to the noise-free measurement Y'_{n-i}.

What we are looking for is an estimate $X^*_{n,n}$ for X_n, which is a linear function of the measurement given by $Y_{(n)}$, that is,

$$X^*_{n,n} = WY_{(n)} \tag{4.1-30}$$

where W is a row matrix of weights, that is, $W = [w_1, w_2, \ldots, w_s]$, where s is the dimension of $Y_{(n)}$; see (4.1-10) and (4.1-11a). For the least-squares estimate

(LSE) we are looking for, we require that the sum of squares of errors be minimized, that is,

$$e(X^*_{n,n}) = e_n = [Y_{(n)} - TX^*_{n,n}]^T [Y_{(n)} - TX^*_{n,n}] \tag{4.1-31}$$

is minimized. As we shall show shortly, it is a straightforward matter to prove using matrix algebra that W of (4.1-30) that minimizes (4.1-31) is given by

$$\hat{W} \equiv (T^T T)^{-1} T^T \tag{4.1-32}$$

It can be shown that this estimate is unbiased [5, p. 182].

Let us get a physical feel for the minimization of (4.1-31). To do this, let us start by using the constant-velocity trajectory example given above with T given by (4.1-18) and $Y_{(N)}$ given by the left-hand side of (4.1-19), that is,

$$Y_{(n)} = \begin{bmatrix} y_n \\ y_{n-1} \\ y_{n-2} \\ \vdots \\ y_0 \end{bmatrix} \tag{4.1-33}$$

and the estimate of the state vector X_n at time n given by

$$X^*_{n,n} = \begin{bmatrix} x^*_{n,n} \\ \dot{x}^*_{n,n} \end{bmatrix} \tag{4.1-34}$$

The $(i + 1)$st row of T transforms the estimate $x^*_{n,n}$ of the state vector at time n back in time to the corresponding estimate of the range coordinate $x^*_{n-i,n}$ at time $n - i$. Specifically,

$$[1 - iT]\begin{bmatrix} x^*_{n,n} \\ \dot{x}^*_{n,n} \end{bmatrix} = x^*_{n,n} - iT\dot{x}^*_{n,n} = x^*_{n-i,n} \tag{4.1-35}$$

as it should. Hence

$$\begin{bmatrix} x^*_{n,n} \\ x^*_{n-1,n} \\ x^*_{n-2,n} \\ \vdots \\ x^*_{0,n} \end{bmatrix} = \begin{bmatrix} 1 & 0 \\ 1 & -T \\ 1 & -2T \\ & \vdots \\ 1 & -nT \end{bmatrix}\begin{bmatrix} x^*_{n,n} \\ \dot{x}^*_{n,n} \end{bmatrix} = TX^*_{n,n} \tag{4.1-36}$$

Substituting (4.1-33) and (4.1-36) into (4.1-31) yields

$$e_n = e(X^*_{n,n}) = \sum_{i=0}^{n} (y_{n-i} - x^*_{n-i,n})^2 \qquad (4.1\text{-}37)$$

Reindexing the above yields

$$e_n = \sum_{j=0}^{n} (y_j - x^*_{j,n})^2 \qquad (4.1\text{-}38)$$

Except for a slight change in notation, (4.1-38) is identical to (1.2-33) of Section 1.2.6. Here we have replaced x^*_n by $x^*_{j,n}$ and e_T by e_n, but the estimation problem is identical. What we are trying to do in effect is find a least-squares fitting line to the data points as discussed in Section 1.2.6 relative to Figure 1.2-10. Here the line estimate is represented by its ordinate at time n, $x^*_{n,n}$, and its slope at time n, $\dot{x}^*_{n,n}$. In contrast in Section 1.2.6 we represented the line fitting the data by its ordinate and slope at time $n = 0$, that is, by x^*_0 and $v^*_0 = \dot{x}^*_0$, respectively. A line is defined by its ordinate and slope at any time. Hence it does not matter which time we use, time $n = n$ or time $n = 0$. (The covariance of the state vector, however, does depend on what time is used.) The state vector estimate gives the line's ordinate and slope at some time. Hence the state vector at any time defines the estimated line trajectory. At time $n = 0$ the estimated state vector is

$$X^*_{0,n} = \begin{bmatrix} x^*_0 \\ v^*_0 \end{bmatrix} = \begin{bmatrix} x^*_0 \\ \dot{x}^*_0 \end{bmatrix} \qquad (4.1\text{-}39)$$

At time n it is given by (4.1-34). Both define the same line estimate.

To further clarify our flexibility in the choice of the time we choose for the state vector to be used to define the estimating trajectory, let us go back to (4.1-9). In (4.1-9) we reference all the measurements to the state vector X_n at time n. We could have just as well have referenced all the measurements relative to the state vector at any other time $n - i$ designated as X_{n-i}. Let us choose time $n - i = 0$ as done in (4.1-39). Then (4.1-9) becomes

$$\begin{bmatrix} Y_n \\ \text{----} \\ Y_{n-1} \\ \text{----} \\ \vdots \\ \text{----} \\ Y_1 \\ \text{----} \\ Y_0 \end{bmatrix} = \begin{bmatrix} M\Phi^n X_0 \\ \text{--------} \\ M\Phi^{n-1} X_0 \\ \text{--------} \\ \vdots \\ \text{--------} \\ M\Phi X_0 \\ \text{--------} \\ M X_0 \end{bmatrix} + \begin{bmatrix} N_n \\ \text{----} \\ N_{n-1} \\ \text{----} \\ \vdots \\ \text{----} \\ N_1 \\ \text{----} \\ N_0 \end{bmatrix} \qquad (4.1\text{-}40)$$

This in turn becomes

$$
\begin{bmatrix} Y_n \\ \hdashline Y_{n-1} \\ \hdashline \vdots \\ \hdashline Y_1 \\ \hdashline Y_0 \end{bmatrix} = \begin{bmatrix} M\Phi^n \\ \hdashline M\Phi^{n-1} \\ \hdashline \vdots \\ \hdashline M\Phi \\ \hdashline M \end{bmatrix} X_0 + \begin{bmatrix} N_n \\ \hdashline N_{n-1} \\ \hdashline \vdots \\ \hdashline N_1 \\ \hdashline N_0 \end{bmatrix}
\tag{4.1-41}
$$

which can be written as

$$
Y_{(n)} = T X_0 + N_{(n)}
\tag{4.1-42}
$$

where $Y_{(n)}$ and $N_{(n)}$ are given by (4.1-11a) with $L = n$ and T is now defined by

$$
T = \begin{bmatrix} M\Phi^n \\ \hdashline M\Phi^{n-1} \\ \hdashline \vdots \\ \hdashline M\Phi \\ \hdashline M \end{bmatrix}
\tag{4.1-43}
$$

In Section 1.2.10 it was indicated that the least-squares fitting line to the data of Figure 1.2-10 is given by the recursive g–h growing-memory (expanding-memory) filter whose weights g and h are given by (1.2-38a and 1.2-38b). The g–h filter itself is defined by (1.2-8a) and (1.2-8b). In Chapters 5 and 6 an indication is given as to how the recursive least-squares g–h filter is obtained from the least-squares filter results of (4.1-30) and (4.1-32). The results are also given for higher order filters, that is, when a polynominal in time of arbitrary degree m is used to fit the data. Specifically the target trajectory $x(t)$ is approximated by

$$
x(t) \doteq p^*(t) = \sum_{k=0}^{m} \bar{a}_k t^k
\tag{4.1-44}
$$

For the example of Figure 1.2-10, $m = 1$ and a straight line (constant-velocity) trajectory is being fitted to the data. For this case the transition–observation matrix is given by (4.1-18). If a constant-accelerating target trajectory is fitted to the data, then, in (4.1-2) and (4.1-44), $m = 2$, and T is given by (4.1-28). In this case, a best-fitting quadratic is being found for the data of Figure 1.2-10. The recursive least-square filter solutions are given in Chapter 6 for $m = 0, 1, 2, 3$; see Table 6.3-1. The solution for arbitrary m is also given in general form; see (5.3-11) and (5.3-13).

The solution for the least-squares estimate $X_{n,n}^*$ given above by (4.1-30) and (4.1-32) requires a matrix inversion in the calculation of the weights. In Section 5.3 it is shown how the least-squares polynomial fit can be obtained without a matrix inversion. This is done by the use of the powerful discrete-time orthogonal Legendre polynomials. What is done is that the polynomial fit of degree m of (4.1-44) is expressed in terms of the powerful discrete-time orthogonal Legendre polynomials (DOLP) having degree m. Specifically (4.1-44) is written as

$$x(r) \doteq p^*(r) = \sum_{k=0}^{m} \beta_k \phi_k(r) \qquad (4.1\text{-}45)$$

where $\phi_k(r)$ is the normalized discrete Legendre polynomial (to be defined in Section 5.3) of degree k and r is an integer time index, specifically, $t = rT$, and the β_k are constants that specify the fit to $x(r)$. Briefly $\phi_k(r)$ is a polynomial in r of degree k with $\phi_k(r)$ orthogonal to $\phi_j(r)$ for $k \neq j$; see (5.3-2). Using this orthogonal polynomial form yields the least-squares solution directly as a linear weighted sum of the $y_n, y_{n-1}, \ldots, y_{n-L}$ without any matrix inversion being required; see (5.3-10) and (5.3-11) for the least-squares polynomial fit, designated there as $[p^*(r)]_n = x^*(r)$. In Section 4.3 another approach, the voltage-processing method, is presented, which also avoids the need to do a matrix inversion. Finally, it is shown in Section 14.4 that when a polynomial fit to the data is being made, the alternate voltage-processing method is equivalent to using the orthogonal discrete Legendre polynomial approach.

In Sections 7.1 and 7.2 the above least-squares polynomial fit results are extended to the case where the measurements consist of the semi-infinite set y_n, $y_{n-1}, \ldots$ instead of $L + 1$ measurements. In this case, the discounted least-squares weighted sum is minimized as was done in (1.2-34) [see (7.1-2)] to yield the fading-memory filter. Again the best-fitting polynomial of the form, given by (4.1-45) is found to the data. In Section 1.2.6, for the constant-velocity target, that is $m = 1$ in (4.1–44), the best-fitting polynomial, which is a straight line in this case, was indicated to be given by the fading memory g–h filter, whose weights g and h are given by (1.2-35a) and (1.2-35b). To find the best-fitting polynomial, in general the estimating polynomial is again approximated by a sum of discrete-time orthogonal polynomials, in this case the orthonormal discrete Laguerre polynomials, which allow the discounted weightings for the semi-infinite set of data. The resulting best-fitting discounted least-squares polynomial fit is given by (7.2-5) in recursive form for the case where the polynomial is of arbitrary degree m. For $m = 1$, this result yields the fading-memory g–h filter of Section 1.2.6. Corresponding convenient explicit results for this recursive fading-memory filter for $m = 0, \ldots, 4$ are given in Table 7.2-2.

In reference 5 (4.1-32) is given for the case of a time-varying trajectory model. In this case M, T, and Φ all become a function of time (or equivalently n) and are replaced by M_n and T_n and $\Phi(t_n, t_{n-1})$, respectively; see pages 172, 173, and 182 of reference 5 and Chapter 15 of this book, in which the time-varying case is discussed.

From (4.1-1) we see that the results developed so far in Section 4.1, and that form the basis for the remaining results here and in Chapters 5 to 15, apply for the case where the measurements are linear related to the state vector through the observation matrix M. In Section 16.2 we extend the results of this chapter and Chapters 5 to 15 for the linear case to the case where Y_n is not linearly related to X_n. This involves using the Taylor series expansion to linearize the nonlinear observation scheme. The case where the measurements are made by a three-dimensional radar in spherical coordinates while the state vector is in rectangular coordinates is a case of a nonlinear observation scheme; see (1.5-2a) to (1.5-2c). Similarly, (4.1-6) implies that the target dynamics, for which the results are developed here and in Chapters 5 to 15, are described by a linear time differential equation; see Chapter 8, specifically (8.1-10). In Section 16.3, we extend the results to the case where the target dynamics are described by a nonlinear differential equation. In this case, a Taylor series expansion is applied to the nonlinear differential equation to linearize it so that the linear results developed in Chapter 4 can be applied.

There are a number of straightforward proofs that the least-squares weight is given by (4.1-32). One is simply to differentiate (4.1-31) with respect to $X_{n,n}^*$ and set the result equal to zero to obtain

$$\frac{de_n}{dX_{n,n}^*} = T^T[Y_{(n)} - TX_{n,n}^*] = 0 \qquad (4.1\text{-}46)$$

Solving for $X_{n,n}^*$ yields (4.1-32) as we desired to show.

In reference 5 (pp. 181, 182) the LSE weight given by (4.1-32) is derived by simply putting (4.1-31) into another form analogous to "completing the squares" and noting that $e(X_{n,n}^*)$ is minimized by making the only term depending on W zero, with this being achieved by having W be given by (4.1-32). To give physical insight into the LSE, it is useful to derive it using a geometric development. We shall give this derivation in the next section. This derivation is often the one given in the literature [75–77]. In Section 4.3 (and Chapter 10) it is this geometric interpretation that we use to develop what is called the voltage-processing method for obtaining a LSE without the use of the matrix inversion of (4.1-32).

4.2 GEOMETRIC DERIVATION OF LEAST-SQUARES SOLUTION

We start by interpreting the columns of the matrix T as vectors in an s-dimensional hyperspace, each column having s entries. There are m' such columns. We will designate these as $t_1, \ldots, t_{m'}$. For simplicity and definiteness assume that $s = 3$, $m' = 2$, and $n = 3$; then

$$T = \begin{bmatrix} t_{11} & t_{12} \\ t_{21} & t_{22} \\ t_{31} & t_{32} \end{bmatrix} \tag{4.2-1}$$

so that

$$t_1 = \begin{bmatrix} t_{11} \\ t_{21} \\ t_{31} \end{bmatrix} \quad \text{and} \quad t_2 = \begin{bmatrix} t_{12} \\ t_{22} \\ t_{32} \end{bmatrix} \tag{4.2-2}$$

$$X_n = X_3 = \begin{bmatrix} x_1 \\ x_2 \end{bmatrix} \tag{4.2-3}$$

and

$$Y_{(n)} = Y_{(3)} = \begin{bmatrix} y_1 \\ y_2 \\ y_3 \end{bmatrix} \tag{4.2-4}$$

Moreover, if we assume the constant-velocity trajectory discussed above, T of (4.1-18) becomes, for $n = 3$,

$$T = \begin{bmatrix} 1 & 0 \\ 1 & -T \\ 1 & -2T \end{bmatrix} \tag{4.2-5}$$

and

$$t_1 = \begin{bmatrix} 1 \\ 1 \\ 1 \end{bmatrix} \quad t_2 = \begin{bmatrix} 0 \\ -T \\ -2T \end{bmatrix} \tag{4.2-6}$$

and

$$X_3 = \begin{bmatrix} x_3 \\ \dot{x}_3 \end{bmatrix} \tag{4.2-7}$$

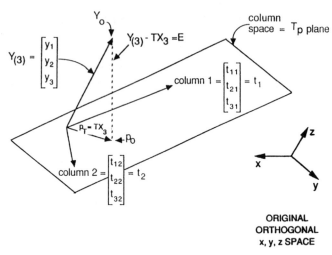

Figure 4.2-1 Projection of data vector $Y_{(3)}$ onto column space of 3×2 T matrix. Used to obtain least-squares solution in three-dimensional space. (After Strang [76].)

In Figure 4.2-1 we show the vectors t_1, t_2, and $Y_{(3)}$. The two vectors t_1 and t_2 define a plane. Designate this plane as T_p. (In general T_p is an m'-dimensional space determined by the m' column vectors of T). Typically $Y_{(3)}$ is not in this plane due to the measurement noise error $N_{(n)}$; see (4.1-11).

Let us go back to the case of arbitrary dimension s for the column space of T and consider the vector

$$
p_T = \begin{bmatrix} p_1 \\ p_2 \\ \vdots \\ p_s \end{bmatrix} = T X_n \tag{4.2-8}
$$

From (4.2-8) we see that the vector p_T is a linear combination of the column vectors of T. Hence the vector p_T is in the space defined by T_p. Now the least-squares estimate picks the X_n that minimizes $e(X_n)$, defined by (4.1-31). That is, it picks the X_n that minimizes

$$
e(X_n) = (Y_{(n)} - T X_n)^T (Y_{(n)} - T X_n) \tag{4.2-9}
$$

Applying (4.2-8) to (4.2-9) gives, for the three-dimensional case being considered,

$$
e(X_n) = \sum_{i=1}^{3} (y_i - p_i)^2 \tag{4.2-10}
$$

But this is nothing more than the Euclidean distance between the endpoints of the vectors p_T and $Y_{(3)}$, these endpoints being designated respectively by p_0 and Y_0 in Figure 4.2-1.

The point p_0 can be placed anywhere in the plane T_p by varying X_n. From simple geometry we know that the distance between the points Y_0 and a point p_0 in the plane T_p is minimized when the vector joining these two points is made to be perpendicular to the plane T_p (at the point p_0 on the plane T_p). That is, the error vector

$$Y_{(3)} - TX_3 = Y_{(3)} - p_T \qquad (4.2\text{-}11)$$

is perpendicular to the plane T_p when the error term $e(X_n)$ is minimized. Then $X_3 = X_{3,3}^*$, where $X_{3,3}^*$ is such that

$$(Y_{(3)} - TX_{3,3}^*) \perp T_p \qquad (4.2\text{-}12)$$

We now obtain an expression for $X_{3,3}^*$. Consider an arbitrary vector in the plane T_p defined by a linear combination of the columns of T, that is, by Tz, where z is an arbitrary $m' \times 1$ column vector that for the example being considered here is a (2×1) dimensional vector. If two vectors represented by the column matrices a and b are perpendicular, then $a^T b = 0$. Hence

$$(Tz)^T (Y_{(3)} - TX_{3,3}^*) = 0 \qquad (4.2\text{-}13)$$

or equivalently, since $(Tz)^T = z^T T^T$

$$z^T (T^T Y_{(3)} - T^T TX_{3,3}^*) = 0 \qquad (4.2\text{-}14)$$

Because (4.2-14) must be true for all z, it follows that it is necessary that

$$T^T TX_{3,3}^* = T^T Y_{(3)} \qquad (4.2\text{-}15)$$

The above in turn yields

$$X_{3,3}^* = (T^T T)^{-1} T^T Y_{(3)} \qquad (4.2\text{-}16)$$

from which it follows that

$$\hat{W} = (T^T T)^{-1} T^T \qquad (4.2\text{-}17)$$

which is the expression for the optimum LSE weight given previously by (4.1-32), as we wanted to show.

Although the above was developed for $m' = 2$ and $s = 3$, it is easy to see that it applies for arbitrary m' and s. In the literature the quantity $(T^T T)^{-1} T^T$ is often referred to as a pseudoinverse operator [78]. This because it provides the solution of $Y_{(n)} = TX_n$ (in the least-squares sense) when T is nonsingular, as it is when $s > m'$, so that T^{-1} does not exist and $X_n = T^{-1}Y_{(n)}$ does not provide a solution for (4.1-31). The case where $s > m'$ is called the overdetermined case. It is the situation where we have more measurements s than unknowns m in our state vector. Also the LSE given by (4.2-16), or equivalently (4.1-30) with W given by (4.1-32), is referred to as the normal-equation solution [75, 76, 79–82]. Actually, to be precise, the normal equation are given by a general form of (4.2-15) given by

$$T^T T X_{n,n}^* = T^T Y_{(n)} \tag{4.2-18}$$

which leads to (4.1-30) with W given by (4.1-32).

   A special case is where T consists of just one column vector t. For this case

$$\begin{aligned}\hat{W} &= (t^T t)^{-1} t^T \\ &= \frac{t^T}{(t^T t)}\end{aligned} \tag{4.2-19}$$

and

$$X_{n,n}^* = \frac{t^T Y_{(n)}}{t^T t} \tag{4.2-20}$$

By way of example consider the case where

$$Y_{n-i} = MX_{n-i} + N_{n-i} \tag{4.2-21}$$

with each term of the above being 1×1 matrices given by

$$Y_{n-1} = [y_{n-i}] \tag{4.2-21a}$$
$$M = [1] \tag{4.2-21b}$$
$$X_{n-i} = [x_{n-i}] \tag{4.2-21c}$$
$$N_{n-i} = [\nu_{n-i}] \tag{4.2-21d}$$

so that

$$y_{n-i} = x_{n-i} + \nu_{n-i} \tag{4.2-21e}$$

This equivalent to only having multiple measurements of the target range for a target modeled as being stationary. For this example

$$t = \begin{bmatrix} 1 \\ 1 \\ \vdots \\ 1 \end{bmatrix} \tag{4.2-22}$$

then

$$X^*_{n,n} = \frac{1}{s} \sum_{i=1}^{s} y_i \tag{4.2-23}$$

which is the sample mean of the y_i's, as expected.

Before proceeding let us digress for a moment to point out some other interesting properties relating to the geometric development of the LSE. We start by calculating the vector p_T for the case $X_3 = X^*_{3,3}$. Specifically, substituting $X^*_{3,3}$ given by (4.2-16) into (4.2-8) yields

$$p_T = T(T^T T)^{-1} T^T Y_{(3)} \tag{4.2-24}$$

Physically p_T given by (4.2-24) is the projection of $Y_{(3)}$ onto the plane T_p; see Figure 4.2-1. Designate this projection vector as p^*_T. The matrix

$$P = T(T^T T)^{-1} T^T \tag{4.2-25}$$

of (4.2-24) that projects $Y_{(3)}$ onto the two-dimensional plane T_p is known as the projection matrix [76]. [Note that for the projection matrix of (4.2-25) a capital P is used whereas for the column matrix p_T of (4.2-8), which represents a vector in the space being projected onto, a lowercase p is used and the subscript T is added to indicate the space projected onto.]

The matrix $I - P$, where I is the identity matrix (diagonal matrix whose entries equal one), is also a projection matrix. It projects $Y_{(3)}$ onto the space perpendicular to T_p. In the case of Figure 4.2-1 it would project the vector $Y_{(3)}$ onto the line perpendicular to the plane T_p forming the vector $Y_{(3)} - T X_3 = Y_{(3)} - p_T$.

The projection matrix P has two important properties. First it is symmetric [76], which means that

$$P^T = P \tag{4.2-26}$$

Second it is idempotent [76], that is,

$$PP = P^2 = P \tag{4.2-27}$$

Conversely, any matrix having these two properties is a projection matrix. For the general form given by (4.2-24) it projects $Y_{(n)}$ onto the column space of T [76].

A special case of interest is that where the column vectors t_i of T are orthogonal and have unit magnitude; such a matrix is called orthonormal. To indicate that the t_i have unit magnitude, that is, are unitary, we here rewrite t_i as $\hat{t}_i$. Then

$$\hat{t}_i^T \hat{t}_j = \begin{cases} 1 & \text{for } i = j \\ 0 & \text{for } i \neq j \end{cases} \tag{4.2-28}$$

Generally the t_i are not unitary and orthogonal; see, for example, (4.1-28) and (4.1-18). However, we shall show in Section 4.3 how to transform T so that the t_i are orthonormal. For an orthonormal matrix

$$T^T T = I \tag{4.2-29}$$

where I is the identity matrix. When T is orthonormal (4.2-25) becomes, for arbitrary m

$$\begin{aligned} P &= T T^T \\ &= \hat{t}_1 \hat{t}_1^T + \hat{t}_2 \hat{t}_2^T + \cdots + \hat{t}_{m'} \hat{t}_{m'}^T \end{aligned} \tag{4.2-30}$$

For the case where $m' = 1$

$$P = \hat{t}_1 \hat{t}_1^T \tag{4.2-31}$$

and

$$p_t = \hat{t}_1 \hat{t}_1^T Y_{(n)} \tag{4.2-32}$$

Here p_t is the projection of $Y_{(n)}$ onto the one-dimensional space T_p, that is, onto the unit vector $\hat{t}_1$.

When T is composed of m orthonormal vectors $\hat{t}_i$, we get

$$p_T = P Y_{(n)} = \hat{t}_1 \hat{t}_1^T Y_{(n)} + \hat{t}_2 \hat{t}_2^T Y_{(n)} + \cdots + \hat{t}_{m'} \hat{t}_{m'}^T Y_{(n)} \tag{4.2-33}$$

that is, p_T is the sum of the projections of $Y_{(n)}$ onto the orthonormal vectors $t_1, \ldots, t_{m'}$. Finally when T is orthonormal so that (4.2-29) applies, (4.2-16) becomes, for arbitrary m',

$$X_{n,n}^* = T^T Y_{(n)} \tag{4.2-34}$$

A better feel for the projection matrix P and its projection p_T is obtained by first considering the case $m' = 1$ above for which (4.2-31) and (4.2-32) apply.

Equation (4.2-32) can be written as

$$p_t = \hat{t}_1(\hat{t}_1^T Y_{(n)})$$
$$= (\hat{t}_1^T Y_{(n)})\hat{t}_1 \tag{4.2-35}$$

As implied above with respect to the discussion relative to Figure 4.2-1, $Y_{(n)}$ and $\hat{t}_1$ can be interpreted as s-dimensional vectors in hyperspace. Physically, in the above, $\hat{t}_1^T Y_{(n)}$ represents the amplitude of the projection of $Y_{(n)}$ onto the unit vector $\hat{t}_1$. The direction of the projection of $Y_{(n)}$ onto $\hat{t}_1$ is $\hat{t}_1$ itself. Hence the projection is the vector $\hat{t}_1$ with an amplitude $\hat{t}_1^T Y_{(n)}$ as given by (4.2-35).

Physically the amplitude of the projection of $Y_{(n)}$ onto the unitary vector $\hat{t}_1$ is given by the vector dot product of $Y_{(n)}$ with $\hat{t}_1$. This is given by

$$\hat{t}_1 \cdot Y_{(n)} = \|\hat{t}_1\| \cdot \|Y_{(n)}\| \cos\theta$$
$$= \|Y_{(n)}\| \cos\theta \tag{4.2-36}$$

where use was made in the above of the fact that $\hat{t}_1$ is unitary so that $\|\hat{t}_1\| = 1$, $\|A\|$ implies the magnitude of vector A, and θ is the angle between the vectors $\hat{t}_1$ and $Y_{(n)}$. If $\hat{t}_1$ is given by the three-dimensional t_1 of (4.2-2) and $Y_{(n)}$ by (4.2-4), then the dot product (4.2-36) becomes, from basic vector analysis,

$$\hat{t}_1 \cdot Y_n = t_{11}y_1 + t_{21}y_2 + t_{31}y_3 \tag{4.2-37}$$

For this case t_{i1} of (4.2-2) is the ith coordinate of the unit vector $\hat{t}_1$ in some three-dimensional orthogonal space; let us say x, y, z. In this space the coordinates x, y, z themselves have directions defined by respectively the unit vectors i, j, k given by

$$i = \begin{bmatrix} 1 \\ 0 \\ 0 \end{bmatrix} \qquad j = \begin{bmatrix} 0 \\ 1 \\ 0 \end{bmatrix} \qquad k = \begin{bmatrix} 0 \\ 0 \\ 1 \end{bmatrix} \tag{4.2-38}$$

Figure 4.2-2 illustrates this dot product for the two-dimensional situation. In this figure i and j are the unit vectors along respectively the x and y axes.

Let us now assume that t_1 is not unitary. In this case we can obtain the projection of $Y_{(n)}$ onto the direction of t_1 by making t_1 unitary. To make t_1 unitary we divide by its magnitude:

$$\hat{t}_1 = \frac{t_1}{\|t_1\|} \tag{4.2-39}$$

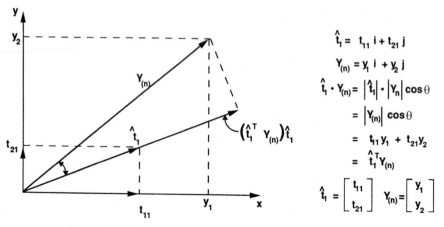

Figure 4.2-2   Projection of vector $Y_{(n)}$ onto unit vector t_1.

But the magnitude of t_1, also called its Euclidean norm, is given by

$$\| t_1 \| = \sqrt{t_1^T t_1} \tag{4.2-40}$$

Hence (4.2-35) becomes, for t_1 not unitary,

$$
p_t = \left(\frac{t_1^T}{\sqrt{t_1^T t_1}} Y_{(n)} \right) \frac{t_1}{\sqrt{t_1^T t_1}}
$$

$$
= \frac{(t_1^T Y_{(n)}) t_1}{t_1^T t_1} = \frac{t_1^T Y_{(n)} t_1}{\|t_1\|^2} \tag{4.2-41}
$$

This is the situation we had in (4.2-20). Thus we again see physically that the least-squares estimate $\hat{X}_{n,n}^*$ of (4.2-20) is the projection of $Y_{(n)}$ onto the direction of the nonunitary vector t_1 as it should be based on the discussion relative to Figure 4.2-1.

4.3   ORTHONORMAL TRANSFORMATION AND VOLTAGE-PROCESSING (SQUARE-ROOT) METHOD FOR LSE

We will now further develop our geometric interpretation of the LSE. We shall show how the projection of $Y_{(n)}$ onto the T_p space can be achieved without the need for the matrix inversion in (4.2-24). This involves expressing the column vectors of T in a new orthonormal space, not the original x, y, z space. We will then show how in this new space the least-squares estimate $X_{n,n}^*$ can in turn easily be obtained without the need for a matrix inversion. This approach is called the voltage-processing (square-root) method for obtaining the least-squares solution. Such approaches are less sensitive to computer round-off errors. Hence these methods should be used where computer round-off errors are a problem. With the rapid development of microcomputer chips that are more accurate (e.g., 32- and 64-bit floating-point computation chips), this problem is being diminished. Two voltage-processing algorithms, the Givens and Gram–Schmidt offer the significant advantage of enabling a high-throughput parallel architecture to be used. The voltage-processing methods will be discussed in much greater detail in Chapters 10 to 14. Here we introduce the method. Specifically, we introduce the Gram–Schmidt method. This method is elaborated on more in Chapter 13. Chapters 11 and 12 respectively cover the Givens and Householder voltage-processing methods.

As done in Figure 4.2-1, we shall for simplicity initially assume t_1, t_2, X_n, T, and $Y_{(n)}$ are given by (4.2-1) through (4.2-4). If the column space of T given by t_1 and t_2 were orthogonal and had unit magnitudes, that is, if they were orthonormal, then we could easily project $Y_{(3)}$ onto the T_p plane, it being given by (4.2-33), with $m' = 2$ and $n = 3$. In general t_1 and t_2 will not be orthogonal. However, we can still obtain the desired projection by finding from t_1 and t_2 an orthonormal pair of vectors in the T_p plane. Designate these unit vectors in the T_p plane as q_1 and q_2. We now show one way we can obtain q_1 and q_2. Pick q_1 along t_1. Hence

$$q_1 = \hat{t}_1 \tag{4.3-1}$$

where $\hat{t}_1$ is the unit vector along t_1 given by (4.2-39) when t_1 is not a unit vector, see Figure 4.3-1. In turn we pick q_2 perpendicular to $\hat{t}_1$ but in the T_p

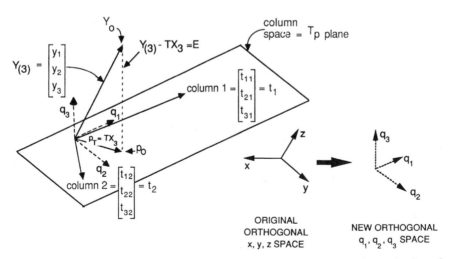

Figure 4.3-1 New orthonormal three-dimensional coordinate system for projection of data vector $Y_{(3)}$ onto column space of 3×2 T matrix. Used to obtain least-squares solution in three-dimensional space. (After Strang [76].)

plane; see Figure 4.3-1. Then the projection of $Y_{(3)}$ onto the T_p plane is given by the sum of the projection of $Y_{(3)}$ onto q_1 and q_2 as in (4.2-33), that is,

$$p_T = q_1^T Y_{(3)} q_1 + q_2^T Y_{(3)} q_2 \qquad (4.3\text{-}2)$$

By adding a third coordinate perpendicular to the T_p-plane whose direction is given by the unit vector q_3, we form a new orthogonal coordinate system in place of the original x, y, z coordinate system in which $Y_{(3)}$ and the t_i were originally defined; see Figure 4.3-1. The directions for x, y, z were given by the unit vectors i, j, k defined by (4.2-38). Those for the new orthogonal coordinate system are given by the unit vectors q_1, q_2, q_3; see Figure 4.3-1. In this new coordinate system, by using (4.2-33) and (4.2-35), we find that the vector $Y_{(3)}$ can be simply written as

$$Y_{(3)} = q_1^T Y_{(3)} q_1 + q_2^T Y_{(3)} q_2 + q_3^T Y_{(3)} q_3 \qquad (4.3\text{-}3)$$

The above can be written as a column matrix in this new q_1, q_2, q_3 orthonormal coordinate system given by

$$Y_{(3)} = \begin{bmatrix} q_1^T Y_{(3)} \\ q_2^T Y_{(3)} \\ q_3^T Y_{(3)} \end{bmatrix} = \begin{bmatrix} y_1' \\ y_2' \\ y_3' \end{bmatrix} \qquad (4.3\text{-}4)$$

where

$$y_1' = q_1^T Y_{(3)} \qquad (4.3\text{-}4a)$$
$$y_2' = q_2^T Y_{(3)} \qquad (4.3\text{-}4b)$$
$$y_3' = q_3^T Y_{(3)} \qquad (4.3\text{-}4c)$$

and where y_i' is the amplitude of the ith coordinate of $Y_{(3)}$ when expressed in the new orthonormal coordinate system defined by q_1, q_2, q_3. Expressed another way, y_i' is the amplitude of the vector component of $Y_{(3)}$ along the ith coordinate direction q_i in the new coordinate system. Here, $Y_{(3)}$ of (4.3-4) replaces $Y_{(3)}$ of (4.2-4), where $Y_{(3)}$ is expressed in x, y, z orthonormal coordinate system. To obtain $Y_{(3)}$ in the new coordinate system as given by (4.3-4), in (4.3-4a) to (4.3-4c) the unit vectors q_1, q_2, q_3 and the vector $Y_{(3)}$ are expressed in the original x, y, z coordinate system. Hence in (4.3-4a) to (4.3-4c)

$$q_j = \begin{bmatrix} q_{1j} \\ q_{2j} \\ q_{3j} \end{bmatrix} \qquad (4.3\text{-}5)$$

where q_{1j} is the amplitude of the component of q_j along the x coordinate, q_{2j}

that along the y coordinate, and q_{3j} that along the z coordinate. In the new q_1, q_2, q_3 coordinate system

$$q_1 = \begin{bmatrix} 1 \\ 0 \\ 0 \end{bmatrix} \quad q_2 = \begin{bmatrix} 0 \\ 1 \\ 0 \end{bmatrix} \quad q_3 = \begin{bmatrix} 0 \\ 0 \\ 1 \end{bmatrix} \tag{4.3-6}$$

We are now in a position to show how we can transform from $Y_{(3)}$ of (4.2-4) given in the x, y, z orthonormal coordinate system to the $Y_{(3)}$ of (4.3-4) in the q_1, q_2, q_3 coordinate system by the use of a matrix orthonormal transformation. Form a matrix Q from the three unit column vectors q_1, q_2, q_3, that is,

$$Q = [q_1 \quad q_2 \quad q_3] \tag{4.3-7}$$

or

$$Q = \begin{bmatrix} q_{11} & q_{12} & q_{13} \\ q_{21} & q_{22} & q_{23} \\ q_{31} & q_{32} & q_{33} \end{bmatrix} \tag{4.3-8}$$

where the coordinates of q_1, q_2, q_3 are expressed in the x, y, z coordinate system. Note that because the columns of (4.3-8) represent orthogonal unit vectors

$$Q^T Q = I \tag{4.3-9}$$

Hence

$$Q^T = Q^{-1} \tag{4.3-10}$$

and also

$$QQ^T = I \tag{4.3-11}$$

A matrix Q having the properties given by (4.3-9) to (4.3-11) is called an orthonormal transformation matrix.

On examining (4.3-4) to (4.3-4c) we see that

$$\begin{bmatrix} y_1' \\ y_2' \\ y_3' \end{bmatrix} = \begin{bmatrix} q_1^T \\ q_2^T \\ q_3^T \end{bmatrix} Y_{(3)} = Q^T Y_{(3)} = Q^T \begin{bmatrix} y_1 \\ y_2 \\ y_3 \end{bmatrix} \tag{4.3-12}$$

For convenience and clarity let $Y_{(3)}$ expressed as a column matrix in the new q_1,

q_2, q_3 coordinate system be written as

$$Y'_{(3)} = \begin{bmatrix} y'_1 \\ y'_2 \\ y'_3 \end{bmatrix} \tag{4.3-13}$$

and $Y_{(3)}$ as expressed in the original x, y, z coordinate system [as given by (4.2-4)] be written as $Y_{(3)}$ without the prime. Then (4.3-12) becomes

$$\begin{aligned} Y'_{(3)} &= Q^T Y_{(3)} \\ &= F Y_{(3)} \end{aligned} \tag{4.3-14}$$

where F defined by

$$F = Q^T \tag{4.3-15}$$

is the sought-after orthonormal transformation matrix that transforms $Y_{(3)}$ from its representation in the x, y, z orthonormal system to $Y'_{(3)}$ given in the q_1, q_2, q_3 orthonormal coordinate system. The rows of F are the columns of Q and hence satisfy the properties given by (4.3-9) to (4.3-11) for an orthonormal transformation matrix, that is,

$$F^T F = F F^T = I \tag{4.3-16}$$
$$F^T = F^{-1} \tag{4.3-17}$$

Now let us use this transformation to obtain the least-squares estimate $X^*_{n,n}$. This LSE is given by the $X^*_{n,n}$ that minimizes (4.1-31), or for the special case where range-only measurements y_i are being made by (4.1-37). Let

$$E = T X^*_{n,n} - Y_{(n)} \tag{4.3-18}$$

From (4.1-31) and (4.2-40) if follows that

$$\| E \|^2 = \| T X^*_{n,n} - Y_{(n)} \|^2 \tag{4.3-19}$$

We want to find the $X^*_{n,n}$ that minimizes (4.3-19). Let us now apply the orthonormal transformation F of (4.3-15) to E to obtain FE and determine the magnitude squared of FE. From (4.2-40)

$$\| FE \|^2 = (FE)^T FE = E^T F^T FE \tag{4.3-20}$$

Applying (4.3-16) to (4.3-20) yields

$$\| FE \|^2 = E^T E = \| E \|^2 \tag{4.3-21}$$

Hence applying an orthonormal transformation to E does not change its magnitude. Thus finding the $X_{n,n}^*$ that minimizes $\|FE\|$ is the same as finding the $X_{n,n}^*$ that minimizes $\|E\|$. From (4.3-18)

$$
\begin{aligned}
FE &= FTX_{n,n}^* - FY_{(n)} \\
&= (FT)X_{n,n}^* - FY_{(n)}
\end{aligned}
\tag{4.3-22}
$$

For simplicity let us again assume $s = 3$ and $m' = 2$ with (4.1-1) to (4.1-2a), (2.4-1a) and (2.4-1b) applying. For this case E is a 3×1 matrix. Let

$$
E = \begin{bmatrix} \varepsilon_1 \\ \varepsilon_2 \\ \varepsilon_3 \end{bmatrix}
\tag{4.3-22a}
$$

Then (4.3-18) becomes

$$
E = \begin{bmatrix} \varepsilon_1 \\ \varepsilon_2 \\ \varepsilon_3 \end{bmatrix} = \begin{bmatrix} t_{11} & t_{12} \\ t_{21} & t_{22} \\ t_{31} & t_{32} \end{bmatrix} \begin{bmatrix} x_1 \\ x_2 \end{bmatrix} - \begin{bmatrix} y_1 \\ y_2 \\ y_3 \end{bmatrix}
\tag{4.3-23}
$$

We now apply F to (4.3-23). First we apply F to the vectors t_1 and t_2 of T. In preparation for doing this note that, due to our choice of q_1, q_2, and q_3, t_1 is only composed of a component along q_1. Thus t_1 can be written as

$$
t_1 = u_{11}q_1 + 0 \cdot q_2 + 0 \cdot q_3
\tag{4.3-24}
$$

In turn t_2 consists of components along q_1 and q_2 so that it can be written as

$$
t_2 = u_{12}q_1 + u_{22}q_2 + 0 \cdot q_3
\tag{4.3-25}
$$

Finally $Y_{(3)}$ consists of the components y_1', y_2', y_3' along respectively q_1, q_2, q_3 so that it can be written as

$$
Y_{(3)} = y_1'q_1 + y_2'q_2 + y_3'q_3
\tag{4.3-26}
$$

The values of u_{ij} are the amplitudes of the unit vectors q_1, q_2, q_3 of which t_j is composed. These amplitudes can be easily obtained by applying an expression such as (4.2-36). This is done in Chapter 13. Now from (4.3-14) the transformation F applied to the column matrices t_1 and t_2 transforms these column vectors from being expressed in the x, y, z coordinate system to the q_1, q_2, q_3 orthogonal coordinate space. On examining (4.3-24) and (4.3-25) we thus see that the column matrices for t_1 and t_2 in the q_1, q_2, q_3 space are given by

$$
Ft_1 = \begin{bmatrix} u_{11} \\ 0 \\ 0 \end{bmatrix}
\tag{4.3-27}
$$

and

$$Ft_2 = \begin{bmatrix} u_{12} \\ u_{22} \\ 0 \end{bmatrix} \tag{4.3-28}$$

Hence

$$FT = \begin{bmatrix} u_{11} & u_{12} \\ 0 & u_{22} \\ --- & --- \\ 0 & 0 \end{bmatrix} = \left[\begin{matrix} U \\ --- \\ 0 \end{matrix}\right]\begin{matrix} \}2 \\ \\ \}1 \end{matrix} \tag{4.3-29}$$

where

$$U = \begin{bmatrix} u_{11} & u_{12} \\ 0 & u_{22} \end{bmatrix} \tag{4.3-29a}$$

Let

$$FE = \begin{bmatrix} \varepsilon_1' \\ \varepsilon_2' \\ \varepsilon_3' \end{bmatrix} \tag{4.3-30}$$

Substituting (4.3-13), (4.3-29), and (4.3-30) into (4.3-22) yields

$$\begin{bmatrix} \varepsilon_1' \\ \varepsilon_2' \\ \varepsilon_3' \end{bmatrix} = \begin{bmatrix} u_{11} & u_{12} \\ 0 & u_{22} \\ 0 & 0 \end{bmatrix} \begin{bmatrix} x_1 \\ x_2 \end{bmatrix} - \begin{bmatrix} y_1' \\ y' \\ y_3' \end{bmatrix} \tag{4.3-31}$$

Writing the above out yields

$$\varepsilon_1' = u_{11}x_1 + u_{12}x_2 - y_1' \tag{4.3-32a}$$
$$\varepsilon_2' = 0 \qquad\quad + u_{22}x_2 - y_2' \tag{4.3-32b}$$
$$\varepsilon_3' = 0 \qquad\quad + 0 \qquad - y_3' \tag{4.3-32c}$$

Examining (4.3-32a) to (4.3-32c), we see that the bottom equation does not contain any component of $X_{n,n}^*$. Hence it does not play a role in the determination of the $X_{n,n}^*$ that minimizes $\| E \|$ or equivalently $\| FE \|$. Only the top two equations enter into the determination of x_1 and x_2. As a result our determination of $X_{n,n}^*$ degenerates into finding the x_1 and x_2 that minimize ε_1' and ε_2'; specifically the x_1 and x_2 that minimizes

$$(\varepsilon_1')^2 + (\varepsilon_2')^2$$

Because we have only two equations [(4.3-32a) and (4.3-32b)] with only two unknowns (x_1 and x_2) to solve for, ε_1' and ε_2' can be forced to be zero. When doing this, (4.3-32a) and (4.3-32b) become

$$y_1' = u_{11}x_1^* + u_{12}x_2^* \qquad (4.3\text{-}33a)$$

$$y_2' = \qquad\quad +u_{22}x_2^* \qquad (4.3\text{-}33b)$$

where we have replaced x_1 and x_2 by their LSE values x_1^* and x_2^* because the solution of (4.3-33a) and (4.3-33b) gives us our desired least-squares solution. Equations (4.3-33a,b) can be written in matrix form as

$$Y_1' = UX_{3,3}^* \qquad (4.3\text{-}34)$$

where

$$Y_1' = \begin{bmatrix} y_1' \\ y_2' \end{bmatrix} \qquad (4.3\text{-}34a)$$

$$X_{3,3}^* = \begin{bmatrix} x_1^* \\ x_2^* \end{bmatrix} \qquad (4.3\text{-}34b)$$

and U is given by (4.3-29a). Physically, Y_1' is the projection of $Y_{(3)}$ onto the T_p plane that is designated as p_T in Figures 4.2-1 and 4.3-1.

Equation (4.3-34) consists of two equations with two unknowns. Hence it is not overdetermined, as is the case for (4.3-23). Thus we can solve (4.3-34) for $X_{3,3}^*$. This can be done by multiplying both sides by the inverse of U to obtain

$$X_{3,3}^* = U^{-1}Y_1' \qquad (4.3\text{-}35)$$

However, on examining (4.3-33a) and (4.3-33b) we see that obtaining the inverse of U to solve for $X_{3,3}^*$ is not necessary. The bottom equation only contains one unknown, x_2^*, which can readily be solved for. Having solved for x_2^*, x_1^* can be solved for using the top equation (4.3-33a). The forms of the equations given by (4.3-33a) and (4.3-33b) are like the forms obtained using the Gauss elimination procedure for solving simultaneous equations. Applying the Gauss elimination procedure to (4.3-23) would yield the Gauss elimination forms similar to those given by (4.3-32a) to (4.3-32c). However, the equations given by (4.3-32a) to (4.3-32c) are not identical to those obtained using the Gauss elimination procedure. Each equation of (4.3-32a) to (4.3-32c) will generally differ from the corresponding one obtained by the Gauss elimination procedure by a constant multiplier, which does not change the solution. How they differ will become apparent when the Givens orthonormal transformation F is further described in Chapters 10 and 11. Having the equations in the Gauss elimination form allows us to solve the simultaneous equation by starting with the bottom equation to solve for x_2^* and then using the top equation to solve for x_1^*. This process of starting with the bottom equation, which only has x_2^* in it, and then going to the top equation, which has x_1^* and x_2^*, to solve for x_1^* is called the back-substitution method; see Chapter 10, specifically (10.2-16) to (10.2-22) and the text relating to these equations.

In the above development we assumed $s = 3$ and $m' = 2$. As shall be seen shortly in this section and as elaborated on more fully in Chapter 10, for the general case where s and m' are arbitrary, on applying the orthonormal transformation F to the general form of E given by (4.3-18), one again obtains the Gauss elimination form of (4.3-32a) to (4.3-32c) with only the top m' equations containing the variables $x_1, \ldots, x_{m'}$. Moreover, the bottom equation of these m' equations contains only the variable $x_{m'}$, the next one up $x_{m'}$ and $x_{m'-1}$, and so on. Because these top m' equations form m' unknowns, $\varepsilon_1' \ldots, \varepsilon_{m'}'$ can be set equal to zero and $x_1, \ldots, x_{m'}$ replaced by $x_1^*, \ldots, x_{m'}$ which can be solved for by the back-substitution method discussed above.

Let us return to the above $s = 3$, $m' = 2$ case and determine the minimum $\| E \|^2$ when $X_{n,n}^*$ of (4.3-19) is the sought-after least-squares solution. From (4.3-21) we know that the minimum of $\| E \|^2$ is equal to the minimum of $\| FE \|^2$. Using the definition of Euclidean norm given by (4.2-40) and the matrix form of FE given by (4.3-30) yields

$$\| FE \|^2 = \varepsilon_1'^2 + \varepsilon_2'^2 + \varepsilon_3'^2 \qquad (4.3-36)$$

But from the above discussion we know that $\| FE \|$ is minimized by setting ε_1' and ε_2' to zero. Hence when the LSE is obtained,

$$\| FE \|^2 = \varepsilon_3'^2 \qquad (4.3-37)$$

Applying (4.3-32c) to (4.3-37) yields

$$\min \| E \| = \min \| FE \|^2 = y_3'^2 \qquad (4.3-38)$$

Physically y_3' is the projection of $Y_{(3)}$ onto the axis perpendicular to the T_p plane, or equivalently onto the q_3 unit vector; see Figure 4.3-1.

We will now extend the detailed results obtained above for the case $s = 3$, $m' = 2$ to the case of arbitrary s and m'. To lead up to the general results, we start by putting the results for the case $s = 3$, $m' = 2$ into a more general form. This we do by breaking F up for the case $s = 3$, $m' = 2$ above into two parts so that it is written as

$$F = \begin{bmatrix} F_1 \\ --- \\ F_2 \end{bmatrix} \begin{matrix} \} 2 \\ \\ \} 1 \end{matrix} \qquad (4.3-39)$$

$$\underbrace{\phantom{F = \begin{bmatrix} F_1 \end{bmatrix}}}_{3}$$

The first part of F, designated as F_1, consists of the first two rows of F, which are the unit row vectors q_1^T and q_2^T. Hence

$$F_1 = \begin{bmatrix} q_1^T \\ q_2^T \end{bmatrix} = \underbrace{\begin{bmatrix} f_1 \\ f_2 \end{bmatrix} \} 2}_{3} \qquad (4.3\text{-}40)$$

where

$$f_1 = q_1^T \qquad (4.3\text{-}40a)$$
$$f_2 = q_2^T \qquad (4.3\text{-}40b)$$

Physically when F_1 multiplies $Y_{(3)}$, it projects $Y_{(3)}$ onto the T_p plane. The second part of F, F_2, consists of the third row of F, which is the row unit vector q_3^T, the unit vector perpendicular to the T_p plane. Hence

$$F = \begin{bmatrix} q_3^T \end{bmatrix} = \underbrace{[f_3]}_{3} \} 1 \qquad (4.3\text{-}41)$$

where

$$f_3 = q_3^T \qquad (4.3\text{-}41a)$$

Applying F of (4.3-39) to E as given by (4.3-18) yields

$$FE = \underbrace{\begin{bmatrix} F_1 T X_{n,n}^* \\ \text{----------} \\ F_2 T X_{n,n}^* \end{bmatrix}}_{1} - \underbrace{\begin{bmatrix} Y_1' \\ \text{----} \\ Y_2' \end{bmatrix}}_{1} \begin{matrix} \} m' = 2 \\ \} 1 \end{matrix} \qquad (4.3\text{-}42)$$

where

$$Y_1' = F_1 Y_{(3)} = \underbrace{\begin{bmatrix} y_1' \\ y_2' \end{bmatrix}}_{1} \} m' = 2 \qquad (4.3\text{-}42a)$$

$$Y_2' = F_2 Y_{(3)} = \underbrace{\begin{bmatrix} y_{(3)}' \end{bmatrix}}_{1} \} 1 \qquad (4.3\text{-}42b)$$

But from (4.3-29) it follows that

$$FT = \begin{bmatrix} F_1 \\ \text{---} \\ F_2 \end{bmatrix} T = \begin{bmatrix} F_1 T \\ \text{---} \\ F_2 T \end{bmatrix} = \underbrace{\begin{bmatrix} U \\ \text{---} \\ 0 \end{bmatrix}}_{m' = 2} \begin{matrix} \} m' = 2 \\ \} 1 \end{matrix} \qquad (4.3\text{-}43)$$

Applying (4.3-43) to (4.3-42) with $X^*_{n,n}$ written as X_3 yields

$$FE = \begin{bmatrix} UX_3 \\ \text{---} \\ 0 \end{bmatrix} - \begin{bmatrix} Y'_1 \\ \text{---} \\ Y'_2 \end{bmatrix} \tag{4.3-44}$$

or

$$E' = \begin{bmatrix} UX_3 - Y'_1 \\ \text{-------------} \\ -Y'_2 \end{bmatrix} \tag{4.3-45}$$

where

$$E' = FE \tag{4.3-45a}$$

Physically E' is E in the transformed coordinate system, the q_1, q_2, q_3 coordinate system.

From (4.3-30) and (4.2-40) we know that

$$\| FE \| = \sum_{i=1}^{3} \varepsilon_i'^2 \tag{4.3-46}$$

$$= \sum_{i=1}^{2} \varepsilon_i'^2 + \varepsilon_3'^2 \tag{4.3-46a}$$

But

$$\| UX_3 - Y'_1 \|^2 = \sum_{i=1}^{2} \varepsilon_i'^2 \tag{4.3-47}$$

$$\| Y'_2 \| = \varepsilon_3'^2 \tag{4.3-48}$$

Therefore

$$\| FE \|^2 = \| UX_3 - Y'_1 \|^2 + \| Y'_2 \|^2 \tag{4.3-49}$$

Although (4.3-49) was developed for the special case $s = 3$, $m' = 2$, it applies for arbitrary s and m', in which case Y'_1 has dimension $m' \times 1$ and Y'_2 has dimension 1×1; see Chapters 10 to 12. For the general case, physically Y'_1 is the projection of $Y_{(n)}$ onto the m'-dimensional space spanned by the m' columns of T. Physically Y'_2 is projection of $Y_{(n)}$ onto a coordinate perpendicular to m'-dimensional space spanned by the m' columns of T. Equivalently Y'_2

is the component of $Y_{(n)}$ perpendicular to Y_1'. Hence

$$Y_2' = Y_{(n)} - Y_1' \tag{4.3-50}$$

Let us relate the above general results to those obtained for the special three-dimensional case of Figures 4.2-1 and 4.3-1. The term Y_1' of (4.3-50) is equivalent to

$$p_T = TX_3 \tag{4.3-51}$$

[see (4.2-8)] while Y_2' is equivalent to

$$E = Y_{(3)} - TX_3 \tag{4.3-52}$$

Here, Y_1' is the only part of $Y_{(n)}$ that plays a role in determining the least-squares estimate $X_{n,n}^*$. The part perpendicular to Y_1', which is Y_2', plays no part in the determination of the least-squares estimate $X_{n,n}^*$, it not being a function of Y_2'; see (4.3-32c) and (4.3-44). The m'-dimensional vector Y_1' is a sufficient statistic; that is, Y_1' is a sufficient statistic of $Y_{(n)}$ for finding the least-squares solution. The transformation F transforms the original set of s equations with m' unknowns to m' equations with m' unknowns. Moreover, as indicated earlier (and to be further discussed in Chapters 10 to 13), these m' equations are in the Gauss elimination form and can be solved easily by the back-substitution method without resorting to a matrix inversion. As discussed before, physically the transformation F transforms the representation of the vectors t_i and $Y_{(n)}$ from the original $x, y, z, \ldots$ orthogonal coordinate system to the new orthogonal coordinate system represented by the orthonormal vectors $q_1, q_2, q_3, \ldots$. In general, the column space of T is formed by the column vectors $t_1, t_2, \ldots, t_{m'}$, or equivalently, by the new set of orthonormal unit vectors $q_1, q_2, \ldots, q_{m'}$. The term T can be augmented to include $Y_{(n)}$; specifically, to T is added the column vector $t_{m'+1}$, where

$$t_{m'+1} = Y_{(n)} \tag{4.3-53}$$

so that the augmented T becomes

$$T_0 = [t_1 \quad t_2 \quad \cdots \quad t_{m'} \quad t_{m'+1}] \tag{4.3-54}$$

Now the $(m' + 1)$-dimensional space of T_0 is represented by the orthonormal unit vectors $q_1, q_2, \ldots, q_{m'+1}$. These unit vectors are chosen in a similar manner to that used for the $(m' + 1 = 3)$-dimensional case discussed above; see Figures 4.2-1 and 4.3-1. Specifically, q_1 is chosen to line up with t_1; q_2 to be in the plane of t_1 and t_2 but orthogonal to t_1; q_3 to be in the three-dimensional space of t_1, t_2, and t_3 but orthogonal to t_1 and t_2; and so on, to $q_{m'+1}$ to be orthogonal to the m'-dimensional space of T. Moreover $q_{m'+1}$ lines up with the part of $Y_{(n)}$ not in the space of T, that is, with

$$Y_{(n)} - Y_1' = Y_2' \tag{4.3-55}$$

Also

$$\| Y_2' \| = \min \| E \| \tag{4.3-56}$$

In summary, for a general s-dimensional $Y_{(n)}$, Y_1' is the projection of $Y_{(n)}$ onto the m'-dimensional column space of T represented by $q_1, q_2, \ldots, q_{m'}$, while Y_2' is the projection of $Y_{(n)}$ onto the coordinate $q_{m'+1}$ that lines up with $Y_{(n)} - Y_1'$ and is perpendicular to the m'-dimensional column space of T. The remaining coordinates of the s-dimensional row space of T and $Y_{(n)}$ are defined by unit vectors $q_{m'+2}, \ldots, q_s$, which are orthogonal to the $(m' + 1)$-dimensional column space of T_0 represented by $q_1, q_2, \ldots, q_{m'+1}$. Hence $Y_{(n)}$ when represented in this space has no components along $q_{m'+2}, \ldots, q_s$ Thus $Y'_{(n)}$, which represents $Y_{(n)}$ expressed in the new orthonormal space $q_1, \ldots, q_s$ is given by

$$
Y'_{(n)} =
\begin{bmatrix}
Y_1' \\
\hline
Y_2' \\
\hline
0
\end{bmatrix}
\begin{array}{l}
\}m' \\
\}1 \\
\}s - m' - 1
\end{array}
\tag{4.3-57}
$$
$$\underbrace{}_{1}$$

Furthermore for this general case (4.3-39) becomes

$$
F =
\begin{bmatrix}
F_1 \\
\hline
F_2 \\
\hline
F_3
\end{bmatrix}
\begin{array}{l}
\}m' \\
\}1 \\
\}s - m' - 1
\end{array}
\tag{4.3-58}
$$
$$\underbrace{}_{s}$$

where

$$
F_1 =
\left.
\begin{bmatrix}
q_1^T \\
q^T \\
\vdots \\
q_{m'}^T
\end{bmatrix}
\right\} m'
\tag{4.3-58a}
$$
$$\underbrace{}_{s}$$

$$
F_2 = \underbrace{\left[q_{m'+1}^T \right]}_{s} \} 1
\tag{4.3-58b}
$$

and

$$
F_3 = \left.\begin{bmatrix} q_{m'+2}^T \\ q_{m'+3}^T \\ \vdots \\ q_s^T \end{bmatrix}\right\} s - m' - 1 \qquad (4.3\text{-}58c)
$$

$$
\underbrace{\phantom{\begin{bmatrix} q_{m'+2}^T \end{bmatrix}}}_{s}
$$

Physically F_1 projects $Y_{(n)}$ onto the m'-dimensional column space of T; F_2 projects $Y_{(n)}$ onto the coordinate aligned along the unit $q_{m'+1}$, which is in turn lined up with $Y_{(n)} - Y_1'$ [see (4.3-50)]; and finally F_3 projects $Y_{(n)}$ onto the space $q_{m'+2}, \ldots, q_s$ orthogonal to $q_1, \ldots, q_{m'+1}$ forming a null vector, as indicated in (4.3-57).

Using (4.3-58) the general form of (4.3-43) becomes

$$
FT = \begin{bmatrix} F_1 \\ \text{---} \\ F_2 \\ \text{---} \\ F_2 \end{bmatrix} T = \begin{bmatrix} F_1 T \\ \text{---} \\ F_2 T \\ \text{---} \\ F_3 T \end{bmatrix} = \left.\begin{bmatrix} U \\ \text{---} \\ 0 \\ \text{---} \\ 0 \end{bmatrix}\right\} \begin{matrix} m' \\ \\ 1 \\ \\ s - m' - 1 \end{matrix} \qquad (4.3\text{-}59)
$$

$$
\underbrace{\phantom{\begin{bmatrix} U \end{bmatrix}}}_{m'}
$$

In turn, if F is applied to the augmented matrix T_0 given by (4.3-54) with $t_{m'+1}$ given by (4.3-53), we obtain, for arbitrary s and m',

$$
FT_0 = \left.\begin{bmatrix} U & | & Y_1' \\ \text{---} & - & \text{--} \\ 0 & | & Y_2' \\ \text{---} & - & \text{--} \\ 0 & | & 0 \\ \text{---} & - & \text{--} \end{bmatrix}\right\} \begin{matrix} m' \\ \\ 1 \\ \\ s - m' - 1 \end{matrix} \qquad (4.3\text{-}60)
$$

$$
\underbrace{}_{m'} \underbrace{}_{1}
$$

This follows from (4.3-57) and (4.3-59).

The computation of the orthonormal vectors $q_1, \ldots, q_{m'}$ could be done off-line in advance if the transition–observation T were known in advance. For example, if the time T between measurements was fixed for the transition–observation matrix T of (4.1-18) or (4.1-28), then the matrix T would be known in advance for arbitrary s and hence $q_1, \ldots, q_{m'}$, could be calculated off-line in advance. However, implicit in having the matrix T be given by (4.1-18) or (4.1-28) is that a track update measurement y_i is obtained at every observation time. In real-world trackers, this is not usually the case, the target echo fading at random times so that at these times no observation is made of y_i at these times i. In this practical case T is not known in advance.

More details on the above Gram-Schmidt method for generation of the orthonormal set of unit vectors $q_1, q_2, \ldots, q_s$ are given in Chapter 13. Further details on the transformation F in general and on two other forms of this transformation, the Givens and Householder, are given in respectively Chapters 10 and 12.

4.4 ADAPTIVE NULLING, THE ORTHONORMAL TRANSFORMATION, AND THE LSE

The orthonormal transformation transcribed above can be applied to obtain a solution to the adaptive sidelobe cancellation problem in a manner paralleling closely that in which the least-squares solution was obtained using the orthonormal transformation. Figure 4.4-1 shows a typical adaptive sidelobe canceler. The sidelobe canceler is used to cancel out jammer (or other)

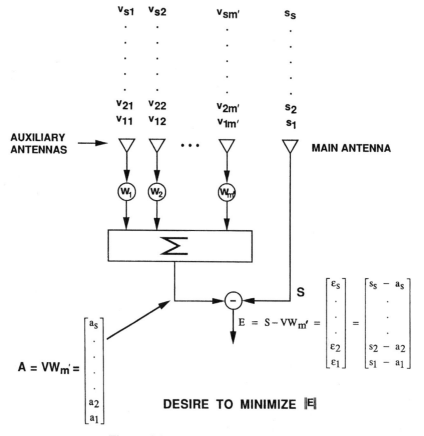

Figure 4.4-1   Sidelobe canceler (SLC).

interference coming in through the antenna sidelobes of the radar's (or communication system's) main antenna [1, 83–87]. Auxiliary antennas are used to generate a replica of the jammer signal in the main antenna output. By substracting this replica from the output of the main antenna, the jammer interference in the main channel is removed. We will elaborate more fully on this sidelobe canceler problem and how the auxiliary channels are used to generate the jammer replica in the remainder of this section. Those not interested in the sidelobe canceler problem may skip the remainder of this section.

For a radar system usually the main antenna has a high-gain, narrow-beam pattern. This beam is pointed in a direction of space where the target is being looked for. To be specific, assume the main beam is formed by a horizontally oriented linear array antenna consisting of P radiating elements [88]. This array electronically scans a vertically oriented fan beam in azimuth in order to locate targets. Figure 4.4-2 shows the main-beam antenna pattern in azimuth when the peak of the main-antenna main lobe is pointed at an angle θ_0. Shown in Figure 4.4-2 is the main-antenna main lobe and its sidelobes. As indicated, the radar is

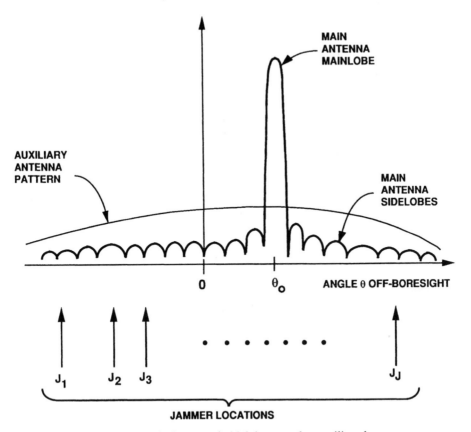

Figure 4.4-2 Main beam and sidelobe canceler auxiliary beam.

looking for the possible presence of a target in the main-antenna main lobe pointed in the direction θ_0.

For Figure 4.4-1 the voltage s_i represents the ith time sample of the main-antenna output. This signal consists of the weak echo from the target when a target is present in the antenna main lobe plus interference. The interference is composed of the radar receiver noise and, if jammers are present, the jammer signal. The jammers are assumed to be located at azimuth angles other than the one where the antenna main lobe is located; see Figure 4.4-2. Hence the jammers are located so as to be coming in through the sidelobes of the main antenna, as indicated in Figure 4.4-2. Here, J jammers are assumed to be present.

The jammer signals coming in through the main-antenna sidelobes can be large in amplitude. Specifically the total jammer interference in the main-antenna output can be 10's of decibels larger than the receiver thermal noise in the main-channel output. As a consequence the interference can reduce the radar system sensitivity by 10's of decibels. Hence, it is desired to remove the interference jammer signals from the main-channel output. If we could generate using another antenna, or antennas, a replica of the jammer signal in the main-channel output, we could subtract it from the main-channel output and as a result eliminate the jammer interference present in the main channel. It is the m' auxiliary antennas of Figure 4.4-1 that are used to generate this replica of the jammer signal in the main channel.

The voltage v_{ij} of Figure 4.4-1 represents the voltage signal output form the jth auxiliary antenna at time i. This signal consists of J-jammer signals plus the thermal noise at the output of the auxiliary-antenna receiver. The echo signal in the auxiliary-antenna outputs is assumed negligible. The auxiliary antenna could be obtained using a random set m' of the P element linear main array antenna. These J elements are shared with the main antenna. Each such radiating element has a low-gain, broad-beam antenna pattern. The pattern formed from one of the auxiliary radar elements is shown in Figure 4.4-2. The other auxiliary radiating elements will have the same antenna pattern. It is possible to find a set of weightings w_j for the auxiliary antenna outputs that when added form the desired replica of the total jammer signal in the main-antenna output.

In order to find such weights, it is necessary that the number of auxiliary antenna be greater than or equal to the number J of jammers present. The question remains as to how to find the weights w_j, $j = 1,,\ldots, m'$, that will generate a replica of the total jammer interference. We shall now show how these weights are obtained by using an orthonormal transformation as done above for the least-squares problems.

Represent the set of signal voltages s_i, $i = 1, 2, \ldots, s$, by the column matrix

$$ s = \begin{bmatrix} s_1 \\ s_2 \\ \vdots \\ s_s \end{bmatrix} \tag{4.4-1} $$

the auxiliary channel voltages $v_{ij}, i = 1, 2, \ldots, s, j = 1, 2, \ldots, m'$, by the matrix

$$V = \begin{bmatrix} v_{11} & v_{12} & \cdots & v_{1m'} \\ v_{21} & v_{22} & \cdots & v_{2m'} \\ \vdots & & & \vdots \\ v_{s1} & v_{s2} & \cdots & v_{sm'} \end{bmatrix} \qquad (4.4\text{-}2)$$

and the weights $w_j, j = 1, 2, \ldots, m'$, by the column matrix

$$W_{m'} = \begin{bmatrix} w_1 \\ w_2 \\ \vdots \\ w_{m'} \end{bmatrix} \qquad (4.4\text{-}3)$$

Let a_i be the ith time sample output of the auxiliary channel summer of Figure 4.4-1. These a_i can be represented by the column matrix

$$A = \begin{bmatrix} a_1 \\ a_2 \\ \vdots \\ a_s \end{bmatrix} \qquad (4.4\text{-}4)$$

It follows from the above that

$$A = V W_{m'} \qquad (4.4\text{-}5)$$

Note that

$$a_i = V_{im'} W_{m'} \qquad (4.4\text{-}6)$$

where $V_{im'}$ is the ith row of V, that is,

$$V_{im'} = \begin{bmatrix} v_{i1} & v_{i2} & \cdots & v_{im'} \end{bmatrix} \qquad (4.4\text{-}6a)$$

Let ε_i be the difference between the main-channel output voltage s_i and the auxiliary-antenna summer output voltage a_i, that is,

$$\varepsilon_i = s_i - a_i \qquad (4.4\text{-}7)$$

Let the column matrix of these differences be given by

$$E = \begin{bmatrix} \varepsilon_1 \\ \varepsilon_2 \\ \vdots \\ \varepsilon_s \end{bmatrix} \qquad (4.4\text{-}8)$$

Then from (4.4-1) to (4.4-8)

$$E = S - V W_{m'} \qquad (4.4\text{-}9)$$

The sidelobe canceler problem can be stated as finding the weights w_i, $i = 1, \ldots, m'$, that cause the ε_i to be minimum. The voltages a_i do not contain the target echo signal (because of the low gain of the auxiliary antenna). Hence the weak echo signal in the main channel will not be canceled out. Thus minimizing ε_i will result in a minimum interference and maximum signal-to-interference ratio at the output of the main channel. To find the w_i that minimize the ε_i, we proceed just as was done for the least-squares problem above in Section 4.3. Specifically, we minimize ε_i by minimizing the sum of the squares of the ε_i, that is, by minimizing

$$\sum_{i=1}^{s} \varepsilon_1^2 = \| E \|^2 = E^T E \tag{4.4-10}$$

Minimizing $\| E \|^2$ maximizes the output signal-to-interference ratio as long as the signal in the auxiliary channels do not effectively contain any signal component [94].

To obtain the minimum of $\| E \|^2$, we parallel the minimization procedure used in Section 4.3 above. Specificaly, first, we apply an orthonormal transformation to (4.4-9) F to E to obtain

$$FE = - \underbrace{\begin{bmatrix} U \\ -- \\ 0 \end{bmatrix}}_{m'} W_{m'} + \underbrace{\begin{bmatrix} S_1' \\ -- \\ S_2' \end{bmatrix}}_{1} \begin{matrix} \} m' \\ \\ \} s - m' \end{matrix} \tag{4.4-11}$$

which becomes

$$FE = - \begin{bmatrix} UW_{m'} \\ \text{-------} \\ 0 \end{bmatrix} + \begin{bmatrix} S_1' \\ --- \\ S_2' \end{bmatrix} \begin{matrix} \} m' \\ \\ \} s - m' \end{matrix}$$

$$= - \begin{bmatrix} UW_{m'} - S_1' \\ \text{-------------} \\ -S_2' \end{bmatrix} \tag{4.4-12}$$

where

$$U = F_1 V \tag{4.4-12a}$$

$$F_2 V = [0] \tag{4.4-12b}$$

$$S_1' = F_1 S \tag{4.4-12c}$$

$$S_2' = F_2 S \tag{4.4-12d}$$

$$F = \begin{bmatrix} F_1 \\ -- \\ F_2 \end{bmatrix} \begin{matrix} \} m' \\ \\ \} s - m' \end{matrix} \tag{4.4-12e}$$

and where U is an upper triangular matrix as was the case for the least-squares problem [see, e.g., (4.3-29a)] and F_1 is the first m' rows of F and F_2 the remaining $s - m'$.

Here F_1 and F_2 have parallel physical meanings to the F_1 and F_2 of the least-squares problem above. As in the case for the least-squares problem, F transforms the s-dimensional column vectors of the matrix V to a new s-dimensional orthonormal coordinate system whose unit vector directions are defined by the rows of F. The first m' orthonormal row vectors of F, designated as F_1, define a subspace of the s-dimensional space. Specifically they define the m'-dimensional space spanned and defined by the m' column vectors of V. The remaining $s - m'$ orthonormal row vectors of F_1, designated as F_2, are orthogonal to the space spanned by F_1 and form the remaining $(s - m')$-dimensional space of the s-dimsensional space. Thus, physically S_1' given by (4.4-12c) is the projection of S onto the space spanned by F_1, or equivalently, the column space of V. In turn S_2' is the projection of S onto the $(s - m')$-dimensional space orthogonal to F_1.

From (4.3-20), (4.3-45) and (4.3-49) the magnitude squared of (4.4-12) becomes

$$\| FE \|^2 = \| E \|^2 = \| UW_{m'} - S_1' \|^2 + \| S_2' \|^2 \tag{4.4-13}$$

We can now determine the w_i that minimizes $\| E \|^2$ above. The term $\| S_2' \|^2$ is independent of the w_i and hence plays no role in the minimization of $\| E \|^2$. Only the term $\| UW_{m'} - S_1' \|^2$ plays a role in this minimization. Because this term consists of m' equations with m' unknown w_i, it can be set equal to zero [just as is done to obtain (4.3-33) from (4.3-32)]. Thus

$$\| UW_{m'} - S_1' \|^2 = 0 \tag{4.4-14}$$

The above is true when

$$UW_{m'} = S_1' \tag{4.4-15}$$

Here we have m' equations to solve for the m' w_i. Because U is upper triangular, the equations of (4.4-15) are in the Gauss elimination forms and hence the back-substitution method can be used to solve for the w_i as done for (4.3-33a) and (4.3-33b) above for the least-squares problem. Thus, we have found the w_i that minimize the main-channel output interference for Figure 4.4-1.

The procedure described above for obtaining the weights of the sidelobe canceler is known as the voltage-processing method, just as it was when used for the least-squares problem of Section 4.3. This is opposed to the power methods, which obtain the weights w_i using the equation [83–86, 94]

$$W_{m'} = \hat{M}^{-1} \hat{\rho} \tag{4.4-16}$$

where the $m' \times n$ matrix $\hat{M}$ is the estimate of the spatial covariance matrix of the voltages v_{ij} across the array and $\hat{\rho}$ is a column matrix of the estimates of the cross correlations between the main-antenna output and the m' auxiliary-channel outputs. Specifically, the p, q element of $\hat{M}$ designated as $\hat{m}_{pq}$ is an estimate of

$$m_{pq} = \overline{v_{ip}v_{iq}^{*}} \tag{4.4-17}$$

where the overbar means time average, that is, average over the time index i and the asterisk signifies complex conjugate, the voltage samples being complex numbers. In (4.4-17) the mean of v_{ij} is assumed to be zero. Typically $\hat{m}_{pq}$ can be obtained from

$$\hat{m}_{pq} = \frac{1}{r}\sum_{i=1}^{r} v_{ip}v_{iq}^{*} \tag{4.4-18}$$

for r large enough. When $p = q$, $\hat{m}_{pq}$ is a power term. For this reason the method using (4.4-16) to obtain $W_{m'}$ is referred to as the power method. For the voltage-processing method the power in the signal is not calculated in order to determine $W_{m'}$. Instead the signal terms are used to determine $W_{m'}$; hence this method is referred to as the voltage method. The above voltage-processing method for doing sidelobe canceling is described in references 83 and 89.

The jth element of the s-dimensional column matrix $\hat{\rho}$ is an estimate of

$$\rho_j = \overline{v_{ij}s_i^{*}} \tag{4.4-19}$$

This estimate can be obtained using

$$\hat{\rho}_j = \frac{1}{r}\sum_{i=1}^{r} v_{ij}s_i^{*} \tag{4.4-20}$$

Here again we are calculating power terms, that is, the product of voltages.

A dedicated circuit implementation of the voltage-processing method for a sidelobe canceler is described in references 83 and 89–92. This implementation uses a parallel-processor architecture called a systolic array. This type of processor can be used to solve the least-squares problem via the voltage-processing method as described above. A description of the use of the systolic array to solve the least-squares problem is given in Section 11.3.

Although the sidelobe canceler was applied to a linear array above, it applies to two-dimensional arrays as well and to the case where the main antenna is a reflector with the auxiliary antennas being other antennas, usually of low gain.

In the above, the sidelobe canceler of Figure 4.4-1 was physically viewed as generating by the use of the auxiliary antenna, a replica of the jammer interference in the main antenna with this replica being subtracted from the

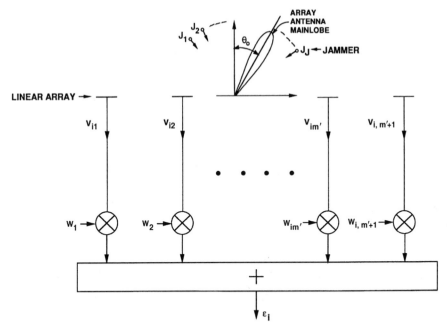

Figure 4.4-3 Fully adaptive phased array.

main-antenna output to cancel out this interference in the main channel. It is also useful to view the sidelobe canceler from another point of view to obtain further insight into it. Specifically, the auxiliary antenna can be viewed as putting nulls in the sidelobes of the main antenna in the directions that the jammers are located with the consequence that they are prevented from entering the main antenna.

The least-squares problem described above can be applied to a fully adaptive array antenna. An example of such an array antenna is the linear array of $m' + 1$ elements shown in Figure 4.4-3. Here the array main beam is set to be steered to an off-boresight angle θ_0. This is the angle at which a target is to be looked for by the radar. Assume, however, that J jammers are present, as shown in Figure 4.4-3. The jammers are located at angles other than the angle θ_0 of the main beam so that the jammer interference signals come in through the sidelobes of the array antenna. These interference signals are assumed to be strong enough so as to be larger in total than the radar receiver noise. What is desired then is to adaptively change the weights w_i of the array so as to remove the interfering signals while still maintaining the main-beam lobe in the direction θ_0 at which one is looking for a target.

McWhirter [93; see also 83, 94] has shown that the fully adaptive array problem is exactly equivalent to be sidelobe canceler problem discussed earlier. He showed that the fully adaptive array problem is transformed to a sidelobe canceler problem if the fully adaptive array is followed by the preprocessor

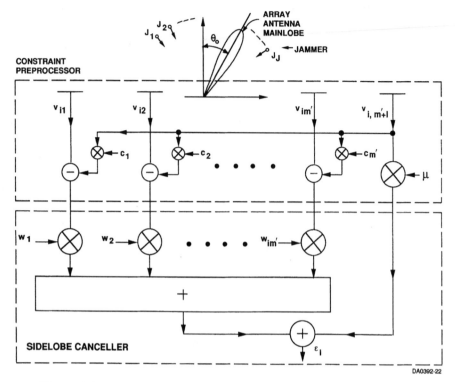

DA0392-22

Figure 4.4-4 Transformation of fully adaptive array to sidelobe canceler.

shown in Figure 4.4-4. At the output of the preprocessor the $(m' + 1)$st array element output times a constant μ becomes the main-channel input for the sidelobe canceler that follows the preprocessor while the remaining m' outputs of the preprocessor becomes the auxiliary outputs for the sidelobe canceler. Physically the constants $c_1, c_2, \ldots, c_{m'}$ of Figure 4.4-4 together with $c_{m'+1} = 1$ for the $(m' + 1)$st array element represent the weights to be used to steer the array to the off-boresight angle θ_0 if no jammer is present. Let this set of $m' + 1$ constants be represented by the $(m' + 1)$-dimensional steering vector $C_{m'+1}$. This steering vector is given by [88].

$$
C_{m'+1} = \begin{bmatrix} c_1 \\ c_2 \\ \vdots \\ c_r \\ \vdots \\ c_{m'} \\ c_{m'+1} \end{bmatrix} = \begin{bmatrix} \varepsilon^{jm'\phi_0} \\ \varepsilon^{j(m'-1)\phi_0} \\ \vdots \\ \varepsilon^{j(m'-r+1)\phi_0} \\ \vdots \\ \varepsilon^{j\phi_0} \\ 1 \end{bmatrix} \qquad (4.4\text{-}21)
$$

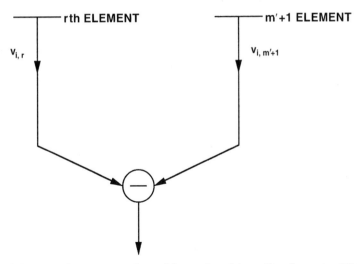

Figure 4.4-5   Two-element array formed from rth and $(m+1)$st elements of full array.

where

$$\phi_0 = \frac{2\pi d \sin \theta_0}{\lambda} \tag{4.4-21a}$$

and where d is the separation between adjacent radiating elements and λ is the propagation wavelength of the radar transmitted signal as in (3.5-3).

In Figure 4.4-4 the $(m'+1)$st element forms a two-element antenna array with each of the other elements. Consider the two-element array formed by the $(m'+1)$st element and rth element. Figure 4.4-5 shows such a two-element array. This two-element array has an antenna gain pattern that has a null in the direction θ_0, the direction in which we want to look for a target and hence the direction where we want the adapted array to have a high-gain beam formed. It is because of this null that the sidelobe canceler following the preprocessor of Figure 4.4-4 does not corrupt the formation of the main lobe in the direction θ_0. The magnitude of the voltage gain pattern of the two-element array antenna is given by [88]

$$G(\theta) = |\sin\left[\tfrac{1}{2}(m' - r + 1)\delta\right]| \tag{4.4-22}$$

where

$$\delta = \frac{2\pi d(\sin \theta - \sin \theta_0)}{\lambda} \tag{4.4-22a}$$

Examining (4.4-22) we see that, as we desired to show, the two-element array antenna has a null in the direction θ_0, the direction at which we want to look for a target.

We will now develop the circuit of Figure 4.4-4 and prove that it transforms the fully adaptive array problem to the sidelobe canceler problem. For the fully adaptive array of Figure 4.4-3 we want to minimize the output power as was done for the sidelobe canceler problem of Figure 4.4-1. Specifically one wants to minimize

$$e_i = V_{i,m'+1} W_{m'+1} \tag{4.4-23}$$

where

$$V_{i,m'+1} = \begin{bmatrix} v_{i1} & v_{i2} & \cdots & v_{i,m'+1} \end{bmatrix} \tag{4.4-23a}$$

$$W_{m'+1} = \begin{bmatrix} w_1 \\ w_2 \\ \vdots \\ w_{m'+1} \end{bmatrix} \tag{4.4-23b}$$

However, in order to prevent the deterioration of the array main-lobe beam in the direction θ_0, where one seeks a target, the following constraint must be added:

$$C_{m'+1}^T W_{m'+1} = \mu \tag{4.4-24}$$

where $C_{m'+1}$ is the steering vector given by (4.4-21) and that is used to steer the array main lobe in the direction θ_0 if no jammers were present. The constraint (4.4-24) ensures that the adapted array will maintain its gain in the direction θ_0 when adaptation jammers are present.

Let $C_{m'}$, $W_{m'}$, and $V_{i,m'}$ be respectively the m'-dimensional matrices obtained if the $(m'+1)$st element is dropped from the matrices $C_{m'+1}$, $W_{m'+1}$, and $V_{i,m'+1}$. Then we can write $C_{m'+1}$, $W_{m'+1}$, and $V_{i,m'+1}$ as

$$C_{m'+1} = \begin{bmatrix} C_{m'} \\ -- \\ c_{m'+1} \end{bmatrix} = \begin{bmatrix} C_{m'} \\ -- \\ 1 \end{bmatrix} \tag{4.4-25}$$

$$W_{m'+1} = \begin{bmatrix} W_{m'} \\ ------ \\ w_{m'+1} \end{bmatrix} \tag{4.4-26}$$

$$V_{i,m'+1} = \begin{bmatrix} V_{i,m'} \vdots v_{i,m'+1} \end{bmatrix} \tag{4.4-27}$$

Substituting (4.4-25) and (4.4-26) into (4.4-24) yields

$$w_{m'+1} = \mu - C_{m'}^T W_{m'} \tag{4.4-28}$$

Now substituting (4.4-26), (4.4-27), and (4.4-28) into (4.4-23) yields

$$\varepsilon_i = V_{i,m'}W_{m'} + v_{i,m'+1}w_{m'+1}$$
$$= V_{i,m'}W_{m'} + v_{i,m'+1}(\mu - C_{m'}^T W_{m'})$$
$$= (V_{i,m'} - v_{i,m'+1}C_{m'}^T)W_{m'} + \mu v_{i,m'+1}$$
$$= \mu v_{i,m'+1} - a_i' \tag{4.4-29}$$

where

$$a_i' = -(V_{i,m'} - v_{i,m'+1}C_{m'}^T)W_{m'} \tag{4.4-29a}$$

On comparing (4.4-29) with (4.4-7) for the sidelobe canceler, we see, after some reflection, that they are identical with $\mu v_{i,m'+1}$ replacing s_i and a_i' replacing a_i. Furthermore, on comparing (4.4-29a) with (4.4-6), we see that a_i' and a_i would be identical if it were not for the term $-v_{i.m'+1}C_{m'}^T$ in a_i' and the negative sign on the right side of (4.4-29a). It is the term $-v_{i,m'+1}C_{m'}^T$ that forms the preprocessor transformation in Figure 4.4-4. The difference in the sign of the right-hand sides of a_i' and a_i results from a choice of convention. Specifically, it results from choosing a subtractor to combine the auxiliary channel sum with the main channel for the sidelobe canceler in Figure 4.4-1. If we chose to add the auxiliary channel to the sum channel, then the expressions for both a_i' and a_i would have a negative sign on the right-hand side. In this case the sign of the weight $W_{m'}$ in the sidelobe canceler of Figure 4.4-1 would reverse.

Because (4.4-29) is in the form of the sidelobe canceler problem, we have transformed our fully adaptive array problem to the sidelobe canceler problem as we set out to do. This completes our proof.

As in the case for the sidelobe canceler, the weights of the adapted full array of Figure 4.4-3 can be thought of as producing nulls in the array antenna pattern in the direction that the jammers are located [84–87, 96] while maintaining the main-lobe peak in the direction where a target is being looked for or tracked.

Another completely different method exists for transforming the fully adaptive array problem of Figure 4.4-3 to the sidelobe canceler problem of Figure 4.4-1. This alternate method involves the use of what is called beam space. In this approach the output of the linear array antenna of $m' + 1$ elements are input into a beamformer network that simultaneously generates $m' + 1$ contiguous beams, as shown in Figure 4.4-6. These $m' + 1$ beams can be formed using a microwave lens [88] (like a Rotman lens [95]), a Butler matrix [88], or a digital Fourier transformation. The output of these beams are applied to a sidelobe canceler. The beam pointing at the angle θ_0, the angle direction at which a target is being looked for, becomes the main antenna beam for the sidelobe canceler. The other m' beam output become the auxiliary-antenna beam inputs for the sidelobe canceler.

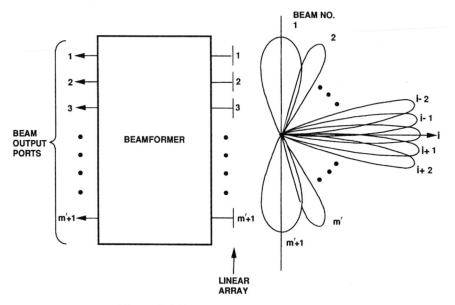

Figure 4.4-6   Array antenna beamformer.

An improvement of this approach called the adaptive-adaptive array was developed by Brookner and Howell [1, 96]. They noted that not all m auxiliary beams are needed in the sidelobe canceler. Only those beam that contain a jammer are needed. Hence, if there are J jammers, only J beams are needed.

4.5   MINIMUM-VARIANCE ESTIMATE

The least-squares estimate developed in the last section does not require any knowledge of the statistics of the measurement noise $N_{(n)}$. The weight vector W of (4.1-30) is only dependent on M and Φ. If we knew in addition the covariance matrix of $N_{(n)}$, then an estimate, which in some sense is better, can be developed. Specifically, an estimate that minimizes the variance of the error of the estimate of X_n can be obtained. This is the estimate that minimizes the diagonal elements of the covariance matrix of the estimate. Let $S^*_{n,n}$ be the covariance matrix of any arbitrary linear estimate $X^*_{n,n}$ of X_n. Then the minimum-variance estimate covariance matrix $\overset{\circ}{S}{}^*_{n,n}$ has the properties

$$(\overset{\circ}{S}{}^*_{n,n})_{i,i} \le (S^*_{n,n})_{i,i} \tag{4.5-1}$$

for all i, where $(\overset{\circ}{S}{}^*_{n,n})_{i,i}$ is the ith diagonal element of $\overset{\circ}{S}{}^*_{n,n}$. The covariance matrix for any linear estimate given by (4.1-30) is given by [5, p. 186]

$$S_{n,n}^* = WR_{(n)}W^T \tag{4.5-2}$$

where $R_{(n)}$ is the covariance matrix of the measurement errors, that is,

$$R_{(n)} = E\{N_{(n)}N_{(n)}^T\} \quad \text{for} \quad E\{N_{(n)}\} = 0 \tag{4.5-3}$$

The linear estimate (4.1-30) for which (4.5-1) is true, and hence is a minimum variance estimate among the class of linear estimates, has weight W given by

$$\mathring{W} = (T^T R_{(n)}^{-1} T)^{-1} T^T R_{(n)}^{-1} \tag{4.5-4}$$

The covariance matrix for the minimum-variance estimate is readily shown to be given by [5, p. 191] (problem 4.5-4)

$$\mathring{S}_{n,n}^* = (T^T R_{(n)}^{-1} T)^{-1} \tag{4.5-5}$$

The minimum-variance estimate weight given by (4.5-4) is unbiased and unique [5, p. 190]. When the measurement error covariance matrix $R_{(n)}$ is equal to a constant times the identity matrix I, that is,

$$R_{(n)} = \sigma_x^2 I \tag{4.5-6}$$

then the minimum-variance estimate weight given by (4.5-4) becomes the least-squares estimate weight given by (4.1-32). If $Y_{(n)}$ has a multidimensional Gaussian distribution, then the minimum-variance estimate weight of (4.5-4) also provides the maximum-likelihood estimate [5, p. 203].

The minimum-variance estimate, in addition to being an estimate for which the values of the diagonal elements of its covariance matrix are minimized, also has the properties that if $S_{n,n}^*$ is the covariance matrix of any other estimate of X_n, then

$$C = S_{n,n}^* - \mathring{S}_{n,n}^* \tag{4.5-7}$$

is positive semidefinite (or equivalently nonnegative definite) [5, p. 193]; that is, for any column matrix $V = [V_1, \dots, V_m]^T$ [9]

$$V_T C V \geq 0 \tag{4.5-8}$$

Equation (4.5-1) follows directly from (4.5-8) and (4.5-7). To show this, simply let $V_i = 0$ for all i except one.

A prediction-ahead estimate $\overset{\circ}{X}{}^{*}_{n+h,n}$ can be obtained from the minimum-variance estimate $\overset{\circ}{X}{}^{*}_{n,n}$ at time n by using

$$\overset{\circ}{X}{}^{*}_{n+h,n} = \Phi^{h}\overset{\circ}{X}{}^{*}_{n,n} \tag{4.5-9}$$

The question arises as to whether the predicted-ahead estimate obtained using the minimum-variance estimate $\overset{\circ}{X}{}^{*}_{n,n}$ is also a minimum-variance estimate. The answer is yes [5, p. 197]. Moreover (4.5-8) and (4.5-7) hold for the covariance of the predicted-ahead minimum-variance estimate. In addition the results also hold if h is negative, that is, for rectrodiction (prediction to a past time prior to when the measurement were obtained). The covariance matrix for the prediction minimum-variance estimate is easily shown to be given by [5, p. 197]

$$\overset{\circ}{S}{}^{*}_{n+h,n} = \Phi^{h}\overset{\circ}{S}{}^{*}_{n,n}\left(\Phi^{h}\right)^{T} \tag{4.5-10}$$

In reference 5 the minimum-variance estimate is given for the case where the target trajectory equation of motion is time varying. The same changes to make the time-invariant least-squares estimate a time-varying estimate apply for the minimum-variance estimate, that is, M, T, and Φ, are replaced by their time-varying counterparts M_n, T_n, and $\Phi(t_{n,n-1})$.

The minimum-variance estimate is a generalization of an estimate obtained by minimizing the weighted least-squares error given by

$$e(X^{*}_{n,n}) = \sum_{i} [(Y_{(n)})_i - (TX^{*}_{n,n})_i]^2 \frac{1}{\sigma_i^2} \tag{4.5-11}$$

or

$$e(X^{*}_{n,n}) = [Y_{(n)} - TX^{*}_{n,n}]^{T}[\text{Diag } R_{(n)}]^{-1}[Y_{(n)} - TX^{*}_{n,n}] \tag{4.5-12}$$

where Diag $R_{(n)}$ is a matrix consisting of the diagonal elements of the measurement noise covariance matrix $R_{(n)}$ (whose ith diagonal element is given by σ_i^2), with all off-diagonal elements being 0 and $(Y_{(n)})_i$ the ith element of the column matrix $Y_{(n)}$. In contrast, the minimum-variance estimate of X_n minimizes

$$e(X^{*}_{n,n}) = (Y_{(n)} - TX^{*}_{n,n})^{T}R_{(n)}^{-1}(Y_{(n)} - TX^{*}_{n,n}) \tag{4.5-13}$$

It was Aitken who suggested the use of the minimization of (4.5-13). This was in 1934 [97].

The minimization of (4.5-13) gives the classical weighted least-squares estimate. Gauss used the minimization of (4.5-12); this was in 1795 (when he was just 18 years old) but first published in 1809 [98]. We will call the estimate based on (4.5-12) the "quasi"-weighted least-squares estimate. Note that when

$R_{(n)}$ is diagonal, the weighted least-squares estimate and the quasi-weighted least squares estimate are identical.

From (4.5-11) one sees an interesting property of the quasi-weighted least-squares estimate that also applies to the classical weighted least-squares estimate. Specifically, the quasi-weighted least-squares estimate weights the errors according to their importance. [Using the classical weighted least-squares error criteria of (4.5-13) does this in a more general way.] As a result the more accurate the measurement $(Y_{(n)})_i$ of $Y_{(n)}$ is, the closer to it the estimate $X_{n,n}^*$ is placed. The weighting consists of dividing each squared error by its own variance. We saw this kind of weighted least-squared estimate being used in Sections 2.3, 2.5, and 2.6 for the Kalman filter; see (2.3-1), (2.5-9), and (2.6-7).

As indicated above, when $R_{(n)}$ is diagonal, the classical least-squares estimate is identical to the quasi-weighted least-squares estimate that minimizes (4.5-11). Let us examine the physical significance of this further. When we are fitting an mth-degree polynomial, as given by (4.1-44), to the data such that (4.5-11) is minimized, we are forcing the data points that have the highest accuracy to have the smallest errors, that is, the smallest deviations from the best-fitting polynomial. For the constant-velocity target radar example of Section 4.1, the sum of the errors given by (4.1-38) becomes the following weighted sum:

$$e_n = \sum_{j=0}^{n} \frac{1}{\sigma_j^2} (y_j - x_{j,n}^*)^2 \qquad (4.5\text{-}14)$$

with $x_{j,n}^*$ still defined by (4.1-35) and σ_j^2 being the variance of the jth radar range measurement.

The range accuracy can vary from look to look due to a number of factors. The power SNR could be varying from one look to the next. This could be due to any number of factors—target cross-sectional variations from look to look, planned or adaptive variations in the radar transmitter energy from look to look, or planned variations of the radar signal bandwidth from look to look. For a pulse compression radar the radar range accuracy for the ith look is given by [99]

$$\sigma_i = \frac{\Delta r}{k_M \sqrt{2SNR_i}} \qquad (4.5\text{-}15)$$

where Δr is the radar range resolution defined as the echo 3 dB down (half-power) width after pulse compression, SNR_i is the single look receiver output SNR for the ith look, and k_M is a constant equal to about 1.6. The above results apply for the chirp waveform pulse compression system described in Section 3.5.1. The results apply for the chirp waveform system using a matched receiver as long as the time sidelobes of the compressed signal are no more than about

50 dB down. (Typically the results also apply for pulse compression systems using nonuniform spectrum transmitter waveforms as long as the compressed pulse sidelobes are no more than 50 dB down.) The range resolution Δr depends on the signal bandwidth. For a radar using a chirp waveform (see Section 3.5.1)

$$\Delta r = \frac{k_0}{2cB_s} \tag{4.5-16}$$

where B_s is the chirp waveform bandwidth, c is the velocity of light (3×10^8 m/sec), and k_0 is a constant typically between about 0.9 and 1.3 depending on the receiver mismatch weighting. For 15-dB sidelobes $k_0 \doteq 0.9$; for 30- to 50-dB sidelobes $k_0 \doteq 1.3$ [88].

In a radar the values of B_s, k_0, and k_M are known for a given look and the echo SNR can be measured. Thus σ_i is known (or at least can be estimated) for the ith look, and hence the minimum-variance estimate polynomial fit to the target trajectory can be obtained for radar systems.

As a by-product of the above weighting, an important property for both the quasi- and classical weighted least-squares estimate is that the estimate is independent of the units used for X_n; that is, the same performance is obtained independent of the units used for the coordinates of X_n. The units can all be different from each other in fact. This arises from the normalization of each squared error term by its own variance with the same units used for both. This invariance property does not apply for the least-squares error estimate that minimizes (4.1-31).

To prove that $\overset{\circ}{W}$ given by (4.5-4) is the minimum-variance weight, we simply differentiate (4.5-13) with rspect to $X_{n,n}^*$, set the result equal to zero, and solve for $X_{n,n}^*$ as done in (4.1-46); see problem 4.5-1.

In the next few chapters we apply the results developed in this chapter to develop the fixed-memory polynomial g–h, g–h–k, and higher order filters. This is followed by the development of the expanding-memory and fading-memory filters, which are put in recursive form.

5

FIXED-MEMORY POLYNOMIAL FILTER

5.1 INTRODUCTION

In Section 1.2.10 we presented the growing-memory g–h filter. For n fixed this filter becomes a fixed-memory filter with the n most recent samples of data being processed by the filter, sliding-window fashion. In this chapter we derive a higher order form of this filter. We develop this higher order fixed-memory polynomial filter by applying the least-squares results given by (4.1-32). As in Section 1.2.10 we assume that only measurements of the target range, designated as $x(t)$, are available, that is, the measurements are one-dimensional, hence $r = 0$ in (4.1-1a). The state vector is given by (4.1-2). We first use a direct approach that involves representing $x(t)$ by an arbitrary mth degree polynomial and applying (4.1-32) [5, pp. 225–228]. This approach is given in Section 5.2. This direct approach unfortunately requires a matrix inversion. In Section 4.3 we developed the voltage-processing approach, which did not require a matrix inversion. In Section 5.3 we present another approach that does not require a matrix inversion. This approach also has the advantage of leading to the development of a recursive form, to be given in Section 6.3, for the growing-memory filter. The approach of Section 5.3 involves using the discrete-time orthogonal Legendre polynomial (DOLP) representation for the polynomial fit. As indicated, previously the approach using the Legendre orthogonal polynomial representation is equivalent to the voltage-processing approach. We shall prove this equivalence in Section 14.4. In so doing, better insight into the Legendre orthogonal polynomial fit approach will be obtained.

5.2   DIRECT APPROACH (USING NONORTHOGONAL mTH-DEGREE POLYNOMIAL FIT)

Assume a sequence of $L + 1$ one-dimensional measurements given by

$$Y_{(n)} = [y_n, y_{n-1}, \ldots, y_{n-L}]^T \tag{5.2-1}$$

with n being the last time a measurement was made. We assume that the underlying process $x(t)$ that generated these data can be approximated by a polynomial of degree m as indicated by (4.1-44), which we rewrite here as

$$x(t) = p^*(t) = [p^*(t)]_n = \sum_{j=0}^{m} (\bar{a}_j)_n t^j \tag{5.2-2}$$

What we want is a least-squares estimate for the coefficients $(\bar{a}_j)_n$ of this polynomial t. The subscript n on the coefficient $(\bar{a}_j)_n$ for the jth polynomial term is used because the estimate of these coefficients will depend on n, the last observation time at which a measurement was made. The subscript n on $[p^*(t)]_n$ similarly is used to indicate that n is the last time a measurement was made.

Let $t = rT$. Then (5.2-2) becomes

$$p^* = p^*(rT) = [p^*(r)]_n = \sum_{j=0}^{m} (\bar{a}_j)_n r^j T^j \tag{5.2-3}$$

or

$$p^* = [p^*(r)]_n = \sum_{j=0}^{m} (z_j^*)_n r^j \tag{5.2-4}$$

where

$$(z_j^*)_n = (\bar{a}_j)_n T^j \tag{5.2-4a}$$

where r becomes a new integer time index for the polynomial p^*. Physically r represents the measurement time index just as n does; it is just referenced to a different starting time. The origin for r is the time at which the first measurement Y_{n-L} is made for the fixed-memory filter; see Figure 5.2-1.

We want the above polynomial to provide a least-square fit to the measured data. This can be achieved by applying the results of Section 4.1 directly. To do this, we must find T of (4.1-32). We can choose for the state vector X_n the

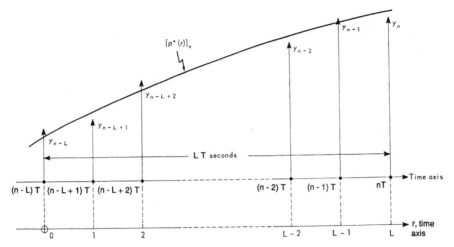

Figure 5.2-1 Polynominal fit $p^* = [p^*(r)]_n$ to range measurements y_i. (From Morrison [5, p. 226].)

coefficients of the polynomial fit given by

$$X_n \equiv X_n' = \begin{bmatrix} (z_0^*)_n \\ (z_1^*)_n \\ \vdots \\ (z_m^*)_n \end{bmatrix} \tag{5.2-5}$$

[Note that here X_n is given by an $(m+1)$-state matrix (as was the case for (4.1-2) instead of m, as in (4.1-1b).] It then follows that the matrix T is given by

$$T \equiv T' = \begin{bmatrix} L^0 & L^1 & L^2 & \cdots & L^m \\ (L-1)^0 & (L-1)^1 & (L-1)^2 & \cdots & (L-1)^m \\ \vdots & \vdots & \vdots & \ddots & \vdots \\ 0^0 = 1 & 0^1 & 0^2 & \vdots & 0^m \end{bmatrix} \tag{5.2-6}$$

(The prime is used on the matrices T and X_n above because we shall shortly develop an alternate, more standard form for the process state vector that uses different expressions for T and X_n.)

It is now a straightforward matter to substitute (5.2-5) and (5.2-6) into (4.1-32) to obtain the least-squares estimate weight W^*. Substituting this value for the weight into (4.2-30) then yields the least-squares estimate $X_{n,n}^*$ in terms of the coefficients $(z_i^*)_n$ of (5.2-5). Knowing these coefficients, we can use (5.2-3) to estimate $X_{n,n}^*$ at time n by choosing $r = L$. By choosing $r = L + h$, we obtain the prediction estimate $X^*_{n+h,n}$. This approach has the disadvantage of requiring a matrix inversion in evaluating $(T^T T)^{-1}$ of (4.1-32). Except for when $T^t T$ is 2×2, this matrix inversion has to be done numerically on the computer, an algebraic solution not being conveniently obtained [5, p. 228].

An approach is developed in the next section that uses an orthogonal polynomial representation for the polynomial fit given by (5.2-3) and as a result does not require a matrix inversion. This new approach also gives further insight into the polynomial least-squares fit. This new approach, as we indicated, is the same as the voltage-processing method described in Section 4.3, which also does not require a matrix inversion. We shall prove this equivalence in Section 14.4.

Before proceeding we will relate the coefficients $\bar{a}_j$ and z_j^* to $D^j x(t)$. The second coefficients of these parameters have been dropped for simplicity. By differentiating the jth term of (5.2-2), we obtain

$$\bar{a}_j = \frac{1}{j!} D^j x(t) \bigg|_{t=0} = \frac{1}{j!} \frac{d^j}{dt^j} x(t) \bigg|_{t=0} \tag{5.2-7}$$

Hence

$$z_j^* = \bar{a}_j T^j = \frac{T^j}{j!} D^j x(t) \bigg|_{t=0} = \frac{T^j}{j!} \frac{d^j}{dt^j} x(t) \bigg|_{t=0} \tag{5.2-8}$$

The parameter z_j^* is a constant times $D^j x(t)$. Hence in the literature it is called the scaled jth-state derivative [5]. We shall discuss it further shortly.

5.3 DISCRETE ORTHOGONAL LEGENDRE POLYNOMIAL APPROACH

As indicated above, an approach is developed in this section that leads to a simple analytical expression for the least-squares polynomial that does not require a matrix inversion. It involves expressing the polynomial fit of (5.2-3) in terms of the discrete orthonormal Legendre polynomials [5, pp. 228–235]. Specifically, the estimating polynomial is expressed, as done in (4.1-45), as

$$[p^*(r)]_n = \sum_{j=0}^{m} (\beta_j)_n \phi_j(r) \tag{5.3-1}$$

where $\phi_j(r)$ is the normalized discrete Legendre polynomial of degree j. Specifically $\phi_j(r)$ is a polynomial in r of degree j. It is given by

$$\phi_j(r) = \frac{1}{c_j} p_j(r) \tag{5.3-1a}$$

where $p_j(r)$ is the unnormalized discrete Legendre polynomial of degree j and the $\phi_i(r)$ and $\phi_j(r)$ for $r = 0, \ldots, L$ are orthogonal for $i \neq j$, that is, they obey the relation

$$\sum_{r=0}^{L} \phi_i(r)\phi_j(r) = \delta_{ij} \qquad (5.3\text{-}2)$$

where δ_{ij} is the Kronecker delta function defined by

$$\delta_{ij} = \left\{ \begin{matrix} 1 & \text{for } i = j \\ 0 & \text{for } i \neq j \end{matrix} \right\} \qquad (5.3\text{-}2a)$$

Because (5.3-2) equals 1 for $i = j$, the $\phi_i(r)$ and $\phi_j(r)$ are called orthonormal. A least-squares estimate for the coefficients $(\beta_j)_n$ will be shortly determined from which $[p^*(r)]_n$ is determined.

In turn the discrete Legendre polynomial is given by

$$p_j(r) = p(r; j, L) \qquad (5.3\text{-}3)$$

with

$$p(r; j, L) = \sum_{\nu=0}^{j} (-1)^\nu \binom{j}{\nu} \binom{j+\nu}{\nu} \frac{r^{(\nu)}}{L^{(\nu)}} \qquad (5.3\text{-}4)$$

where

$$x^{(m)} = x(x-1)(x-2)\ldots(x-m+1) \qquad (5.3\text{-}4a)$$

The normalizing constant of (5.3-1a) is given by

$$c_j = c(j, L) \qquad (5.3\text{-}5)$$

where

$$[c(j, L)]^2 = \sum_{r=0}^{L} [p(r; j, L)]^2$$
$$= \frac{(L+j+1)^{(j+1)}}{(2j+1)L^{(j)}} \qquad (5.3\text{-}5a)$$

From (5.3-3), (5.3-1a), and (5.3-2) it follows that the discrete Legendre polynomials satisfy the orthogonality condition

$$\sum_{r=0}^{L} p(r; i, L)p(r; j, L) = 0 \qquad i \neq j \qquad (5.3\text{-}6)$$

TABLE 5.3-1. First Four Discrete Orthogonal Legendre Polynomials

$$p(x; 0, L) = 1$$

$$p(x; 1, L) = 1 - 2\frac{x}{L}$$

$$p(x; 2, L) = 1 - 6\frac{x}{L} + 6\frac{x(x-1)}{L(L-1)}$$

$$p(x; 3, L) = 1 - 12\frac{x}{L} + 30\frac{x(x-1)}{L(L-1)} - 20\frac{x(x-1)(x-2)}{L(L-1)(L-2)}$$

$$p(x; 4, L) = 1 - 20\frac{x}{L} + 90\frac{x(x-1)}{L(L-1)} - 140\frac{x(x-1)(x-2)}{L(L-1)(L-2)}$$
$$+ 70\frac{x(x-1)(x-2)(x-3)}{L(L-1)(L-2)(L-3)}$$

Table 5.3-1 gives the first four discrete Legendre polynomials. Note that $p_j(r)$ and $p(r; i, L)$ are used here to represent the Legendre polynomial, whereas $p, p^*, p(t)$, and $p(r)$ are used to represent a general polynomial. The presence of the subscript or the three variables in the argument of the Legendre polynomial make it distinguishable from the general polynomials, such as given by (4.1-44), (4.1-45), and (5.3-1).

The error given by (4.1-31) becomes

$$e_n = \sum_{r=0}^{L}\{y_{n-L+r} - [p^*(r)]_n\}^2 \tag{5.3-7}$$

Substituting (5.3-1) into (5.3-7), differentiating with respect to $(\beta_i)_n$, and setting the result equal to 0 yield, after some straightforward manipulation,

$$\sum_{k=0}^{L}\sum_{j=0}^{m}(\beta_j)_n\phi_i(k)\phi_j(k) = \sum_{k=0}^{L}y_{n-L+k}\phi_i(k) \quad i = 0, 1, \ldots, m \tag{5.3-8}$$

where for convenience r is replaced by k. Changing the order of the summation yields in turn

$$\sum_{j=0}^{m}(\beta_j)_n\sum_{k=0}^{L}\phi_i(k)\phi_j(k) = \sum_{k=0}^{L}y_{n-L+k}\phi_i(k) \tag{5.3-9}$$

Using (5.3-2) yields the least-squares estimate for $(\beta_i)_n$ given by

$$(\beta_j)_n^* = \sum_{k=0}^{L}y_{n-L+k}\phi_j(k) \quad j = 0, 1, \ldots, m \tag{5.3-10}$$

Substituting the above in (5.3-1) yields finally

$$[p^*(r)]_n = \sum_{j=0}^{m} \left[\sum_{k=0}^{L} y_{n-L+k} \phi_j(k) \right] \phi_j(r) \qquad (5.3\text{-}11)$$

The above final answer requires no matrix inversion for the least-squares estimate solution. This results from the use of the orthogonal polynomial representation. Equation (5.3-11) gives an explicit general functional expression for the least-squares estimate polynomial fit directly in terms of the measurements y_i. This was not the case when using the direct approach of Section 5.2, which did not involve the orthogonal polynomial representation. There a matrix inversion was required. If the entries in this matrix are in algebraic or functional form, then, as mentioned before, this matrix inversion cannot be carried out algebraically easily except when the matrix to be inverted is 2×2. For large matrices the matrix entries need to be specific numerical values with the inverse obtained on a computer; the inverse then is not in algebraic or functional form.

The least-squares estimate polynomial solution given by (5.3-11) can be used to estimate the process values at times in the future $(r > L)$ or past $(r < L)$. Estimates of the derivatives of the process can also be obtained by differentiating (5.3-11) [5]. To do this, we let $t = rT$, where T is the time between the measurements y_i. Then $dt = Tdr$ and

$$D = \frac{d}{dt} = \frac{1}{T}\frac{d}{dr} \qquad (5.3\text{-}12)$$

Applying this to (5.3-11) yields the following expression for the least-squares polynomial fit estimate p^* and its derivatives [5, p. 231]

$$[D^i p^*(r)]_n = \frac{1}{T^i} \sum_{j=0}^{m} \left[\sum_{k=0}^{L} y_{n-L+k} \phi_j(k) \right] \frac{d^i}{dr^i} \phi_j(r) \qquad (5.3\text{-}13)$$

At time $n + 1$ when the measurement y_{n+1} is received, the $L + 1$ measurements $Y_{(n)}$ of (5.2-1) can be replaced by the $L + 1$ measurements of

$$Y_{(n+1)} = (y_{n+1}, y_n, \ldots, y_{n-L+1})^T \qquad (5.3\text{-}14)$$

A new fixed-memory polynomial filter least-squares estimate is now obtained. Equation (5.3-11) is again used to obtain the estimating polynomial $[p^*(i)]_{n+1}$. This new polynomial estimate is based on the latest $L + 1$ measurements. But now time has moved one interval T forward, and the measurements used are those made at times $n - L + 1$ to $n + 1$ to give the new coefficient estimates $(\beta_j)^*_{n+1}$ in (5.3-10) for the time-shifted data. This process is then repeated at time $n + 2$ when Y_{n+2} is obtained, and so on.

5.4   REPRESENTATION OF POLYNOMIAL FIT IN TERMS OF ITS DERIVATIVES (STATE VARIABLE REPRESENTATION OF POLYNOMIAL FIT IN TERMS OF PROCESS DERIVATIVES)

We begin by developing in this section a very useful alternate representation for the polynomial function of time. This representation lets us obtain the transformation matrix for the process that provides an alternate way to obtain the process state variable estimate at other times. Instead of expressing the polynomial process in terms of the mth-degree polynomial $[p^*(r)]_n$, it is possible to express the process in terms of its first m derivatives at any time, as shall now be shown.

For a process given by an mth-degree polynomial, its state vector at any time n can be expressed in terms of its first m derivatives by

$$X(t_n) = X_n = \begin{bmatrix} x \\ Dx \\ \vdots \\ D^m x \end{bmatrix} = \begin{bmatrix} x_n \\ Dx_n \\ \vdots \\ D^m x_n \end{bmatrix} \tag{5.4-1}$$

where D is defined by (5.3-12). Let

$$y_n = x_n + \nu_n \tag{5.4-2}$$

If we assume that only x_n is measured and not its derivatives, then

$$Y_n = MX_n + N_n \tag{5.4-3}$$

where because range is the only measurement

$$M = \begin{bmatrix} 1 & 0 & 0 & \cdots & 0 \end{bmatrix} \tag{5.4-3a}$$

and Y_n and N_n are 1×1 matrices given by

$$Y_n = [y_n] \quad \text{and} \quad N_n = [\nu_n] \tag{5.4-3b}$$

For definiteness and convenience in the ensuing discussion let us assume that the polynomial fit p^* is of degree 2, that is, $m = 2$ and is given by

$$x_n = (\bar{a}_0)_n + (\bar{a}_1)_n rT + (1/2!)(\bar{a}_2)_n r^2 T^2 \tag{5.4-4}$$

where we have now shifted the origin of r so that $r = 0$ at time n, the time at which the estimate $X_{n,n}^*$ is to be obtained. Then

$$Dx_n = \frac{1}{T}(\bar{a}_1 T + \bar{a}_2 r T^2) = \bar{a}_1 \quad \text{for } r = 0 \tag{5.4-5}$$

$$D^2 x_n = \bar{a}_2 \tag{5.4-6}$$

where for simplicity the subscript n on $\bar{a}_i$ has been dropped. It is next easy to show for $m = 2$ that the transition matrix Φ that goes from the state X_n to X_{n+1} is given by

$$
\Phi = \begin{bmatrix} 1 & T & \dfrac{1}{2}T^2 \\ 0 & 1 & T \\ 0 & 0 & 1 \end{bmatrix}
\tag{5.4-7}
$$

The reader can verify this by substituting (5.4-4) to (5.4-6) into (5.4-1) and multiplying by (5.4-7). The transition matrix that goes from X_n to X_{n+h} is $\Phi_h = \Phi^h$.

For the case where measurements are available at $L + 1$ times as given by (5.2-1),

$$
Y_{(n)} = \begin{bmatrix} MX_n \\ \hline MX_{n-1} \\ \hline \vdots \\ \hline MX_{n-L} \end{bmatrix} + \begin{bmatrix} \nu_n \\ \hline \nu_{n-1} \\ \hline \vdots \\ \hline \nu_{n-L} \end{bmatrix}
\tag{5.4-8}
$$

which in turn can be written as

$$
Y_{(n)} = \begin{bmatrix} MX_n \\ \hline M\Phi^{-1}X_n \\ \hline \vdots \\ \hline M\Phi^{-L}X_n \end{bmatrix} + \begin{bmatrix} \nu_n \\ \hline \nu_{n-1} \\ \hline \vdots \\ \hline \nu_{n-L} \end{bmatrix}
\tag{5.4-9}
$$

or

$$
Y_{(n)} = \begin{bmatrix} M \\ \hline M\Phi^{-1} \\ \hline \vdots \\ \hline M\Phi^{-L} \end{bmatrix} X_n + \begin{bmatrix} \nu_n \\ \hline \nu_{n-1} \\ \hline \vdots \\ \hline \nu_{n-L} \end{bmatrix}
\tag{5.4-10}
$$

or

$$Y_{(n)} = TX_n + N_{(n)} \tag{5.4-11}$$

where

$$T = \begin{bmatrix} M \\ \text{----} \\ M\Phi^{-1} \\ \text{-------} \\ \vdots \\ \text{-------} \\ M\Phi^{-L} \end{bmatrix} \qquad N_{(n)} = \begin{bmatrix} \nu_n \\ \text{-----} \\ \nu_{n-1} \\ \text{-----} \\ \vdots \\ \text{-----} \\ \nu_{n-L} \end{bmatrix} \tag{5.4-11a}$$

Note that the above results [(5.4-8) to (5.4-11a) with (5.2-1) are a special case of (4.1-5) to (4.1-11b). Here the Y_i and N_i are 1×1 matrices [see (5.4-3b)] instead of $(r + 1) \times 1$ column matrices [as in (4.1-1a) and (4.1-1c)].

Sometimes in the literature, instead of representing the state vector X_n by its derivatives, as done in (5.4-1) for a process represented by an mth-degree polynomial, the scaled derivatives z_j of (5.2-8) are used to form the scaled state vector Z_n given by

$$Z_n = \begin{bmatrix} z_0 \\ z_1 \\ z_2 \\ \vdots \\ z_m \end{bmatrix} = \begin{bmatrix} z \\ \dot{z} \\ \ddot{z} \\ \vdots \\ D^m z \end{bmatrix} = \begin{pmatrix} x \\ TDx \\ \dfrac{T^2}{2!} D^2 x \\ \vdots \\ \dfrac{T^m}{m!} D^m x \end{pmatrix}_n \tag{5.4-12}$$

where, as before T is the time between the measurements. The transition matrix for the scaled state vector Z_n when $m = 2$ is given by

$$\Phi_z = \begin{bmatrix} 1 & 1 & 1 \\ 0 & 1 & 2 \\ 0 & 0 & 1 \end{bmatrix} \tag{5.4-13}$$

where the subscript z is used here to emphasize that Φ is the transition matrix for the scaled state vector z_n.

5.5   REPRESENTATION OF LEAST-SQUARES ESTIMATE IN TERMS OF DERIVATIVE STATE VECTOR

The least-squares estimate of the process state vector given by (5.4-1) can be written at time $r = L + h$ as [with now the origin of r again at $n - L$ as given in

Figure (5.2-1)]

$$X^*_{n+h,n} = \begin{bmatrix} p^*(L+h) \\ Dp^*(L+h) \\ \vdots \\ D^m p^*(L+h) \end{bmatrix}_n \tag{5.5-1}$$

For simplicity the subscript n has been dropped from $p^*(r)$. Sometimes we shall drop the r variable as well, just leaving p^*.

In terms of the scaled state vector one obtains

$$Z^*_{n+h,n} = \begin{bmatrix} p^*(L+h) \\ TDp^*(L+h) \\ \dfrac{T^2}{2!}D^2 p^*(L+h) \\ \vdots \\ \dfrac{T^m}{m!}D^m p^*(L+h) \end{bmatrix}_n \tag{5.5-2}$$

Using the orthogonal polynomial least-squares solution given by (5.3-11) and (5.3-13), it can be shown that the least-squares solution for Z_{n+h} is [5, p. 237]

$$Z^*_{n+h,n} = W(h)_z Y_{(n)} \tag{5.5-3}$$

TABLE 5.5-1. Elements of Associate Stirling Matrix S of First Kind for $i,j = 0, \ldots, 10$

i \ j	0	1	2	3	4	5	6	7	8	9	10
0	1	0	0	0	0	0	0	0	0	0	0
1	0	1	1	2	6	24	120	720	5,040	40,320	362,880
2	0	0	1	3	11	50	274	1,764	13,068	109,584	1,026,576
3	0	0	0	1	6	35	225	1,624	13,132	118,124	1,172,700
4	0	0	0	0	1	10	85	735	6,769	67,284	723,680
5	0	0	0	0	0	1	15	175	1,960	22,449	269,325
6	0	0	0	0	0	0	1	21	322	4,536	63,273
7	0	0	0	0	0	0	0	1	28	546	9,450
8	0	0	0	0	0	0	0	0	1	36	870
9	0	0	0	0	0	0	0	0	0	1	45
10	0	0	0	0	0	0	0	0	0	0	1

Note: Used to obtain weights of optimum least-squares fixed-memory filter.

Source: From Morrison [5, p. 85].

where [5, p. 239]

$$W(h)_z = \Phi(h)_z SGCB \qquad (5.5\text{-}3a)$$

where S is the associate Stirling matrix of the first kind. Values for S are given in Table 5.5-1 for i,j up to 10. The recursive expression for the i,j element of S, $[S]_{i,j}$ is given by [5, p. 116]

$$[S]_{i,j} = [S]_{i-1,j-1} + (j-1)[S]_{i,j-1} \qquad (5.5\text{-}4)$$

where initially

$$[S]_{0,0} = 1 \text{ and } [S]_{0,j} = 0 = [S]_{i,0} \text{ for } i,j \geq 1 \qquad (5.5\text{-}4a)$$

The i,jth term of the matrix G is given by

$$[G]_{i,j} = (-1)^j \binom{j}{i} \binom{j+1}{i} \frac{1}{L^{(i)}} \qquad (5.5\text{-}5)$$

TABLE 5.5-2. Factor of Elements of G Matrix for $i,j = 1, \ldots, 10$

$i \backslash j$	0	1	2	3	4	5	6	7	8	9	10
0	1	−1	1	−1	1	−1	1	−1	1	−1	1
1		−2	6	−12	20	−30	42	−56	72	−90	110
2			6	−30	90	−210	420	−756	1,260	−1,980	2,970
3				−20	140	−560	1,680	−4,200	9,240	−18,480	34,320
4					70	−630	3,150	−11,550	34,650	−90,090	210,210
5						−252	2,772	−16,632	72,072	−252,252	756,756
6							924	−12,012	84,084	−420,420	1,681,680
7								−3,432	51,480	−411,840	2,333,760
8									12,870	−218,790	1,969,110
9										−48,620	923,780
10											184,756

Notes: Used to obtain weights of optimum least-squares fixed-memory filter. The G matrix is defined by

$$[G]_{ij} = (-1)^j \binom{j}{i} \binom{j+i}{i} \frac{1}{L^{(i)}}.$$

Displayed above is

$$(-1)^j \binom{j}{i} \binom{j+i}{i}$$

Example: $[G]_{3,5} = -560/L^{(3)}$, $[G]_{7,8} = 51,480/L^{(7)}$.

With appropriate adjustment of signs, the above matrix also gives the coefficients of the discrete Legendre polynomials [see (5.3-4)].

Source: From Morrison [5, p. 241].

Values of $[G]_{ij}$ for $0 \le i,j \le 10$ are given in Table 5.5-2. Those of the matrix B are given by

$$[B]_{ij} = p_i(r)|_{r=L-j} \quad \text{for} \quad \begin{array}{c} 0 \le i \le m \\ 0 \le j \le L \end{array} \tag{5.5-6}$$

The matrix C is diagonal, with the diagonal elements given by

$$[C]_{ij} = \frac{1}{c_j^2} \delta_{ij} \qquad 0 \le i,j \le m \tag{5.5-7}$$

where δ_{ij} is the Kronecker delta function defined by (5.3-2a). Finally the i,j element of $\Phi(h)_z$ is defined by

$$[\Phi(h)_z]_{ij} = \binom{j}{i} h^{j-1} \qquad 0 \le i,j \le m \tag{5.5-8}$$

where $\Phi(0) = I$. Using (5.5-4) to (5.5-8) and or Tables 5.5-1 and 5.5-2, (5.5-3a) can be programmed on a computer to provide optimum weight $W(h)$, and by the use of (5.5-3), least-squares estimate of the scaled state vector. Note that $i = j = 0$ is first element of the above matrices.

5.6 VARIANCE OF LEAST-SQUARES POLYNOMIAL ESTIMATE

Substituting (5.4-11) into (5.5-3) yields

$$Z^*_{n+h,n} = W(h)_z TX_n + W(h)_z N_{(n)} \tag{5.6-1}$$

It then directly follows that the covariance matrix of the scaled least-squares estimate $Z^*_{n+h,n}$ is given by

$$_sS^*_{n+h,n} = W(h)_z R_{(n)} W(h)_z^T \tag{5.6-2}$$

where $R_{(n)}$ is the covariance matrix of $N_{(n)}$. Often the measurements have zero mean and are uncorrelated with equal variance σ_x^2. In this case $R_{(n)}$ is given by (4.5-6), and (5.6-2) becomes

$$_sS^*_{n+h,n} = \sigma_x^2 W(h)_z W(h)_z^T \tag{5.6-3}$$

which can be calculated numerically on a computer once $W(h)$ is programmed using (5.5-3a).

When the polynomial fit p^* is of degree $m = 1$, reference 5 (p. 243) shows that the above results yield the following algebraic form for the covariance matrix for the scaled least-squares state vector estimate $Z_{n,n}^*$;

$$_sS_{n,n}^* = \sigma_x^2 \begin{bmatrix} \dfrac{2(2L+1)}{(L+2)(L+1)} & \dfrac{6}{(L+2)(L+1)} \\[3mm] \dfrac{6}{(L+2)(L+1)} & \dfrac{12}{(L+2)(L+1)L} \end{bmatrix} \tag{5.6-4}$$

In addition the covariance matrix for the one-step-ahead scaled prediction state vector $Z_{n+1,n}^*$ is given by [5, p. 245]

$$_sS_{n+1,n}^* = \sigma_x^2 \begin{pmatrix} \dfrac{2(2L+3)}{(L+1)L} & \dfrac{6}{(L+1)L} \\[3mm] \dfrac{6}{(L+1)L} & \dfrac{12}{(L+2)(L+1)L} \end{pmatrix} \tag{5.6-5}$$

It is readily shown [5, p. 245] that the covariance matrix for the unscaled prediction rate vector $X_{n+h,n}^*$ is given by

$$S_{n+h,n}^* = D(T)\,_sS_{n+h,n}^*D(T) \tag{5.6-6}$$

where

$$[D(T)]_{ij} = \frac{j!}{T^j}\delta_{ij} \qquad 0 \le i,j \le m \tag{5.6-6a}$$

In the above we have used the unsubscripted S for the covariance matrix of the unscaled state vector and $_sS$ for the covariance matrix of the scaled state vector Z^*. [In reference 5 (p. 246) S is used for the unscaled vector and S for the scaled vector, see Table 5.6-1.]

It can be easily shown [5, p. 245] that

$$\begin{aligned} [S_{n+h,n}^*]_{ij} &= \frac{i!\,j!}{T^{i+j}}[_sS_{n+h,n}^*]_{ij} \\[2mm] &= \frac{\sigma_x^2}{T^{i+j}}\sum_{k=0}^{m}\frac{d^i}{dr^i}\phi_k(r)\frac{d^j}{dr^j}\phi_k(r)\bigg|_{r=L+h} \end{aligned} \tag{5.6-7}$$

TABLE 5.6-1. Covariance Matrix Notation

	Brookner (this book)	Morrison [5]
Unscaled	S (capital S)	S (sans serif capital S)
Scaled	$_sS$ (subscripted capital S)	S (italic capital S)

5.7 SIMPLE EXAMPLE

Assume $R_{(n)}$ is diagonal as given by (4.5-6) with $\sigma_x = 10$ ft. We want to design a first-degree (that is, $m = 1$) fixed-memory smoothing filter whose rms one-step position prediction error is 3 ft. From (5.6-5) the variance of the one-step prediction is given by the 0,0 element of (5.6-5), that is,

$$[_sS^*_{n+1,n}]_{0,0} = \sigma_x^2 \frac{2(2L+3)}{(L+1)L} \tag{5.7-1}$$

(In this chapter 5 and chapter 7, to be consistent with the literature [5], we index the rows and columns of the covariance matrix starting with the first being 0, the second 1, and so on, corresponding with the derivative being estimated). Substituting into the above yields

$$9 = 100 \times \frac{2(2L+3)}{(L+1)L} \tag{5.7-2}$$

Solving yields that $L = 45$ is needed; thus $L + 1 = 46$ measurements are required in $Y_{(n)}$. The variance of the unscaled velocity estimate can be obtained using (5.6-5) and (5.6-6) to yield [5, p. 246]

$$\begin{aligned}
[S^*_{n+1,n}]_{1,1} &= \frac{1}{T^2} [_sS^*_{n+1,n}]_{1,1} \\
&= \frac{1}{T^2} \sigma_x^2 \frac{12}{(L+2)(L+1)L} \\
&= \frac{0.0123}{T^2}
\end{aligned} \tag{5.7-3}$$

Assume it is desired that the rms velocity error be 4 ft/sec. Then from (5.7-3) it follows that $T = 0.028$ sec, or equivalently, about 36 measurements per second must be made.

5.8 DEPENDENCE OF COVARIANCE ON L, T, m, AND h

Using (5.6-7) it can be shown that for large L [5, p. 250]

$$[S^*_{n+h,n}]_{ij} \approx \frac{\lambda_{ij}}{T^{i+j}L^{i+j+1}} \sigma_x^2 \tag{5.8-1}$$

where λ_{ij} is a constant dependent on h and m but not on L. Values for λ_{ii} for $i, m = 0, \ldots, 10$ are given in Table 5.8-1 for $h = 0$ and $h = 1$, that is, for filtering to the present time n and for one-step-ahead prediction. Table 5.8-2 gives λ_{ii} for $h = -\frac{1}{2}L$, that is, smoothing to the center of the observation

TABLE 5.8-1. Values of Constant λ_{ii} in VRF Equation for Fixed-Memory Filter When L Large and $h = 0^a$ and $h = 1^b$

i \ m	0	1	2	3	4	5	6	7	8	9	10
0	1.0(0)										
1	4.0(0)	1.2(1)									
2	9.00(0)	1.92(2)	7.2(2)								
3	1.600(1)	1.200(3)	2.592(4)	1.008(5)							
4	2.500(1)	4.800(3)	3.175(5)	6.451(6)	2.540(7)						
5	3.600(1)	1.470(4)	2.258(6)	1.306(8)	2.540(9)	1.006(10)					
6	4.900(1)	3.763(4)	1.143(7)	1.452(9)	7.684(10)	1.449(12)	5.754(12)				
7	6.400(1)	8.467(4)	4.752(7)	1.098(10)	1.229(12)	6.120(13)	1.128(15)	4.488(15)			
8	8.100(1)	1.728(5)	1.537(8)	6.323(10)	1.299(13)	1.333(15)	6.344(16)	1.149(18)	4.578(18)		
9	1.000(2)	3.267(5)	4.516(8)	2.968(11)	1.018(14)	1.874(16)	1.804(18)	8.301(19)	1.483(21)	5.914(21)	
10	1.210(2)	5.808(5)	1.913(9)	1.187(12)	6.363(14)	1.919(17)	3.259(19)	2.988(21)	1.339(23)	2.366(24)	9.439(24)

Note: 1.449(12) means 1.449×10^{12}.

Example: $i = 2$, $m = 3$, $\mathrm{VRF} = (2.592 \times 10^4)/T^4 L^5$; see (5.8-1).

^{a} Smoothing to endpoint.

^{b} One-step prediction.

Source: from Morrison [5, p. 258].

TABLE 5.8-2. Values of Constant λ_{ii} in VRF Equation for Fixed-Memory Filter When L Large and $h = \frac{1}{2}$ [a]

i / m	0	1	2	3	4	5	6	7	8	9	10
0	1.0(0)	1.0(0)	2.25(0)	2.250(0)	3.516(0)	3.516(0)	4.758(0)	4.785(0)	6.056(0)	6.056(0)	7.328(0)
1		1.2(1)	1.20(1)	7.500(1)	7.500(1)	2.297(2)	2.297(2)	5.168(2)	5.168(2)	9.771(2)	9.771(2)
2			7.20(2)	7.200(2)	8.820(3)	8.820(3)	4.465(4)	4.465(4)	1.501(5)	1.501(5)	3.963(5)
3				1.008(5)	1.008(5)	2.041(6)	2.041(6)	1.544(7)	1.544(7)	7.247(7)	7.247(7)
4					2.540(7)	2.540(7)	7.684(8)	7.684(8)	8.116(9)	8.116(9)	5.073(10)
5						1.0061(10)	1.0061(10)	4.250(11)	4.250(11)	5.977(12)	5.977(12)
6							5.754(12)	5.754(12)	3.237(14)	3.237(14)	5.846(15)
7								4.488(15)	4.488(15)	3.243(17)	3.243(17)
8									4.578(18)	4.578(18)	4.131(20)
9										5.914(21)	5.914(21)
10											9.439(24)

Note: 2.540(7) means 2.540×10^7.

Example: $i = 3$, $m = 5$, VRF $= (2.041 \times 10^6)/T^6 L^7$; see (5.8-1).

[a] Smoothing to center of observation interval.

Source: From Morrison [5, p. 259].

221

interval. Knowing λ_{ii} from these tables and using (5.8-1), the variance $[S^*_{n+h,n}]_{ii}$ can be obtained for $h = 0$, 1, or $-\frac{1}{2}L$ for large L.

As mentioned in Section 1.2.4.4, often in the literature the normalized value of the covariance matrix elements is defined. As before, the normalized covariance matrix elements are normalized relative to the covariance of the measurement error σ_x^2 and referred to as the variance reduction factors (VRFs) [5, p. 256]; see Section 1.2.4.4. Thus

$$\text{VRF}\{[S^*_{n+h,n}]_{ij} = \frac{1}{\sigma_x^2}[S^*_{n+h,n}]_{ij} \qquad (5.8\text{-}2)$$

From (5.8-1) it is apparent that the covariance elements decrease with increasing L. Consider a diagonal element of the covariance matrix defined by (5.8-1). Assume T, L, and σ_x^2 are fixed. The question we want to address is how $[S^*_{n+h,n}]_{ii}$ varies with m. We see that it depends on the variation λ_{ii} with m. Examining Table 5.8-1 indicates that λ_{ii} increases with increasing m; hence, the diagonal elements of the covariance matrix increase with increasing m. Because of this increase with m, it follows that it is desirable to keep m as small as possible. This subject will be discussed further in Sections 5.9 and 5.10.

It is not difficult to show [5, p. 254] from (5.6-7) that, for a fixed m and L, $[S^*_{n+h,n}]_{ii}$ is a polynomial in h of degree $2(m-i)$ with its zeros in the observation interval from $n-L$ to n or equivalently in the interval $h = -L, \ldots, 0$. As a result it increases monotonically outside the observation interval, as shown in Figure 5.8-1. Consequently, predictions or retrodictions far

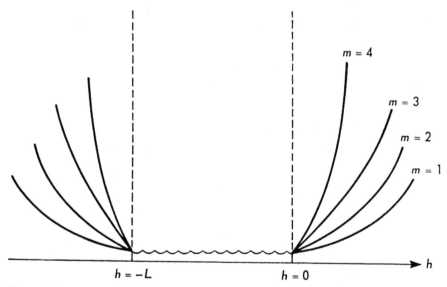

Figure 5.8-1 Functional monotonic increase of variance $[S^*_{n+h,n}]_{ii}$ of h-step prediction and retrodiction outside of data interval for least-squares fixed-memory filter. (From Morrison [5, p. 254].)

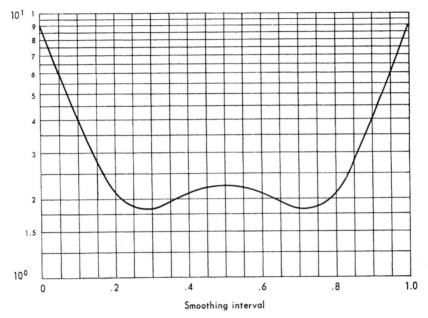

Figure 5.8-2 Plot of $L \cdot [S^*_{n+h,n}]_{0,0}$ for fixed-memory filter for $m = 2$ and $L \to \infty$. (From Morrison [5, p. 255].)

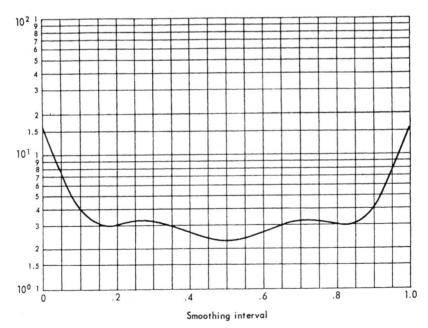

Figure 5.8-3 Plot of $L \cdot [S^*_{n+h,n}]_{0,0}$ for fixed-memory filter for $m = 3$ and $L \to \infty$. (From Morrison [5, p. 255].)

outside the measurement interval should be avoided. Figures 5.8-2 and 5.8-3 give plots of $[S^*_{n+h,n}]_{0,0}$ for $m = 2$ and $m = 3$. These curves are at their minimum value or near their minimum value at the center of the data interval. For this reason, it is desirable when determining the state of the target to smooth to the center of the observation interval if appropriate.

Examination of (5.8-1) indicates that, for L and m fixed, $[S^*_{n+h,n}]_{ij}$ increases as T is reduced whenever i or j or both are greater than 0. For $i = j = 0$, $[S^*_{n+h,n}]_{ij}$ is independent of T. Note that the filter integration time is given by

$$T_f = LT \tag{5.8-3}$$

Thus reducing T while keeping T_f fixed causes $[S^*_{n+h,n}]_{ij}$ to decrease monotonically; see (5.8-1). In this case L increases as T decreases so that an ever-increasing number of measurements is obtained over the fixed interval T_f. As a result $[S^*_{n+h,n}]_{ij}$ will decrease to zero in theory as L increases. However, in practice, as T goes to zero, the measurements will become correlated so that at some point the variance will not decrease as L increases, or equivalently, as T decreases.

Let us use (5.8-1) to obtain the square root of $[S^*_{n+h,n}]_{ii}$ for large L for important special cases. Specifically for $m = 1$ and $L = n$

$$\frac{\sigma_{n,n}}{\sigma_x} = \frac{\sigma_{n+1,n}}{\sigma_x} = \frac{2}{\sqrt{L}} \tag{5.8-4}$$

$$\frac{\dot{\sigma}_{n,n}}{\sigma_x} = \frac{\dot{\sigma}_{n+1,n}}{\sigma_x} = \frac{2\sqrt{3}}{T_f\sqrt{L}} \tag{5.8-5}$$

$$\frac{\sigma_{m,n}}{\sigma_x} = \frac{1}{\sqrt{L}} \tag{5.8-6}$$

$$\frac{\dot{\sigma}_{m,n}}{\sigma_x} = \frac{2\sqrt{3}}{T_f\sqrt{L}} \tag{5.8-7}$$

where m as a subscript here is used to represent the midpoint time index, that is, $m = \frac{1}{2}n$.

The above are extremely useful equations for quick back-of-the-envelope designs for determining sensitivity to system parameters for the aid-in-system design and for checks for detailed simulations. For $m = 2$ we obtain

$$\frac{\sigma_{n,n}}{\sigma_x} = \frac{\sigma_{n+1,n}}{\sigma_x} = \frac{3}{\sqrt{L}} \tag{5.8-8}$$

$$\frac{\dot{\sigma}_{n,n}}{\sigma_x} = \frac{\dot{\sigma}_{n+1,n}}{\sigma_x} = \frac{8\sqrt{3}}{T_f\sqrt{L}} \tag{5.8-9}$$

$$\frac{\ddot{\sigma}_{n,n}}{\sigma_x} = \frac{\ddot{\sigma}_{n+1,n}}{\sigma_x} = \frac{12\sqrt{5}}{T_f^2\sqrt{L}} \tag{5.8-10}$$

$$\frac{\sigma_{m,n}}{\sigma_x} = \frac{1.5}{\sqrt{L}} \tag{5.8-11}$$

$$\frac{\dot{\sigma}_{m,n}}{\sigma_x} = \frac{2\sqrt{3}}{T_f\sqrt{L}} \tag{5.8-12}$$

$$\frac{\ddot{\sigma}_{m,n}}{\sigma_x} = \frac{12\sqrt{5}}{T_f^2\sqrt{L}} \tag{5.8-13}$$

The above clearly indicates the penalty in the estimate in going from an $m = 1$ to an $m = 2$ filter if it is not needed. Specifically,

$$\frac{\left[\sigma_{n,n}/\sigma_x\right]_{m=2}}{\left[\sigma_{n,n}/\sigma_x\right]_{m=1}} = \frac{\left[\sigma_{n+1,n}/\sigma_x\right]_{m=2}}{\left[\sigma_{n+1,n}/\sigma_x\right]_{m=1}} = 1.5 \tag{5.8-14}$$

$$\frac{\left[\dot{\sigma}_{n,n}/\sigma_x\right]_{m=2}}{\left[\dot{\sigma}_{n,n}/\sigma_x\right]_{m=1}} = \frac{\left[\dot{\sigma}_{n+1,n}/\sigma_x\right]_{m=2}}{\left[\dot{\sigma}_{n+1,n}/\sigma_x\right]_{m=1}} = 4 \tag{5.8-15}$$

$$\frac{\left[\sigma_{m,n}/\sigma_x\right]_{m=2}}{\left[\sigma_{m,n}/\sigma_x\right]_{m=1}} = 1.5 \tag{5.8-16}$$

$$\frac{\left[\dot{\sigma}_{m,n}/\sigma_x\right]_{m=2}}{\left[\dot{\sigma}_{m,n}/\sigma_x\right]_{m=1}} = 1 \tag{5.8-17}$$

Of course if the target has an acceleration and an $m = 1$ filter is used, then a bias error would be suffered if an $m = 1$ g–h filter is used instead of an $m = 2$ g–h–k filter; see problem 5.8-1.

5.9   SYSTEMATIC ERRORS (BIAS, LAG, OR DYNAMIC ERROR)

In Sections 5.2 and 5.3 we used a polynomial of degree m to fit to $L + 1$ measurements given by the vector $Y_{(n)}$ of (5.2-1) when obtaining our least-squares estimate of $x_{n,n}^*$. The question arises as to what errors we incur if the target trajectory is a polynomial of degree larger than m or is not a polynomial at all. The estimate $x_{n,n}^*$ will then have a systematic error as well as a random error due to the measurement errors of (5.4-2) or more generally (4.1-1c). The magnitude of this systematic error will be developed in this section. Also, in Section 1.2.4.5 we pointed out that for the constant g–h filter it did not pay to have the rms of the predicted position estimate $\sigma_{n+1,n}$ much smaller than the bias error b^* or vice versa. As a result these errors were balanced. We somewhat arbitrarily chose b^* to equal $3\sigma_{n+1,n}$. We shall in this section show that the same type of situation prevails for the general least-squares polynomial estimate. In the next section we shall balance these errors by choosing $b^* = \sigma_{n+1,n}$.

First let us recall that even if the trajectory is not actually given by a polynomial, there always exists a polynomial of sufficiently high degree that can approximate the trajectory as accurately as desired. Let this degree be d and let the polynomial be given by

$$p(t) = \sum_{k=0}^{d} a_k t^k \qquad (5.9\text{-}1)$$

Then the trajectory state vector at time n can be written as

$$A_n = \begin{bmatrix} p(t) \\ Dp(t) \\ \vdots \\ D^d p(t) \end{bmatrix}_{t=nT} \qquad (5.9\text{-}2)$$

Assume that a polynomial of degree $m < d$, such as (5.2-2), is used to obtain the least-squares estimate (5.3-11) of the trajectory. Then the resultant estimate $X_{n,n}^*$ can now be written in terms of A_n plus the random part $N_{n,n}^*$ of the error in the estimate plus a systematic error term $B_{n,n}$, which we express as $-B_{n,n}$; specifically,

$$X_{n,n}^* = A_n - B_{n,n}^* + N_{n,n}^* \qquad (5.9\text{-}3)$$

where

$$B_{n,n}^* = \begin{bmatrix} b^*(t) \\ Db^*(t) \\ \vdots \\ D^d b^*(t) \end{bmatrix}_{t=nT} \qquad (5.9\text{-}3\text{a})$$

where $b^*(t)$ is the systematic error for the trajectory position estimate. Here, $b^*(t)$ is a polynomial of degree d whose first m terms equal zero (the terms of degree less than or equal to m) because the least-squares estimating polynomials of degree m provide an unbiased estimate of the target trajectory up to degree m; see Section 4.1.

(It should be pointed out that $b^*(t)$ actually depends on the observation interval over which the least-squares estimate is being obtained. As a result it should be written as $[b^*(t)]_n$. For simplicity we have dropped the subscript n, and it should be understood that the bias error is indeed dependent on n.)

Using the orthogonal Legendre polynomial representation of Section 5.3 for $b^*(t)$ it can be shown that [5, p. 268]

$$(b_i^*)_{n+h,n} = \left[\frac{d^i}{dt^i} b^*(r) \right]_n \bigg|_{r=L+h} = \frac{1}{T^i} \sum_{j=m+1}^{d} (\bar{\gamma}_j)_n \frac{d^i}{dr^i} p_j(r) \bigg|_{r=L+h} \qquad (5.9\text{-}4)$$

where

$$(\bar{\gamma}_j)_n = \frac{1}{c_j^2} \sum_{k=0}^{L} x_{n-L+k} \, p_j(k) \quad \text{for } m+1 \le j \le d \qquad (5.9\text{-}4a)$$

where c_j is given by (5.3-5) and r is indexed as in Fig. 5.2-1.

Equation (5.9-4) is the weighted sum of the product of an orthogonal Legendre polynomial and one of its derivatives. Likewise the diagonal covariance elements given by $[S^*_{n+h,n}]_{ii}$ of (5.6-7) are made of such sums. The zeros of these polynomials and in addition their derivatives are inside the observation interval; see Figures 5.9-1 and 5.9-2. It is for this reason that the covariance matrix and their derivatives are small in the observation interval and increase monotonically outside; see Figures 5.8-2 and 5.8-3. Because the systematic errors and their derivatives are made up of a similar sum, the same is thus true for them. Half of the polynomials and their derivatives have a zero at the center of the observation interval.

These results are not surprising since the polynomial fit $[p^*(r)]_n$ was designed to provide a least-squares fit to the data inside the observation interval and nothing was stated as to how well the fit would be outside the observation interval. As a result the point on the trajectory where the target state vector is best known or nearly best known, in the sense of having a minimum bias error,

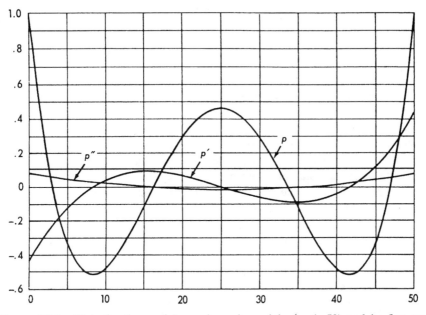

Figure 5.9-1 Plot of orthogonal Legendre polynomial $p(x; 4, 50)$ and its first two derivatives. (From Morrison [5, p. 269].)

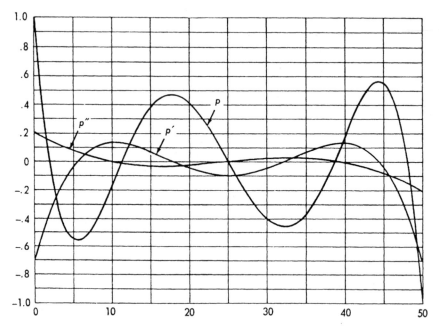

Figure 5.9-2　Plot of orthogonal Legendre polynomial $p(x; 5, 50)$ and its first two derivatives. (From Morrison [5, p. 269].)

is the center of the observation interval. The same was true relative to the covariance of the target estimate, tending to be minimum or near minimum at the center of the observation interval. Hence the time at which the best estimate of the state vector of the target is obtained is at or near the center of the observation interval. It is for this reason that smoothing to the center of the observation interval is often done; see Sections 3.5.2 and 5.8.

From (5.9-4) it follows that the systematic errors decrease as the order m of the polynomial fit increases. This occurs until $m = d$, at which point there is no systematic error. In reference 5 it is also shown that for a given L and m the systematic error decreases as T decreases. This is not surprising since it is well known that a polynomial fit of degree m to a set of data points $L + 1$ provides a better fit when the data points are observed over a small interval rather than a large interval. In reference 5 it is shown that the magnitude of the systematic error of (5.9-4) increases as L increases when T is fixed. Hence, for a fixed T, L (or equivalently T_f) should be chosen as small as possible in order to minimize the systematic error. This is opposite to what is needed to minimize the estimate covariance elements given by (5.8-1). There it is required to have L, or equivalently T_f, as large as possible.

We thus can not simultaneously minimize the systematic error and the estimate noise error using the least-squares polynomial fit described above. Consequently, a compromise choice for L is needed that balances the systematic error and the estimate noise error so that they are of the same order of magnitude as was done in Section 1.2.4.5; see (1.2-22) and (1.2-23). One way of doing this is described in the next section.

5.10 BALANCING SYSTEMATIC AND RANDOM ESTIMATION ERRORS

Assume that when no measurement errors are present the data set given by $Y(n)$ of (5.2-1) can be described at time $t = nT$ with negligible error by the polynomial of degree d given by

$$p(t) = \sum_{k=0}^{d}(a_k)_n t^k \tag{5.10-1}$$

Assume that the power series $p(t)$ given above converges rapidly so that if it is truncated to a polynomial of degree m then the truncation error is dominated by the first neglected term $(a_{m+1})_n t^{m+1}$. Then (to within a sign) the ith derivative of the systematic error can be expressed as [5, p. 275]

$$D^i b^*(r) \approx \frac{1}{T^i} \frac{(a_{m+1})_n T^{m+1} L^{(m+1)}}{\binom{2m+2}{m+1}} \frac{d^i}{dr^i} p_{m+1}(r) \tag{5.10-2}$$

Let us determine the systematic error at the end of the observation interval, that is, at $r = L$; see Figure 5.2-1. For $r = L$ it follows from (5.3-3) and (5.3-4) that $|p_{m+1}(L)| = p(L; m+1, L) = 1$. Hence the systematic error is bounded by [5, p. 276]

$$[|b^*(L)|] \leq \frac{|a_{m+1}| T^{m+1} L^{(m+1)}}{\binom{2m+2}{m+1}} \tag{5.10-3}$$

where $|a_{m+1}|$ is the magnitude of an estimated known bound on the coefficient a_{m+1} of the $(m+1)$st term and

$$\binom{n}{m} = \frac{n!}{m!(n-m)!} \tag{5.10-3a}$$

is the standard binomial coefficient, and $x^{(m)}$ is defined by (5.3-4a).

In contrast, from (5.8-1) the variance of the smoothed position error is [5, p. 276]

$$[S_{n,n}^*]_{0,0} = \frac{\lambda_{0,0}(m)}{L} \sigma_x^2 \tag{5.10-4}$$

where $\lambda_{0,0}$ is written as $\lambda_{0,0}(m)$ to emphasize its dependence on m. The term

$\lambda_{0,0}(m)$ is determined from Table 5.8-1. Equation (5.10-3) clearly shows how the systematic error increases with increasing L, while (5.10-4) shows how the variance of the trajectory position error decreases with increasing L.

To balance the systematic and random prediction errors, $|b^*(r)|$ is set equal to k times the rms prediction error, as done in Section 1.2.4.5. There k was set equal to 3; see (1.2-23). Here let us set $k = 1$. For a given m we find the L that makes the systematic error and the rms of the random error equal. We add the two errors to find the total error. We do this for $m = 0, 1, 2, \ldots$. Then we select the m that results in the lowest total error. This procedure allows the determination of the L and m that gives simultaneously the lowest total error and the best balance between the systematic and random errors.

5.11 TREND REMOVAL

Sometimes the trajectory of the target is known approximately. For example, this could be the case when tracking a satellite whose orbit is approximately known. Generally the trajectory of the satellite is known to be given by some elliptical equation of motion having approximately known parameters. Assume that a set of $L + 1$ new measurements $Y_{(n)}$ given by (5.2-1) is made at the times $n - L, \ldots, n$ of the satellite position. We want to use these $L + 1$ measurements to improve our known estimate of the target trajectory. We cannot directly use our linear least-squares Legendre polynomial filtering described in Section 5.3 for a target having its dynamics described by this elliptical trajectory. This is because the target motion is not described by a polynomial function of time. Instead the motion dynamics of the target is described by an elliptical path that is not nonlinear. This problem can be circumvented because the exact elliptical trajectory of the target can be approximated by a polynomial trajectory of high enough degree, d as given by (5.9-1), such that it has negligible error. The linear least-squares orthogonal Legendre polynomial filtering process described in Section 5.3 can then be applied to the trajectory to obtain an estimating polynomial of degree $m \le d$ that fits the data. An improvement on this polynomial approximation procedure will actually be used. This improvement reduces the degree m of the approximating polynomial fit that has to be used. This in turn is important because we saw in Section 5.8 that the variance of the least-squares estimate decreases with decreasing m. With this technique the polynomial of degree d, which has negligble errors, actually never has to be found. As we shall see, the technique involves a trend removal from the most recent data of the trajectory. The technique is general in that it can be applied to a linear as well as a nonlinear motion dynamics case where either is described by a high-order polynomial of degree d.

This trend removal approach applies to any situation where we have an initial approximate polynomial fit of degree m to the trajectory (the actual dynamics being linear or nonlinear) based on the past data. Assume that $L + 1$ new observations $Y_{(n)}$ of (5.2-1) have been obtained and that we want to improve our estimate of the trajectory. Then we use the existing polynomial fit based on the past data to obtain an estimate of what the $L+1$ new observations should be. We next find the differences between the actual new $L+1$ measurements and their predictions based on the past measurements. Let $\delta Y_{(n)}$ [given by (5.11-3)] represent this set of $L+1$ differences. It is to this set of differences $\delta Y_{(n)}$ that the least-squares estimate procedure is applied in order to update the target trajectory. That is, we find a best-fitting polynomial to the differences $\delta Y_{(n)}$ rather than $Y_{(n)}$. The advantage is that the $\delta Y_{(n)}$ has most of its motion (trend) removed from it and we are left with mostly the rms errors and some part of the trajectory trend that the past data does not allow us to predict. This trend removal technique is now detailed.

Assume that based on past data we know the approximate elliptical orbit can be approximated by a polynomial of degree m given by (4.1-44), that is,

$$p^*(t) = \sum_{k=0}^{m} \overline{a}_k t^k \tag{5.11-1}$$

Next form the differences

$$\delta y_{n-i} = y_{n-i} - p^*_{n-i} \quad \text{for } 0 \le i \le L \tag{5.11-2}$$

where

$$p^*_{n-i} = p^*(t_{n-i}) = p^*[(n-i)T] \tag{5.11-2a}$$

The above set of $L+1$ differences terms can be thought of as forming the new measurement vector $\delta Y_{(n)}$ is given by

$$\delta Y_{(n)} = (\delta y_n, \delta y_{n-1}, \dots, \delta y_{n-L})^T \tag{5.11-3}$$

Assume no measurement noise errors are present in the δy_{n-i}. Then the set of $L+1$ measurements defined by $\delta Y_{(n)}$ is exactly described by the difference polynomial given by

$$\delta p(t) = p(t) - p^*(t) = \sum_{i=0}^{d} (a_i - \overline{a}_i)t^i \tag{5.11-4}$$

where $p(t)$ is the polynomial of degree d that describes with negligible error the true trajectory, $p(t)$ being given by (5.10-1). The least-squares orthogonal Legendre polynomial fit described in Section 5.3 can now be applied to $\delta Y_{(n)}$ to obtain an estimate for this difference polynomial δp. Instead of the estimate $X^*_{n+h,n}$ for the state vector at time $n+h$, the estimate $\delta X^*_{n+h,n}$ given by

$$\delta X^*_{n+h,n} = X^*_{n+h,n} - \overline{X}_{n+h,n} \tag{5.11-5}$$

is obtained, where

$$\bar{X}_{n+h,n} = \begin{bmatrix} p^*(t) \\ Dp^*(t) \\ \vdots \\ D^m p^*(t) \end{bmatrix}_{t=(n+h)T} \tag{5.11-5a}$$

The desired estimate $X^*_{n+h,n}$ is obtained from $\delta X^*_{n+h,n}$ by adding $\bar{X}_{n+h,n}$ to $\delta X^*_{n+h,n}$.

Applying the least-squares filtering to the data set $\delta Y_{(n)}$ instead of $Y_{(n)}$ yields a much smaller systematic error. This is because the mth-degree polynomial fit to the data set $\delta Y_{(n)}$ has coefficients that are much smaller than those for $Y_{(n)}$, the known trend in the data set being removed. Most important, as a result of this trend removal, the dominant truncation error coefficient resulting when an mth-degree polynomial fit is used is now given by $a_{m+1} - \bar{a}_{m+1}$ instead of a_{m+1}; see (5.11-4). In turn this will result in the systematic error given by (5.10-3) being smaller. As a result the degree m of the polynomial fit can be reduced and L can be increased, with both of these leading to a reduction in the covariance matrix of the estimate and in turn a reduction of the sum of the systematic plus random error, that is, the total error.

6

EXPANDING-MEMORY
(GROWING-MEMORY)
POLYNOMIAL FILTERS

6.1 INTRODUCTION

The fixed-memory filter described in Chapter 5 has two important disadvantages. First, all the data obtained over the last $L + 1$ observations have to be stored. This can result in excessive memory requirements in some instances. Second, at each new observation the last $L + 1$ data samples have to be reprocessed to obtain the update estimate with no use being made of the previous estimate calculations. This can lead to a large computer load. When these disadvantages are not a problem, the fixed-memory filter would be used generally. Two filters that do not have these two disadvantages are the expanding-memory filter and the fading memory filter. The expanding memory filter is, as discussed in Section 1.2.10 and later in Section 7.6, suitable for track initiation and will be covered in detail in this chapter. The fading memory filter as discussed in Chapter 1 is used for steady state tracking, as is the fixed-memory filter, and will be covered in detail in Chapter 7.

Before proceeding it is important to highlight the advantages of the fixed-memory filter. First, if bad data is acquired, the effect on the filter will only last for a finite time because the filter has a finite memory of duration $L + 1$; that is, the fixed-memory filter has a finite transient response. Second, fixed-memory filters of short duration have the advantage of allowing simple processor models to be used when the actual process model is complex or even unknown because simple models can be used over short observation intervals. These two advantages are also obtained when using a short memory for the fading-memory filter discussed in Chapter 7.

233

6.2 EXTRAPOLATION FROM FIXED-MEMORY FILTER RESULTS

All the results given in Chapter 5 for the fixed-memory filter apply directly to the expanding-memory filter except now L is increasing with time instead of being fixed. To allow for the variation of L, it is convenient to replace the variable L by n and to have the first observation y_{n-L} be designated as y_0. The measurement vector $Y_{(n)}$ of (5.2-1) becomes

$$Y_{(n)} = [y_n, y_{n-1}, \ldots, y_0]^T \tag{6.2-1}$$

where n is now an increasing variable. The filter estimate is now based on all the $n+1$ past measurements. All the equations developed in Chapter 5 for the fixed-memory state vector estimate covariance matrix [such as (5.6-4), (5.6-7), and (5.8-1)] and systematic error [such as (5.10-2) and (5.10-3)] apply with L replaced by n. The least-squares polynomial fit equations given by (5.3-11), (5.3-13), and (5.5-3) also applies with L again replaced by n.

In this form the smoothing filter has the disadvantage, as already mentioned, of generally not making use of any of the previous estimate calculations in order to come up with the newest estimate calculation based on the latest measurement. An important characteristic of the expanding-memory filter, for which n increases, is that it can be put in a recursive form that allows it to make use of the last estimate plus the newest observation y_n to derive the latest estimate with the past measurements $(y_0, y_1, \ldots, y_{n-1})$ not being needed. This results in a considerable savings in computation and memory requirements because the last n measurements do not have to be stored, only the most recent state vector estimate, $X_{n,n-1}^*$. This estimate contains all the information needed relative to the past measurements to provide the next least-squares estimate. The next section gives the recursive form of the least-squares estimate orthogonal Legendre filter.

6.3 RECURSIVE FORM

It can be shown [5, pp. 348–362] after quite some manipulation that the filter form given by (5.3-13) can be put in the recursive forms of Table 6.3-1 for a one-state predictor when $m = 0, 1, 2, 3$. The results are given in terms of the scaled state vector $Z_{n+1,n}^*$ [see (5.4-12)]. As indicated before only the last one-state update vector $Z_{n,n-1}^*$ has to be remembered to do the update. This is an amazing result. It says that the last one-step update state vector $Z_{n,n-1}^*$ of dimension $m+1$ contains all the information about the previous n observations in order to obtain the linear least-squares polynomial fit to the past data $Y_{(n-1)}$ and the newest measurement y_n. Stated another way, the state vector $Z_{n,n-1}^*$ is a sufficient statistic [8, 9, 100].

TABLE 6.3-1. Expanding-Memory Polynomial Filter

Define

$$
\begin{pmatrix} z_0^* \\ z_1^* \\ z_2^* \\ \\ z_3^* \end{pmatrix}_{n+1,n}
=
\begin{pmatrix} x^* \\ T\dot{x}^* \\ \dfrac{T^2}{2!}\ddot{x}^* \\ \dfrac{T^3}{3!}\dddot{x}^* \end{pmatrix}_{n+1,n}
\qquad
\varepsilon_n \equiv y_n - \left(z_0^*\right)_{n,n-1}
$$

Degree 0 [a]:

$$
\left(z_0^*\right)_{n+1,n} = \left(z_0^*\right)_{n,n-1} + \frac{1}{n+1}\,\varepsilon_n
$$

Degree 1 [a]:

$$
\left(z_1^*\right)_{n+1,n} = \left(z_1^*\right)_{n,n-1} + \frac{6}{(n+2)(n+1)}\,\varepsilon_n
$$

$$
\left(z_0^*\right)_{n+1,n} = \left(z_0^*\right)_{n,n-1} + \left(z_1^*\right)_{n+1,n} + \frac{2(2n+1)}{(n+2)(n+1)}\,\varepsilon_n
$$

Degree 2 [a]:

$$
\left(z_2^*\right)_{n+1,n} = \left(z_2^*\right)_{n,n-1} + \frac{30}{(n+3)(n+2)(n+1)}\,\varepsilon_n
$$

$$
\left(z_1^*\right)_{n+1,n} = \left(z_1^*\right)_{n,n-1} + 2\left(z_2^*\right)_{n+1,n} + \frac{18(2n+1)}{(n+3)(n+2)(n+1)}\,\varepsilon_n
$$

$$
\left(z_0^*\right)_{n+1,n} = \left(z_0^*\right)_{n,n-1} + \left(z_1^*\right)_{n+1,n} - \left(z_2^*\right)_{n+1,n} + \frac{3(3n^2+3n+2)}{(n+3)(n+2)(n+1)}\,\varepsilon_n
$$

Degree 3 [a]:

$$
\left(z_3^*\right)_{n+1,n} = \left(z_3^*\right)_{n,n-1} + \frac{140}{(n+4)(n+3)(n+2)(n+1)}\,\varepsilon_n
$$

$$
\left(z_2^*\right)_{n+1,n} = \left(z_2^*\right)_{n,n-1} + 3\left(z_3^*\right)_{n+1,n} + \frac{120(2n+1)}{(n+4)(n+3)(n+2)(n+1)}\,\varepsilon_n
$$

$$
\left(z_1^*\right)_{n+1,n} = \left(z_1^*\right)_{n,n-1} + 2\left(z_2^*\right)_{n+1,n} - 3\left(z_3^*\right)_{n+1,n} + \frac{20(6n^2+6n+5)}{(n+4)(n+3)(n+2)(n+1)}\,\varepsilon_n
$$

$$
\left(z_0^*\right)_{n+1,n} = \left(z_0^*\right)_{n,n-1} + \left(z_1^*\right)_{n+1,n} - \left(z_2^*\right)_{n+1,n}
$$
$$
+ \left(z_3^*\right)_{n+1,n} + \frac{8(2n^3+3n^2+7n+3)}{(n+4)(n+3)(n+2)(n+1)}\,\varepsilon_n
$$

[a] In all cases, n starts at zero.

Source: From Morrison [5].

The filter equations given in Table 6.3-1 for $m = 1$ are exactly the same as those of the g–h growing-memory filter originally given in Section 1.2.10 for track initiation; compare (1.2-38a) and (1.2-38b) with the expressions for g and h given in Table 6.3-1 for $m = 1$; see problem 6.5-2. The filter of Section 1.2.10 and Table 6.3-1 for $m = 1$ are for a target characterized as having a constant-velocity dynamics model. The equations in Table 6.3-1 for $m = 2$ are for when the target dynamics has a constant acceleration and corresponds to the g–h–k growing-memory filter. The equations for $m = 3$ are the corresponding equations for when the target dynamics have a constant jerk, that is, a constant rate of change of acceleration. Practically, filters of higher order than $m = 3$ are not warranted. Beyond jerk are the yank and snatch, respectively, the fourth and fifth derivatives of position. The equation for $m = 0$ is for a stationary target. In this case the filter estimate of the target position is simply an average of the $n + 1$ measurements as it should be; see (4.2-23) and the discussion immediately before it. Thus we have developed the growing-memory g–h filter and its higher and lower order forms from the theory of least-squares estimation. In the next few sections we shall present results relative to the growing-memory filter with respect to its stability, track initiation, estimate variance, and systematic error.

6.4 STABILITY

Recursive differential equations such as those of Table 6.3-1 are called stable if any transient responses induced into them die out eventually. (Stated more rigorously, a differential equation is stable if its natural modes, when excited, die out eventually.) It can be shown that all the recursive differential expanding-memory filter equations of Table 6.3-1 are stable.

6.5 TRACK INITIATION

The track initiation of the expanding-memory filters of Table 6.3-1 needs an initial estimate of $Z^*_{n,n-1}$ for some starting n. If no a prori estimate is available, then the first $m + 1$ data points could be used to obtain an estimate for $Z^*_{m,m-1}$, where m is the order of the expanding-memory filter being used. This could be done by simply fitting an mth-order polynomial filter through the first $m + 1$ data points, using, for example, the Lagrange interpolation method [5]. However, an easier and better method is available. It turns out that we can pick any arbitrary value for $Z^*_{0,-1}$ and the growing memory filter will yield the right value for the scaled state vector $Z^*_{m+1,m}$ at time m. In fact the estimate $Z^*_{m+1,m}$ will be least-squares mth-order polynomial fit to the first $m + 1$ data samples independent of the value chosen for $Z^*_{0,-1}$; see problems 6.5-1 and 6.5-2. This is what we want. Filters having this property are said to be self-starting.

6.6   VARIANCE REDUCTION FACTOR

For large n the VRF for the expanding-memory filter can be obtained using
(5.8-1) and Tables (5.8-1) and (5.8-2) with L replaced by n. Expressions for the
VRF for arbitrary n are given in Table 6.6-1 for the one-step predictor when
$m = 0, 1, 2, 3$. Comparing the one-step predictor variance of Table 6.6-1 for
$m = 1$ with that given in Section 1.2.10 for the growing-memory filter indicates
that they are identical, as they should be; see (1.2-42). Also note that the same
variance is obtained from (5.6-5) for the least-squares fixed-memory filter.

TABLE 6.6-1. VRF for Expanding-Memory One-Step Predictors[a]
(Diagonal Elements of $S^*_{n+1,n}$)

Degree (m)	Output	VRF
0	$x^*_{n+1,n}$	$\dfrac{1}{(n+1)^{(1)}}$
1	$\dot{x}^*_{n+1,n}$	$\dfrac{12}{T^2(n+2)^{(3)}}$
	$x^*_{n+1,n}$	$\dfrac{2(2n+3)}{(n+1)^{(2)}}$
2	$\ddot{x}^*_{n+1,n}$	$\dfrac{720}{T^4(n+3)^{(5)}}$
	$\dot{x}^*_{n+1,n}$	$\dfrac{192n^2 + 744n + 684}{T^2(n+3)^{(5)}}$
	$x^*_{n+1,n}$	$\dfrac{9n^2 + 27n + 24}{(n+1)^{(3)}}$
3	$\dddot{x}^*_{n+1,n}$	$\dfrac{100,800}{T^6(n+4)^{(7)}}$
	$\ddot{x}^*_{n+1,n}$	$\dfrac{25,920n^2 + 102,240n + 95,040}{T^4(n+4)^{(7)}}$
	$\dot{x}^*_{n+1,n}$	$\dfrac{1200n^4 + 10,200n^3 + 31,800n^2 + 43,800n + 23,200}{T^2(n+4)^{(7)}}$
	$x^*_{n+1,n}$	$\dfrac{16n^3 + 72n^2 + 152n + 120}{(n+1)^{(4)}}$

[a]Recall that $x^{(m)} = x(x-1)(x-2) \cdots (x-m+1)$; see (5.3-4a).
Source: (From Morrison [5].)

6.7 SYSTEMATIC ERRORS

Because the systematic error of the expanding-memory filter grows as n grows (see Section 5.10), this filter cannot be cycled indefinitely. The fixed-memory filter of Chapter 5 and g–h fading-memory filter of Section 1.2.6 do not have this problem. The g–h fading-memory filter and its higher order forms are developed in the next section from the least-squares estimate theory results developed in Section 4.1.

7

FADING-MEMORY (DISCOUNTED LEAST-SQUARES) FILTER

7.1 DISCOUNTED LEAST-SQUARES ESTIMATE

The fading-memory filter introduced in Chapter 1, is similar to the fixed-memory filter in that it has essentially a finite memory and is used for tracking a target in steady state. As indicated in Section 1.2.6, for the fading-memory filter the data vector is semi-infinite and given by

$$Y_{(n)} = [y_n, y_{n-1}, \ldots]^T \tag{7.1-1}$$

The filter realizes essentially finite memory for this semi-infinite data set by having, as indicated in section 1.2.6, a fading-memory. As for the case of the fixed-memory filter in Chapter 5, we now want to fit a polynomial $p^* = [p^*(r)]_n$ [see (5.2-3), e.g.)] to the semi-infinite data set given by (7.1-1). Here, however, it is essential that the old, stale data not play as great a role in determining the polynomial fit to the data, because we now has a semi-infinite set of measurements. For example, if the latest measurement is at time n and the target made a turn at data sample n-10, then we do not want the samples prior to the $n - 10$ affecting the polynomial fit as much. The least-squares polynomial fit for the fixed-memory filter minimized the sum of the squares of the errors given by (5.3-7). If we applied this criteria to our filter, then the same importance (or weight) would be given an error resulting from the most recent measurement as well as one resulting for an old measurement. To circumvent this undesirable feature, we now weight the error due to the old data less than that due to recent data. This is achieved using a discounted, least-squares weighting as done in (1.2-34); that is, we minimize

$$e_n = \sum_{r=0}^{\infty} \{y_{n-r} - \left[p^*(r)\right]_n\}^2 \theta^r \qquad (7.1-2)$$

where here positive r is now running backward in time and

$$0 \leq \theta < 1 \qquad (7.1-2a)$$

The parameter θ here determines the discounting of the old data errors, as done in Section 1.2.6. For the most recent measurement $y_n, r = 0$ in (7.1-2) and $\theta^0 = 1$ with the error based on the most recent measurement at time $r = 0$ being given maximum weight. For the one-time-interval-old data $y_{n-1}, r = 1$ and $\theta^r = \theta$ so that these one-time-interval-old data are not given as much weight (because $0 \leq \theta < 1$), with the result that the error for the polynomial fit to this data point can be greater than it was for the most recent data point in obtaining the best estimating polynomial, which satisfies (7.1-2). For the two-time-interval-old data point given by $y_{n-2}, r = 2$ and $\theta^r = \theta^2$, and the error for this time sample can even be bigger, and so forth. Thus with this weighting the errors relative to the fitting polynomial are discounted more and more as the data gets older and older. The minimum of (7.1-2) gives us what we called in Section 1.2.6 a discounted least-squares fit of the polynomial to the semi-infinite data set. The memory of the resulting filter is dependent on θ. The smaller θ is the shorter the filter memory because the faster the filter discounts the older data. This filter is also called the fading-memory filter. It is a generalization of the fading-memory g–h filter of Section 1.2.6. The g–h filter of Section 1.2.6 is of degree $m = 1$, here we fit a polynomial p^* of arbitrary degree m.

To find the polynomial fit p^* of degree m that minimizes (7.1-2), an orthogonal polynomial representation of p^* is used, just as was done for the fixed-memory filter when minimizing (5.3-7) by the use of the Legendre polynomial; see (5.3-1). Now, however, because the data is semi-infinite and because of the discount weighting by θ^r, a different orthogonal polynomial is needed. The discrete orthogonal polynomial used now is the discrete Laguerre polynomial described in the next section.

7.2 ORTHOGONAL LAGUERRE POLYNOMIAL APPROACH

The discounted least-squares estimate polynomial p^* that minimizes (7.1-2) is represented by the sum of normalized orthogonal polynomials as done in (5.3-1) except that the orthonormal discrete Laguerre polynomial $\phi_j(r)$ of degree j is defined by the equations [5, pp. 500–501]

$$\phi_j(r) = K_j p_j(r) \qquad (7.2-1)$$

where

$$K_j = \frac{1}{c_j} = \left(\frac{1-\theta}{\theta^j}\right)^{\frac{1}{2}} \tag{7.2-1a}$$

$$c_j = c(j,\theta) \tag{7.2-1b}$$

$$[c(j,\theta)]^2 = \frac{\theta^j}{1-\theta} \tag{7.2-1c}$$

$$p_j(r) = p(r;j,\theta)$$

$$= \theta^j \sum_{\nu=0}^{j} (-1)^\nu \binom{j}{\nu} \left(\frac{1-\theta}{\theta}\right)^\nu \binom{r}{\nu} \tag{7.2-1d}$$

where $p(r;j,\theta)$ is the orthogonal discrete Laguerre polynomial, which obeys the following discrete orthogonal relationship:

$$\sum_{r=0}^{\infty} p(r;i,\theta)p(r;j,\theta) = \begin{cases} 0 & j \neq i \\ [c(j,\theta)]^2 & j = i \end{cases} \tag{7.2-2}$$

Tabel 7.2-1 gives the first four discrete Laguerre polynomials. The orthonormal Laguerre polynomial $\phi_j(r)$ obeys the orthonormal relationship

$$\sum_{r=0}^{\infty} \phi_i(r)\phi_j(r)\theta^r = \delta_{ij} \tag{7.2-3}$$

TABLE 7.2-1. First Four Orthogonal Discrete Laguerre Polynomials

$$p(x;0,\theta) = 1$$

$$p(x;1,\theta) = \theta\left[1 - \frac{1-\theta}{\theta}x\right]$$

$$p(x;2,\theta) = \theta^2\left[1 - 2\left(\frac{1-\theta}{\theta}\right)x + \left(\frac{1-\theta}{\theta}\right)^2 \frac{x(x-1)}{2!}\right]$$

$$p(x;3,\theta) = \theta^3\left[1 - 3\left(\frac{1-\theta}{\theta}\right)x + 3\left(\frac{1-\theta}{\theta}\right)^2 \frac{x(x-1)}{2!} - \left(\frac{1-\theta}{\theta}\right)^3 \frac{x(x-1)(x-2)}{3!}\right]$$

$$p(x;4,\theta) = \theta^4\left[1 - 4\left(\frac{1-\theta}{\theta}\right)x + 6\left(\frac{1-\theta}{\theta}\right)^2 \frac{x(x-1)}{2!}\right.$$

$$\left. - 4\left(\frac{1-\theta}{\theta}\right)^3 \frac{x(x-1)(x-2)}{3!} + \left(\frac{1-\theta}{\theta}\right)^4 \frac{x(x-1)(x-2)(x-3)}{4!}\right]$$

Substituting (7.2-1) into (5.3-1), and this in turn into (7.1-2), and performing the minimization yields [5, p. 502]

$$(\beta_j)_n = \sum_{k=0}^{\infty} y_{n-k} \phi_j(k) \theta^k \qquad 0 \le j \le m \qquad (7.2\text{-}4)$$

However, the above solution is not recursive. After some manipulation [5, pp. 504–506], it can be shown that the discounted least-squares mth-degree polynomial filter estimate for the ith derivative of x, designated as $D^i x^*$, is given by the recursive solution

$$(D^i x^*)_{n-r,n} = \left\{ \left(-\frac{1}{T}\right)^i \sum_{j=0}^{m} \left[\frac{d^i}{dr^i} \phi_j(r) \right] \frac{K_j \theta^j (1-q)^j}{(1-q\theta)^{j+1}} \right\} y_n \qquad (7.2\text{-}5)$$

where q is the backward-shifting operator given by

$$q^k y_n = y_{n-k} \qquad (7.2\text{-}5a)$$

for k an integer and q has the following properties:

$$(1-q)^2 = 1 - 2q + q^2 \qquad (7.2\text{-}6)$$

$$(1-q)^{-1} = 1 + q + q^2 + \cdots \qquad (7.2\text{-}7)$$

$$= \sum_{k=0}^{\infty} q^k$$

$$(\beta - q)^{-1} = \beta^{-1} \left(1 - \frac{q}{\beta}\right)^{-1}$$

$$= \beta^{-1} \left(1 + \frac{q}{\beta} + \left(\frac{q}{\beta}\right)^2 + \cdots \right) \qquad (7.2\text{-}8)$$

It is not apparent at first that (7.2-5) provides a recursive solution for $D^i x^*$. To verify this the reader should write out (7.2-5) for $i = 0$ and $m = 1$. Using (7.2-5) the recursive equations of Table 7.2-2 for the fading-memory filters are obtained for $m = 0, \ldots, 4$ [5, pp. 506–507].

The filter equations for $m = 1$ are identical to the fading memory g–h filter of Section 1.2.6. Specifically, compare g and h of (1.2-35a) and (1.2-35b) with those of Table 7.2-2 for $m = 1$. Thus we have developed the fading-memory g–h filter from the least-squares estimate as desired. In the next sections we shall discuss the fading-memory filter stability, variance, track initiation, and systematic error, as well as the issue of balancing systematic and random prediction errors and compare this filter with the fixed-memory filter. Note that the recursive fading-memory filters given by Table 7.2-2 only depend on the

TABLE 7.2-2. Fading-Memory Polynomial Filter

Define

$$
\begin{pmatrix} z_0^* \\ z_1^* \\ z_2^* \\ z_3^* \\ z_4^* \end{pmatrix}_{n+1,n}
=
\begin{pmatrix} x^* \\ TDx^* \\ \dfrac{T^2}{2!}D^2x^* \\ \dfrac{T^3}{3!}D^3x^* \\ \dfrac{T^4}{4!}D^4x^* \end{pmatrix}_{n+1,n}
\qquad
\varepsilon_n = y_n - \left(z_0^*\right)_{n,n-1}
$$

Degree 0:

$$
\left(z_0^*\right)_{n+1,n} = \left(z_0^*\right)_{n,n-1} + (1-\theta)\varepsilon_n
$$

Degree 1:

$$
\left(z_1^*\right)_{n+1,n} = \left(z_1^*\right)_{n,n-1} + (1-\theta)^2\varepsilon_n
$$
$$
\left(z_0^*\right)_{n+1,n} = \left(z_0^*\right)_{n,n-1} + \left(z_1^*\right)_{n+1,n} + (1-\theta^2)\varepsilon_n
$$

Degree 2:

$$
\left(z_2^*\right)_{n+1,n} = \left(z_2^*\right)_{n,n-1} + \tfrac{1}{2}(1-\theta)^3\varepsilon_n
$$
$$
\left(z_1^*\right)_{n+1,n} = \left(z_1^*\right)_{n,n-1} + 2\left(z_2^*\right)_{n+1,n} + \tfrac{3}{2}(1-\theta)^2(1+\theta)\varepsilon_n
$$
$$
\left(z_0^*\right)_{n+1,n} = \left(z_0^*\right)_{n,n-1} + \left(z_1^*\right)_{n+1,n} - \left(z_2^*\right)_{n+1,n} + (1-\theta^3)\varepsilon_n
$$

Degree 3:

$$
\left(z_3^*\right)_{n+1,n} = \left(z_3^*\right)_{n,n-1} + \tfrac{1}{6}(1-\theta)^4\varepsilon_n
$$
$$
\left(z_2^*\right)_{n+1,n} = \left(z_2^*\right)_{n,n-1} + 3\left(z_3^*\right)_{n+1,n} + (1-\theta)^3(1+\theta)\varepsilon_n
$$
$$
\left(z_1^*\right)_{n+1,n} = \left(z_1^*\right)_{n,n-1} + 2\left(z_2^*\right)_{n+1,n} - 3\left(z_3^*\right)_{n+1,n} + \tfrac{1}{6}(1-\theta)^2(11+14\theta+11\theta^2)\varepsilon_n
$$
$$
\left(z_0^*\right)_{n+1,n} = \left(z_0^*\right)_{n,n-1} + \left(z_1^*\right)_{n+1,n} - \left(z_2^*\right)_{n+1,n} + \left(z_3^*\right)_{n+1,n} + (1-\theta^4)\varepsilon_n
$$

Degree 4:

$$
\left(z_4^*\right)_{n+1,n} = \left(z_4^*\right)_{n,n-1} + \tfrac{1}{24}(1-\theta)^5\varepsilon_n
$$
$$
\left(z_3^*\right)_{n+1,n} = \left(z_3^*\right)_{n,n-1} + 4\left(z_4^*\right)_{n+1,n} + \tfrac{5}{12}(1-\theta)^4(1+\theta)\varepsilon_n
$$
$$
\left(z_2^*\right)_{n+1,n} = \left(z_2^*\right)_{n,n-1} + 3\left(z_3^*\right)_{n+1,n} - 6\left(z_4^*\right)_{n+1,n} + \tfrac{5}{24}(1-\theta)^3(7+10\theta+7\theta^2)\varepsilon_n
$$
$$
\left(z_1^*\right)_{n+1,n} = \left(z_1^*\right)_{n,n-1} + 2\left(z_2^*\right)_{n+1,n} - 3\left(z_3^*\right)_{n+1,n} + 4\left(z_4^*\right)_{n+1,n}
$$
$$
+ \tfrac{5}{12}(1-\theta)^2(5+7\theta+7\theta^2+5\theta^3)\varepsilon_n
$$
$$
\left(z_0^*\right)_{n+1,n} = \left(z_0^*\right)_{n,n-1} + \left(z_1^*\right)_{n+1,n} - \left(z_2^*\right)_{n+1,n} + \left(z_3^*\right)_{n+1,n} - \left(z_4^*\right)_{n+1,n} + (1-\theta^5)\varepsilon_n
$$

Source: From Morrison [5].

past infinite set of measurements y_n, y_{n-1}, ... through the past prediction estimate $Z^*_{n,n-1}$ just as was the case for the expanding-memory filter of Section 6.3. Hence as in that case $Z^*_{n,n-1}$ is a sufficient statistic.

7.3 STABILITY

The fading-memory filters described in Section 7.2 are all stable for $|\theta| < 1$. For large n the transient (natural modes) of the fading-memory filter vary as $n^m \theta^n$ [5, p. 508]. As a result the transient error persists longer the larger m is. Thus it is desired to keep m as small as possible in order to keep the transient as short as possible. On the other hand the filter systematic errors increase with decreasing m. Hence a compromise is needed.

Making θ smaller will also cause the transient to die out faster. However, making θ smaller will also reduce the filter memory (as was the case for the discounted least-squares g–h filter of Section 1.2.6), the old data not being weighted as heavily; see (7.1-2). Based on the results obtained for the fixed-memory filter, it follows that the shorter the memory time the smaller the systematic errors but the larger the variance of the filter estimates. This results in another compromise being needed. In Section 7.8 we discuss the balancing of the systematic and random errors.

7.4 VARIANCE REDUCTION FACTORS

The VRF for the fading-memory filter is given in Table 7.4-1 for the one-step predictor when $m = 0, \ldots, 3$. A general expression for the i, j covariance matrix elements is given by [5, p. 528]

$$[S^*_{n-r,n}]_{ij} = \sigma^2_x \frac{(1-\theta)^{i+j+1} \lambda_{ij}(\theta, r; m)}{T^{i+j}} \tag{7.4-1}$$

where values of the diagonal $\lambda_{ii}(\theta, r; m)$ are given in Table 7.4-2 for $i - 0, \ldots, 10$ and $m = 0, \ldots, 10$ when θ is close to unity. For the fading-memory one-step predictor with $m = 1$, the exact covariance matrix is given by [5, p. 532]

$$_s S^*_{n+1,n} = \sigma^2_x \begin{pmatrix} \dfrac{(1-\theta)(5+4\theta+\theta^2)}{(1+\theta)^3} & \dfrac{(1-\theta)^2(3+\theta)}{(1+\theta)^3} \\[4mm] \dfrac{(1-\theta)^2(3+\theta)}{(1+\theta)^3} & \dfrac{2(1-\theta)^3}{(1+\theta)^3} \end{pmatrix} \tag{7.4-2}$$

The variance for the one-step position predictor given by (7.4-2) (the 0,0

TABLE 7.4-1. Fading-Memory Filter VRF for One-Step Predictor

Degree (m)	Output	VRF ($0 < \theta < 1$)
0	$x^*_{n+1,n}$	$\dfrac{1-\theta}{1+\theta}$
1	$Dx_{n+1,n}*$	$\dfrac{2}{T^2}\dfrac{(1-\theta)^3}{(1+\theta)^3}$
	$x^*_{n+1,n}$	$\dfrac{1-\theta}{(1+\theta)^3}(5 + 4\theta + \theta^2)$
2	$D^2x^*_{n+1,n}$	$\dfrac{6}{T^4}\dfrac{(1-\theta)^5}{(1+\theta)^5}$
	$Dx^*_{n+1,n}$	$\dfrac{1}{T^2}\dfrac{(1-\theta)^3}{(1+\theta)^5}\left(\dfrac{49 + 50\theta + 13\theta^2}{2}\right)$
	$x^*_{n+1,n}$	$\dfrac{1-\theta}{(1+\theta)^5}(19 + 24\theta + 16\theta^2 + 6\theta^3 + \theta^4)$
3	$D^3x^*_{n+1,n}$	$\dfrac{20}{T^6}\dfrac{(1-\theta)^7}{(1+\theta)^7}$
	$D^2x^*_{n+1,n}$	$\dfrac{1}{T^4}\dfrac{(1-\theta)^5}{(1+\theta)^7}(126 + 152\theta + 46\theta^2)$
	$Dx^*_{n+1,n}$	$\dfrac{1}{T^2}\dfrac{(1-\theta)^3}{(1+\theta)^7}\left(\dfrac{2797 + 4{,}634\theta + 3{,}810\theta^2 + 1{,}706\theta^3 + 373\theta^4}{18}\right)$
	$x^*_{n+1,n}$	$\dfrac{1-\theta}{(1+\theta)^7}(69 + 104\theta + 97\theta^2 + 64\theta^3 + 29\theta^4 + 8\theta^5 + \theta^6)$

Note: The VRF of $D^i x^*$ is defined as $E\{(D^i x^*_{n+1,n})^2\}/\sigma_x^2$ and is thus the diagonal element of the estimate covariance matrix when the variance of the input errors is unity.

Source: From Morrison [5, pp. 526, 527].

element) agrees with the results given in Section 1.2.10 for the fading-memory g–h filter; see (1.2-41). The results of (7.4-2) also agree with those of Table 7.4-1 for $m = 1$.

7.5 COMPARISON WITH FIXED-MEMORY POLYNOMIAL FILTER

The fading-memory filter is very similar to the fixed-memory filter of Chapter 5 in that (unlike the expanding-memory filter)it has effectively a fixed memory.

TABLE 7.4-2. Values of Constant λ_{ii} in Equation for One-Step Prediction Using Fading-Memory Filter for θ Close to Unity

i \\ m	0	1	2	3	4	5	6	7	8	9	10
0	5.0(-1)	1.25(0)	2.063(0)	2.906(0)	3.770(0)	4.647(0)	5.534(0)	6.429(0)	7.331(0)	8.238(0)	9.150(0)
1		2.5(-1)	1.75(0)	5.781(0)	1.375(0)	2.714(1)	4.748(1)	7.636(1)	1.154(2)	1.661(2)	2.303(2)
2			1.875(-1)	2.531(0)	1.378(1)	4.906(1)	1.358(2)	3.177(2)	6.594(2)	1.251(3)	2.211(3)
3				1.563(-1)	3.438(0)	2.777(1)	1.377(2)	5.070(2)	1.525(3)	3.958(3)	9.184(3)
4					1.367(-1)	4.443(0)	4.982(1)	3.276(2)	1.546(3)	5.800(3)	1.839(4)
5						1.231(-1)	5.537(0)	8.218(1)	6.914(2)	4.065(3)	1.860(4)
6							1.128(-1)	6.711(0)	1.273(2)	1.333(3)	9.552(3)
7								1.047(-1)	7.960(0)	1.878(2)	2.396(3)
8									9.819(-2)	9.279(0)	2.666(2)
9										9.274(-2)	1.067(1)
10											8.810(-2)

Notes: 5.070(2) means 5.070×10^2. VRF $= [(1-\theta)^{2i+1}/T^{2i}] \, \lambda_{ii}(m)$ [see (7.4-1)].
Example: $i = 2$, $m = 3$, VRF $= [(1-\theta)^5/T^4] \times 2.531 \times 10^0$.
Source: From Morrison [5, p. 529].

As indicated previously, this memory depends on θ. The question we address in this section is what is the effective memory of the fading-memory filter. A natural basis is to find the θ for which the variance of the fading-memory filter estimate is identical to that of the fixed-memory filter. We first answer this question for the one-step predictor of first degree $(m = 1)$. Equation (5.6-5) gives the covariance matrix for the fixed-memory filter while (7.4-2) gives it for the fading-memory filter. Equating the 0, 0 elements of these two matrices gives $L \doteq 30$ for $\theta = 0.9$. Thus the fading-memory filter has an effective memory (smoothing time) of $30T$. Note the same procedure of equating variances was used in Section 1.2.10 [see (1.2-40)] for track initiation. There it was used to determine the time to transition from the track initiation growing-memory filter to the g–h steady-state fading-memory filter. From the above we see that the track initiation transition time turns out to be the time when the memory of the growing-memory filter equals that for the fading-memory filter. This is an intuitively satisfying result. It says that for minimum transient in switching from the track initiation filter to the steady state filter, we should transition when the growing-memory filter has processed as much data as the fading-memory filter uses in steady state.

Equating the 1,1 elements yields a slightly different value for L. Specifically for $\theta = 0.9$, $L \doteq 34$ is obtained. Equating the off-diagonal elements yields $L \doteq 32$. Thus equating the three distinct covariance matrix terms yields three different L. They are reasonably close however. Using Table 7.4-2 one can obtain the one-step predictor VRF for the fading-memory filter for θ close to unity.

Similarly, from Table 5.8-1 one can obtain the one-step predictor VRFs for the fixed-memory filter for large L. Equating corresponding covariance matrix elements obtained with these tables will yield the effective memory of the fading-memory filter as a function of θ when θ is close to unity, that is, when the filter memory is large. By way of example, using the position estimate variance terms when $m = 1$, one obtains [5, p. 534].

$$\frac{4}{L} = 1.25(1 - \theta) \tag{7.5-1}$$

or equivalently

$$L = \frac{3.2}{1 - \theta} \tag{7.5-2}$$

In general, equating other elements of the covariance matrices yields an equation of the form

$$L = \frac{\text{const}}{1 - \theta} \tag{7.5-3}$$

TABLE 7.5-1. Memory $L+1$ of Fixed-Memory Filter Equal to That of Fading-Memory Filter with Discounting Parameter θ

$i \backslash m$	0	1	2	3	4	5	6	7	8	9	10
0	2.00	3.20	4.36	5.51	6.63	7.75	8.85	9.96	11.0	12.1	13.2
1		3.63	4.79	5.92	7.04	8.15	9.25	10.4	11.4	12.5	13.6
2			5.21	6.34	7.46	8.56	9.66	10.8	11.8	12.9	14.0
3				6.76	7.87	8.98	10.1	11.2	12.3	13.3	14.4
4					8.29	9.40	10.5	11.6	12.7	13.7	14.8
5						9.82	10.9	12.0	13.1	14.2	15.2
6							11.3	12.4	13.5	14.6	15.7
7								12.8	13.9	15.0	16.1
8									14.4	15.4	16.5
9										15.8	16.9
10											17.4

Note: $L = \text{const.}/(1 - \theta)$.

Example: Let $m = 1$. Equating position VRFs gives $L = 3.2/(1 - \theta)$. Equating velocity VRFs gives $L = 3.63/(1 - \theta)$. Let $\theta = 0.9$. Then $L = 32$ or $L = 36$, respectively.

Source: From Morrison [5, p. 535].

The constants obtained using Tables 7.4-2 and 5.8-1 are tabulated in Table 7.5-1 for $i = 0, \ldots, 10$ and $m = 0, \ldots, 10$.

7.6 TRACK INITIATION

In Sections 1.2.10 and 1.3 track initiation was discussed for the fading-memory g–h and g–h–k filters, which are, respectively, the $m = 1$, 2 fading-memory filters being discussed here in Chapter 7. The basic method for track initiation described Sections 1.2.10 and 1.3 can be extended, in general, to any degree fading-memory filter. We will now review this subject to get further insight into track initiation and to extend the results to higher degree fading-memory filters.

We would like to initiate the fading-memory filter at time $n = 0$ with some scaled vector $Z_{0,-1}$, which is as close as possible to the true state of the target. In general, we do not have the a priori knowledge that permits the initiation of the fading-memory filter at time $n = 0$. In this case the filter could be started at some later time like $n = n_0$ when sufficient information is available to obtain a good estimate of $Z^*_{n_0+1, n_0}$. One possibility is to fit a polynomial through the first $m + 1$ samples so that $n_0 = m$ and use this to provide the estimate for $Z^*_{n+1, n} = Z^*_{m+1, m}$. This is essentially what was done to initiate track for the expanding-memory filter. Using this approach, however, does not give a good estimate for the steady-state fading-memory filter.

To see why the state estimate $Z^*_{n_0+1, n_0}$ so obtained is not a good one, consider a fading-memory filter that has been operating satisfactory in steady state for a while. At a time n it has a scaled one-step prediction state vector $Z^*_{n,n-1}$, which together with the observation y_n yields the updated scaled estimate $Z^*_{n+1,n}$; see Table 7.2-2. What distinguishes $Z^*_{n+1,n}$ from $Z^*_{m+1,m}$ obtained above? The estimate $Z^*_{n+1,n}$ has the steady-state variance and steady-state systematic error expected for the fading-memory filter being used. On the other hand, $Z^*_{m+1,m}$ does not have a variance or systematic error expected in steady state if the effective memory of the fading-memory filter is much larger than $m + 1$. Instead, its variance will be much larger because the amount of data used to estimate $Z^*_{m+1,m}$ is much shorter than that used in the steady state to estimate $Z^*_{n+1,n}$. The systematic error, on the other hand, will be much smaller because a smaller length of data is used to estimate the sealed vector $Z^*_{m+1,m}$. However, the total error, random plus systematic, will be larger for the estimate $Z^*_{m+1,m}$ than for $Z^*_{n+1,n}$.

What is needed is an estimate for $Z^*_{n_0+1,n_0}$ obtained using a larger set of measurements than the first $m + 1$ samples. The way to obtain such an estimate is to use an expanding-memory filter of the same degree m as the fading-memory filter to obtain this estimate at a time $n = n_0$. The time n_0 is that for which the expanding-memory filter has the same variance as the fading-memory

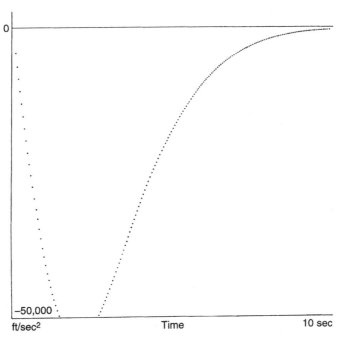

Figure 7.6-1 Starting transient error for critically damped one-step predictor filter of degree $m = 3$ when critically damped filter itself is used for track initiation after first four data points are fitted to third-degree polynomial in order to initialize filter. Parameters $\theta = 0.945$, $T = 0.05$ sec, and $\sigma_x = 5$ ft. (From Morrison [5, p. 541].)

filter in steady state, both filters having the same degree m. At this time the two
filters have the same memory. Also, at this time n_0 the systematic error $Z^*_{n_0+1,n_0}$
will approximately match that of the fading-memory filter in steady state. This
is exactly the procedure used in Section 1.2.10 for track initiation of the fading-
memory g–h filter. Here we are applying the procedure to higher order
(arbitrary m) fading-memory filters.

In Section 7.5 we discussed how to determine the memory $L+1$ of an
expanding-memory filter that has the same variance as that of a fading-memory
filter having the same degree. Thus those results can be used to determine the
memory $n_0 + 1$ of the expanding-memory filter needed to estimate $Z^*_{n_0+1,n_0}$.
The $Z^*_{n_0+1,n_0}$ obtained from the expanding-memory filter at time $n = n_0$ is then
used to start the fading-memory filter at time $n = n_0 + 1$. To obtain the estimate
$Z^*_{n_0+1,n_0}$, the expanding-memory filter is used to track the target for the first
$n_0 + 1$ measurements. As discussed in Section 6.5, the expanding-memory filter
is self-starting and hence needs no special track initiation procedures; it can be
started at time zero with any scaled vector $Z^*_{0,-1}$.

Figure 1.3-7 gave an example of the initiation of a critically damped g–h–k
filter, or equivalently a fading-memory filter of degree 2 (i.e., $m = 2$), using a
polynomial fit to the first $m + 1$ observations to obtain $Z^*_{m+1,m}$. The transient
resulting is clearly very objectional. Figure 1.3-8 gives the result when an

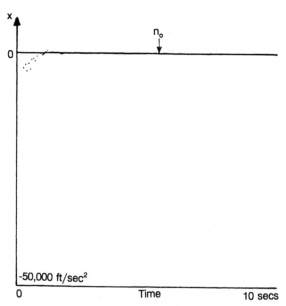

Figure 7.6-2 Starting transient error for critically damped one-step predictor filter of
degree 3 when expanding-memory polynomial filter is used for track initiation. Switch
from expanding-memory polynomial filter to g–h–k critically damped filter occurs after
101st observation, that is, $n_0 = 100$. Parameters: $\theta = 0.945, T = 0.05$ sec, and
$\sigma_x = 5$ ft. (From Morrison [5, p. 542].)

expanding-memory filter is used for track initiation in the manner described above. The transient is now much smaller. A second example is given in Figures 7.6-1 and 7.6-2 for a third degree ($m = 3$) filter. Figure 7.6-1 shows the poor results obtained when a polynomial fit to the first $m + 1$ observations is used for track initiation. In contrast Figure 7.6-2 shows the excellent results obtained when an expanding-memory filter is used for track initiation.

7.7 SYSTEMATIC ERRORS

A procedure similar to that used for the fixed-memory polynomial filter in Sections 5.9 and 5.10 can be used to obtain the systematic error $b^*(r)$ for the fading-memory filter. Instead of using the orthogonal Legendre polynomial, the orthogonal Laguerre polynomial is used for the representation of the systematic error. The resulting equation for the bound on the systematic error, equivalent to (5.10-3), for the fading-memory filter is [5, p. 547]

$$b^*(r) = \frac{(a_{m+1})_n T^{m+1}(m+1)!}{(1-\theta)^{m+1}} p_{m+1}(r) \qquad (7.7\text{-}1)$$

where $p_{m+1}(r)$ is defined by (7.2-1d).

7.8 BALANCING THE SYSTEMATIC AND RANDOM PREDICTION ERROR

As indicated at the end of Section 7.3, the fading-memory filter systematic and random errors have to be balanced in a similar manner to the used for the fixed-memory filter in Section 5.10. The memory for the fading-memory filter is determined by the value of θ. For a given m, just as was done for the fixed-memory filter, the value of θ can be found that results in a balance of the systematic error bounded by (7.7-1) and the random error variance given by (7.4-1) or (7.4-2). This would be done for $m = 0, 1, \ldots$ to find the m that results in the minimum total error, that is, the sum of the systematic and estimate random error, just as was done for the fixed-memory filter; see last paragraph of Section 5.10.

8

GENERAL FORM FOR LINEAR TIME-INVARIANT SYSTEM

8.1 TARGET DYNAMICS DESCRIBED BY POLYNOMIAL AS A FUNCTION OF TIME

8.1.1 Introduction

In Section 1.1 we defined the target dynamics model for target having a constant velocity; see (1.1-1). A constant-velocity target is one whose trajectory can be expressed by a polynomial of degree 1 in time, that is, $d = 1$, in (5.9-1). (In turn, the tracking filter need only be of degree 1, i.e., $m = 1$.) Alternately, it is a target for which the first derivative of its position versus time is a constant. In Section 2.4 we rewrote the target dynamics model in matrix form using the transition matrix Φ; see (2.4-1), (2.4-1a), and (2.4-1b). In Section 1.3 we gave the target dynamics model for a constant accelerating target, that is, a target whose trajectory follows a polynomial of degree 2 so that $d = 2$; see (1.3-1). We saw that this target also can be alternatively expressed in terms of the transition equation as given by (2.4-1) with the state vector by (5.4-1) for $m = 2$ and the transition matrix by (5.4-7); see also (2.9-9). In general, a target whose dynamics are described exactly by a dth-degree polynomial given by (5.9-1) can also have its target dynamics expressed by (2.4-1), which we repeat here for convenience:

$$X_{n+1} = \Phi X_n$$

where the state vector X_n is now defined by (5.4-1) with m replaced by d and the transition matrix is a generalized form of (5.4-7). Note that in this text d represents the true degree of the target dynamics while m is the degree used by the tracking filter to approximate the target dynamics. For the nonlinear dynamics model case, discussed briefly in Section 5.11 when considering the tracking of a satellite, d is the degree of the polynomial that approximates the elliptical motion of the satellite to negligible error.

We shall now give three ways to derive the transition matrix of a target whose dynamics are described by an arbitrary degree polynomial. In the process we give three different methods for describing the target dynamics for a target whose motion is given by a polynomial.

8.1.2   Linear Constant-Coefficient Differential Equation

Assume that the target dynamics is described exactly by the dth-degree polynomial given by (5.9-1). Then its dth derivative equals a constant, that is,

$$D^d x(t) = \text{const} \tag{8.1-1}$$

while its $(d+1)$th derivative equals zero, that is,

$$D^{d+1} x(t) = 0 \tag{8.1-2}$$

As a result the class of all targets described by polynomials of degree d are also described by the simple linear constant-coefficient differential equation given by (8.1-2). Given (8.1-1) or (8.1-2) it is a straightforward manner to obtain the target dynamics model form given by (1.1-1) or (2.4-1) to (2.4-1b) for the case where $d = 1$. Specifically, from (8.1-1) it follows that for this $d = 1$ case

$$Dx(t) = \dot{x}(t) = \text{const} \tag{8.1-3}$$

Thus

$$\dot{x}_{n+1} = \dot{x}_n \tag{8.1-4}$$

Integrating this last equation yields

$$x_{n+1} = x_n + T\dot{x}_n \tag{8.1-5}$$

Equations (8.1-4) and (8.1-5) are the target dynamics equations for the constant-velocity target given by (1.1-1). Putting the above two equations in matrix form yields (2.4-1) with the transition matrix Φ given by (2.4-1b), the desired result. In a similar manner, starting with (8.1-1), one can derive the form of the target dynamics for $d = 2$ given by (1.3-1) with, in turn, Φ given by (5.4-7). Thus for a target whose dynamics are given by a polynomial of degree d, it is possible to obtain from the differential equation form for the target dynamics given by (8.1-1) or (8.1-2), the transition matrix Φ by integration.

8.1.3 Constant-Coefficient Linear Differential Vector Equation for State Vector $X(t)$

A second method for obtaining the transition matrix Φ will now be developed. As indicated above, in general, a target for which

$$D^d x(t) = \text{const} \qquad (8.1\text{-}6)$$

can be expressed by

$$X_{n+1} = \Phi X_n \qquad (8.1\text{-}7)$$

Assume a target described exactly by a polynomial of degree 2, that is, $d = 2$. Its continuous state vector can be written as

$$X(t) = \begin{bmatrix} x(t) \\ \dot{x}(t) \\ \ddot{x}(t) \end{bmatrix} = \begin{bmatrix} x(t) \\ Dx(t) \\ D^2 x(t) \end{bmatrix} \qquad (8.1\text{-}8)$$

It is easily seen that this state vector satisfies the following constant-coefficient linear differential vector equation:

$$\begin{bmatrix} Dx(t) \\ D^2 x(t) \\ D^3 x(t) \end{bmatrix} = \begin{bmatrix} 0 & 1 & 0 \\ 0 & 0 & 1 \\ 0 & 0 & 0 \end{bmatrix} \begin{bmatrix} x(t) \\ Dx(t) \\ D^2 x(t) \end{bmatrix} \qquad (8.1\text{-}9)$$

or

$$\frac{d}{dt} X(t) = AX(t) \qquad (8.1\text{-}10)$$

where

$$A = \begin{bmatrix} 0 & 1 & 0 \\ 0 & 0 & 1 \\ 0 & 0 & 0 \end{bmatrix} \qquad (8.1\text{-}10a)$$

The constant-coefficient linear differential vector equation given by (8.1-9), or more generally by (8.1-10), is a very useful form that is often used in the literature to describe the target dynamics of a time-invariant linear system. As shown in the next section, it applies to a more general class of target dynamics models than given by the polynomial trajectory. Let us proceed, however, for the time being assuming that the target trajectory is described exactly by a polynomial. We shall now show that the transition matrix Φ can be obtained from the matrix A of (8.1-10).

First express $X(t + \varsigma)$ in a vector Taylor expansion as

$$X(t + \varsigma) = X(t) + \varsigma DX(t) + \frac{\varsigma^2}{2!} D^2 X(t) \cdots$$

$$= \sum_{\nu=0}^{\infty} \frac{\varsigma^\nu}{\nu!} D^\nu X(t) \tag{8.1-11}$$

From (8.1-10)

$$D^\nu X(t) = A^\nu X(t) \tag{8.1-12}$$

Therefore (8.1-11) becomes

$$X(t + \varsigma) = \left[\sum_{\nu=0}^{\infty} \frac{(\varsigma A)^\nu}{\nu!} \right] X(t) \tag{8.1-13}$$

We know from simple algebra that

$$e^x = \sum_{\nu=0}^{\infty} \frac{x^\nu}{\nu!} \tag{8.1-14}$$

Comparing (8.1-14) with (8.1-13), one would expect that

$$\sum_{\nu=0}^{\infty} \frac{(\varsigma A)^\nu}{\nu!} = \exp(\varsigma A) = G(\varsigma A) \tag{8.1-15}$$

Although A is now a matrix, (8.1-15) indeed does hold with $\exp = e$ being to a matrix power being defined by (8.1-15). Moreover, the exponent function $G(\varsigma A)$ has the propensities one expects for an exponential. These are [5, p. 95]

$$G(\varsigma_1 A) G(\varsigma_2 A) = G[(\varsigma_1 + \varsigma_2)A] \tag{8.1-16}$$

$$[G(\varsigma_1 A)]^k = G(k\varsigma_1 A) \tag{8.1-17}$$

$$\frac{d}{d\varsigma} G(\varsigma A) = G(\varsigma A) A \tag{8.1-18}$$

We can thus rewrite (8.1-13) as

$$X(t + \varsigma) = \exp(\varsigma A) X(t) \tag{8.1-19}$$

Comparing (8.1-19) with (8.1-7), we see immediately that the transition matrix is

$$\Phi(\varsigma) = \exp(\varsigma A) \tag{8.1-20}$$

for the target whose dynamics are described by the constant-coefficient linear vector differential equation given by (8.1-10). Substituting (8.1-20) into (8.1-19) yields

$$X(t_n + \varsigma) = \Phi(\varsigma)X(t_n) \qquad (8.1\text{-}21)$$

Also from (8.1-15), and (8.1-20) it follows

$$\Phi(\varsigma) = I + \varsigma A + \frac{\varsigma^2}{2!}A^2 + \frac{\varsigma^3}{3!}A^3 + \cdots \qquad (8.1\text{-}22)$$

From (8.1-17) it follows that

$$(\exp \varsigma A)^k = \exp k\varsigma A \qquad (8.1\text{-}23)$$

Therefore

$$[\Phi(\varsigma)]^k = \Phi(k\varsigma) \qquad (8.1\text{-}24)$$

By way of example, assume a target having a polynomial trajectory of degree $d = 2$. From (8.1-10a) we have A. Substituting this value for A into (8.1-22) and letting $\varsigma = T$ yields (5.4-7), the transition matrix for the constant-accelerating target as desired.

8.1.4 Constant-Coefficient Linear Differential Vector Equation for Transition Matrix Φ

A third useful alternate way for obtaining Φ is now developed [5. pp. 96–97]. First, from (8.1-21) we have

$$X(\varsigma) = \Phi(\varsigma)X(0) \qquad (8.1\text{-}25)$$

Differentiating with respect to ς yields

$$\left[\frac{d}{d\varsigma}\Phi(\varsigma)\right]X(0) = \frac{d}{d\varsigma}X(\varsigma) \qquad (8.1\text{-}26)$$

The differentiation of a matrix by ς consists of differentiating each element of the matrix with respect to ς. Applying (8.1-10) and (8.1-25) to (8.1-26) yields

$$\left[\frac{d}{d\varsigma}\Phi(\varsigma)\right]X(0) = AX(\varsigma)$$

$$= A\Phi(\varsigma)X(0) \qquad (8.1\text{-}27)$$

Thus

$$\frac{d}{d\varsigma}\Phi(\varsigma) = A\Phi(\varsigma) \qquad (8.1\text{-}28)$$

On comparing (8.1-28) with (8.1-10) we see that the state vector $X(t)$ and the transition matrix $\Phi(\varsigma)$ both satisfy the same linear, time-invariant differential vector equation. Moreover, given this differential equation, it is possible to obtain $\Phi(\varsigma)$ by numerically integrating it. This provides a third method for obtaining $\Phi(\varsigma)$.

Define the matrix inverse of Φ by Ψ, that is,

$$\Psi(\varsigma) = [\Phi(\varsigma)]^{-1} \qquad (8.1\text{-}29)$$

The inverse Ψ satisfies the associated differential equation [5, p. 97]

$$\frac{d}{d\varsigma}\Psi(\varsigma) = -\Psi(\varsigma)A \qquad (8.1\text{-}30)$$

Thus $\Psi(\varsigma)$ can be obtained by numerically integrating the above equation.

To show that (8.1-30) is true, we first verify that the solution to (8.1-30) is

$$\Psi(\varsigma) = \Psi(0)\exp(-\varsigma A) \qquad (8.1\text{-}31)$$

This we do by differentiating the above to obtain

$$\frac{d}{d\varsigma}\Psi(\varsigma) = -\Psi(0)[\exp(-\varsigma A)]A$$
$$= -\Psi(\varsigma)A \qquad (8.1\text{-}32)$$

Thus (8.1-31) satisfies (8.1-30), as we wished to show. For $\Psi(0)$ let us choose

$$\Psi(0) = I \qquad (8.1\text{-}33)$$

This yields for $\Psi(\varsigma)$ the following:

$$\Psi(\varsigma) = \exp(-\varsigma)A \qquad (8.1\text{-}34)$$

It now only remains to show that the above is the inverse of Φ. To do this, we use (8.1-16), which yields

$$\exp(\varsigma A)\exp(-\varsigma A) = \exp(0)$$
$$= I \qquad (8.1\text{-}35)$$

This completes our proof that $\Phi^{-1} = \Psi$ and Ψ satisfies (8.1-30).

For a target whose trajectory is given by a polynomial, it does not make sense to use the three ways given in this section to obtain Φ. The Φ can easily be obtained by using the straightforward method illustrated in Section 2.4; see (2.4-1), (2.4-1a), and (2.4-1b) and (1.3-1) in Section 1.3. However, as shall be seen later, for more complicated target models, use of the method involving the integration of the differential equation given by (8.1-28) represents the preferred method. In the next section we show that (8.1-10) applies to a more general class of targets than given by a polynomial trajectory.

8.2 MORE GENERAL MODEL CONSISTING OF THE SUM OF THE PRODUCT OF POLYNOMIALS AND EXPONENTIALS

In the preceeding section we showed that the whole class of target dynamics consisting of polynomials of degree d are generated by the differential equation given by (8.1-2). In this section we consider the target whose trajectory is described by the sum of the product of polynomials and exponentials as given by

$$x(t) = \sum_{j=0}^{k} p_j(t)e^{\lambda_j t} \tag{8.2-1}$$

where $p_j(t)$ is a polynomial whose degree shall be specified shortly. The above $x(t)$ is the solution of the more general [than (8.1-2)] linear, constant-coefficient differential vector equation given by [5, pp. 92–94]

$$(D^{d+1} + \gamma_d D^d + \cdots + \gamma_1 D + \gamma_0)x(t) = 0 \tag{8.2-2}$$

We see that (8.1-2) is the special case of (8.2-2) for which $\gamma_0 = \gamma_1 = \cdots = \gamma_d = 0$. The λ_j of (8.2-1) are the k distinct roots of the characteristic equation

$$\lambda^{d+1} + \gamma_d \lambda^d + \cdots + \gamma_1 \lambda + \gamma_0 = 0 \tag{8.2-3}$$

The degree of $p_j(t)$ is 1 less than the multiplicity of the root λ_j of the characteristic equation.

By way of example let $d = 2$. Then

$$(D^3 + \gamma_2 D^2 + \gamma_1 D + \gamma_0)x(t) = 0 \tag{8.2-4}$$

Let the state vector $X(t)$ for this process defined by (8.1-8). Then it follows directly from (8.2-4) that

$$\frac{d}{dt}X(t) = \begin{pmatrix} \dot{x} \\ \ddot{x} \\ \dddot{x} \end{pmatrix}_t = \begin{pmatrix} 0 & 1 & 0 \\ 0 & 0 & 1 \\ -\gamma_0 & -\gamma_1 & -\gamma_2 \end{pmatrix} \begin{pmatrix} x \\ \dot{x} \\ \ddot{x} \end{pmatrix}_t \tag{8.2-5}$$

or

$$\frac{d}{dt}X(t) = AX(t) \qquad (8.2\text{-}6)$$

where

$$A \equiv \begin{pmatrix} 0 & 1 & 0 \\ 0 & 0 & 1 \\ -\gamma_0 & -\gamma_1 & -\gamma_2 \end{pmatrix} \qquad (8.2\text{-}6a)$$

This gives us a more general form for A than obtained for targets following exactly a polynomial trajectory as given in Section 8.1; see (8.1-10a).

The matrix A above can be made even more general. To do this, let

$$\hat{X}(t) = GX(t) \qquad (8.2\text{-}7)$$

where G is an arbitrary constant 3×3 nonsingular matrix. Applying (8.2-7) to (8.2-6) yields

$$\frac{d}{dt}G^{-1}\hat{X}(t) = AG^{-1}\hat{X}(t) \qquad (8.2\text{-}8)$$

Because G is a constant, the above becomes

$$G^{-1}\frac{d}{dt}\hat{X}(t) = AG^{-1}\hat{X}(t) \qquad (8.2\text{-}9)$$

or

$$\frac{d}{dt}\hat{X}(t) = GAG^{-1}\hat{X}(t) \qquad (8.2\text{-}10)$$

or finally

$$\frac{d}{dt}\hat{X}(t) = B\hat{X}(t) \qquad (8.2\text{-}11)$$

where

$$B = GAG^{-1} \qquad (8.2\text{-}11a)$$

Because G is arbitrary, B is arbitrary, but constant. Thus, (8.2-6) applies where A can be an arbitrary matrix and not just (8.2-6a).

9

GENERAL RECURSIVE MINIMUM-VARIANCE GROWING-MEMORY FILTER (BAYES AND KALMAN FILTERS WITHOUT TARGET PROCESS NOISE)

9.1 INTRODUCTION

In Section 6.3 we developed a recursive least-squares growing memory-filter for the case where the target trajectory is approximated by a polynomial. In this chapter we develop a recursive least-squares growing-memory filter that is not restricted to having the target trajectory approximated by a polynomial [5. pp. 461–482]. The only requirement is that Y_{n-i}, the measurement vector at time $n - i$, be linearly related to X_{n-i} in the error-free situation. The Y_{n-i} can be made up to multiple measurements obtained at the time $n - i$ as in (4.1-1a) instead of a single measurement of a single coordinate, as was the case in (4.1-20), where $Y_{n-1} = [y_{n-1}]$. The Y_{n-i} could, for example, be a two-dimensional measurement of the target slant range and Doppler velocity. Extensions to other cases, such as the measurement of three-dimensional polar coordinates of the target, are given in Section 16.2 and Chapter 17.

Assume that at time n we have $L + 1$ observations $Y_n, Y_{n-1}, \ldots, Y_{n-L}$ obtained at, respectively, times $n, n - 1, \ldots, n - L$. These $L + 1$ observations are represented by the matrix $Y_{(n)}$ of (4.1-11a). Next assume that at some later time $n + 1$ we have another observation Y_{n+1} given by

$$Y_{n+1} = M\Phi X_n + N_{n+1} \tag{9.1-1}$$

Assume also that at time n we have a minimum-variance estimate of $X_{n,n}^*$ based on the past $L + 1$ measurements represented by $Y_{(n)}$. This estimate is given by (4.1-30) with W_n given by (4.5-4).

In turn the covariance matrix $\overset{\circ}{S}_{n,n}^*$ is given by (4.5-5). Now to determine the new minimum-variance estimate $X_{n+1,n+1}^*$ from the set of data consisting of $Y_{(n)}$

and Y_{n+1}, one could again use (4.1-30) and (4.5-4) with $Y_{(n)}$ now replaced by $Y_{(n+1)}$, which is $Y_{(n)}$ of (4.1-11a) with Y_{n+1} added to it. Correspondingly the matrices T and $R_{(n)}$ would then be appropriately changed to account for the increase in $Y_{(n)}$ to include Y_{n+1}. This approach, however, has the disadvantage that it does not make use of the extensive computations carried out to compute the previously minimum-variance estimate $X_{n,n}^*$ based on the past data $Y_{(n)}$. Moreover, it turns out that if Y_{n+1} is independent of $Y_{(n)}$, then the minimum-variance estimate of $X_{n+1,n+1}^*$ can be obtained directly from Y_{n+1} and $X_{n,n}^*$ and their respective variances R_{n+1} and $S_{n,n}^*$. This is done by obtaining the minimum-variance estimate of $X_{n+1,n+1}^*$ using Y_{n+1} and $X_{n,n}^*$ together with their variances. No use is made of the original data set $Y_{(n)}$. This says that the estimate $X_{n,n}^*$ and its covariance matrix $S_{n,n}^*$ contain all the information we need about the previous $L+1$ measurements, that is, about $Y_{(n)}$. Here, $X_{n,n}^*$ and its covariance matrix are sufficient statistics for the information contained in the past measurement vector $Y_{(n)}$ together with its covariance matrix $R_{(n)}$. (This is similar to the situation where we developed the recursive equations for the growing- and fading-memory filters in Sections 6.3, 7.2, and 1.2.6.)

9.2 BAYES FILTER

The recursive form of the minimum variance estimate based on Y_{n+1} and $X_{n,n}^*$ is given by [5, p. 464]

$$\mathring{X}_{n+1,n+1}^* = \mathring{X}_{n+1,n}^* + \mathring{H}_{n+1}(Y_{n+1} - M\mathring{X}_{n+1,n}^*) \tag{9.2-1}$$

where

$$\mathring{H}_{n+1} = \mathring{S}_{n+1,n+1}^* M^T R_1^{-1} \tag{9.2-1a}$$

$$\mathring{S}_{n+1,n+1}^* = [(\mathring{S}_{n+1,n}^*)^{-1} + M^T R_1^{-1} M]^{-1} \tag{9.2-1b}$$

$$\mathring{S}_{n+1,n}^* = \Phi \mathring{S}_{n,n}^* \Phi^T \tag{9.2-1c}$$

$$\mathring{X}_{n+1,n}^* = \Phi \mathring{X}_{n,n}^* \tag{9.2-1d}$$

The above recursive filter is referred to in the literature as the Bayes filter by some (this is because it can be derived using the Bayes theorem on conditional probabilities [128].) The only requirement needed for the recursive minimum-variance filter to apply is that Y_{n+1} be independent of $Y_{(n)}$. When another measurement Y_{n+2} is obtained at a later time $n+2$, which is independent of the previous measurements, then the above equations (indexed up one) can be used again to obtain the estimate $\mathring{X}_{n+2,n+2}^*$. If $Y_{(n)}$ and Y_{n+1} are dependent, the Bayes filter could still be used except that it would not now provide the minimum-variance estimate. If the variates are reasonably uncorrelated though, the estimate could possibly still be a good one.

9.3   KALMAN FILTER (WITHOUT PROCESS NOISE)

If we apply the inversion lemma given by (2.6-14) to (9.2-1b), we obtain after some manipulations the following equivalent algebraic equation for the recursive minimum-variance growing-memory filter estimate [5, p. 465]:

$$\overset{\circ}{X}{}^{*}_{n,n} = \overset{\circ}{X}{}^{*}_{n,n-1} + \overset{\circ}{H}_n(Y_n - MX^{*}_{n,n-1}) \tag{9.3-1}$$

where

$$\overset{\circ}{H}_n = \overset{\circ}{S}{}^{*}_{n,n-1}M^T(R_1 + MS^{*}_{n,n-1}M^T)^{-1} \tag{9.3-1a}$$

$$\overset{\circ}{S}{}^{*}_{n,n} = (1 - H_nM)\overset{\circ}{S}{}^{*}_{n,n-1} \tag{9.3-1b}$$

$$\overset{\circ}{S}{}^{*}_{n,n-1} = \Phi\overset{\circ}{S}{}^{*}_{n-1,n-1}\Phi^T \tag{9.3-1c}$$

$$\overset{\circ}{X}{}^{*}_{n,n-1} = \Phi\overset{\circ}{X}{}^{*}_{n-1,n-1} \tag{9.3-1d}$$

The preceding Kalman filter equations are the same as given by (2.4-4a) to (2.4-4j) except that the target model dynamic noise (U_n or equivalently its covariance matrix Q_n) is not included. Not including the target model dynamic noise in the Kalman filter can lead to computational problems for the Kalman filter [5, Section 12.4]. This form of the Kalman filter is not generally used for this reason, and it is not a form proposed by Kalman. The Kalman filter with the target process noise included is revisited in Chapter 18.

9.4   COMPARISON OF BAYES AND KALMAN FILTERS

As discussed in Sections 2.3, 2.5, and 2.6, the recursive minimum-variance growing-memory filter estimate is a weighted sum of the estimates Y_{n+1} and $X^{*}_{n+1,n}$ with the weighting being done according to the importance of the two estimates; see (2.3-1), (2.5-9), and (2.6-7). Specifically, it can be shown that the recursive minimum-variance estimate can be written in the form [5, p. 385]

$$\overset{\circ}{X}{}^{*}_{n+1,n+1} = \overset{\circ}{S}{}^{*}_{n+1,n+1}[(\overset{\circ}{S}{}^{*}_{n+1,n})^{-1}\overset{\circ}{X}{}^{*}_{n+1,n} + MR_1^{-1}Y_{n+1}] \tag{9.4-1}$$

If the covariance matrix of y_{n+1} is dependent on n, then R_1 is replaced by R_{n+1}. The recursive minimum-variance Bayes and Kalman filter estimates are maximum-likelihood estimates and maximum a posteriori estimates when Y_{n+1} and $Y_{(n)}$ are uncorrelated and Gaussian. When $Y_{(n)}$ is Gaussian we can derive (9.4-1) by maximizing the probability of the X_{n+1} given $Y_{(n+1)}$ using Bayes' rule. All other properties given in Section 4.5 for the minimum-variance estimate also apply. The Kalman filter has the advantage over the Bayes filter of eliminating the need for two matrix inversions in (9.2-1b), which have a size equal to the state vector $X_{n,n}^*$ [which can be large, e.g., 10×10 for the example (3.4-6)]. The Kalman filter on the other hand only requires a single matrix inversion in (9.3-1a) of an order equal to the measurement vector Y_{n+1} (which has a dimension 4×4) for the example of Section 2.4, where the target is measured in polar coordinates; see (2.4-7). It is also possible to incorporate these four measurements one at a time if they are independent of each other. In this case no matrix inversion is needed.

9.5 EXTENSION TO MULTIPLE MEASUREMENT CASE

In the Bayes and Kalman filters it is not necessary for Y_{n+1} to be just a single measurement at time t_{n+1}. The term Y_{n+1} could be generalized to consist of $L + 1$ measurements at $L + 1$ times given by

$$Y_{n+1}, Y_n, Y_{n-1}, \ldots, Y_{n-L+1} \qquad (9.4\text{-}2)$$

For this more general case we can express the above $L + 1$ measurements as a vector given by

$$Y_{(n+1)} = \begin{bmatrix} Y_{n+1} \\ \text{----} \\ Y_n \\ \text{----} \\ \vdots \\ \text{----} \\ Y_{n-L+1} \end{bmatrix} \qquad (9.4\text{-}3)$$

Then from (4.1-5) through (4.1-10), (4.1-11) follows. It then immediately follows that (9.2-1) through (9.2-1d) and (9.3-1) through (9.3-1d) apply with M replaced by T of (4.1-11b) and Y_{n+1} replaced by $Y_{(n+1)}$ of (9.4-3).

10

VOLTAGE LEAST-SQUARES
ALGORITHMS REVISITED

10.1 COMPUTATION PROBLEMS

The least-squares estimates and minimum-variance estimates described in Section 4.1 and 4.5 and Chapter 9 all require the inversion of one or more matrices. Computing the inverse of a matrix can lead to computational problems due to standard computer round-offs [5, pp. 314–320]. To illustrate this assume that

$$s = 1 + \varepsilon \qquad (10.1\text{-}1)$$

Assume a six-decimal digit capability in the computer. Thus, if $s = 1.000008$, then the computer would round this off to 1.00000. If, on the other hand, $s = 1.000015$, then the computer would round this off to 1.00001. Hence, although the change in ε is large, a reduction of 33.3% for the second case (i.e., 0.000005/0.000015), the change in s is small, 5 parts in 10^6 (i.e., 0.000005/ 1.000015). This small error in s would seem to produce negligible effects on the computations. However, in carrying out a matrix inversion, it can lead to serious errors as indicated in the example to be given now. Assume the nearly singular matrix [5]

$$A = \begin{pmatrix} s & 1 \\ 1 & 1 \end{pmatrix} \qquad (10.1\text{-}2)$$

where $s = 1 + \varepsilon$. Inverting A algebraically gives

$$A^{-1} = \frac{1}{s - 1} \begin{pmatrix} 1 & -1 \\ -1 & s \end{pmatrix} \qquad (10.1\text{-}3)$$

If $\varepsilon = 0.000015$, then from (10.1-3) we obtain the following value for A^{-1} without truncation errors:

$$A^{-1} = \frac{\begin{pmatrix} 1 & -1 \\ -1 & 1.000015 \end{pmatrix}}{0.000015}$$

$$= 10^4 \begin{pmatrix} 6.66 & -6.66 \\ -6.66 & 6.66 \end{pmatrix} \qquad (10.1\text{-}4)$$

However, if ε is truncated to 0.00001, then (10.1-3) yields

$$A^{-1} = \frac{\begin{pmatrix} 1 & -1 \\ -1 & 1.00001 \end{pmatrix}}{0.00001}$$

$$\approx 10^4 \begin{pmatrix} 10 & -10 \\ -10 & 10 \end{pmatrix} \qquad (10.1\text{-}5)$$

Thus the 5 parts in 10^6 error in s results in a 50% error in each of the elements of A^{-1}.

Increasing the computation precision can help. This, however, can be costly in computer hardware and/or computer time. There are, however, alternative ways to cope with this problem. When doing a LSE problem this involves the use of the voltage least-squares, also called square-root algorithms, which are not as sensitive to computer round-off errors. This method was introduced in Section 4.3 and will be described in greater detail in Section 10.2 and Chapters 11 to 14. Section 10.2.3 discusses a measure, called the condition number, for determining the accuracy needed to invert a matrix.

The inverse of the matrices in (4.1-32) and (4.5-4) will be singular or nearly singular when the time between measurements is very small, that is, when the time between measurements T of (4.1-18) or (4.1-28) is small. Physically, if range measurements are only being made and they are too close together, then the velocity of the state vector $X_{n,n}^*$ cannot be accurately estimated. Mathematically, the rows of T matrix become very dependent when the measurements are too close together in time. When this happens, the matrices of the least-squares and minimum-variance estimates tend to be singular. When the columns of T are dependent, the matrix is said to not have full column rank. Full column rank is required for estimating $X_{n,n}^*$ [5, Section 8.8]. The matrix T has full column rank when its columns are independent. It does not have full rank if one of its columns is equal to zero.

The examples of matrix T given by (4.1-18) for a constant-velocity target and (4.1-28) for a constant-accelerating target show that the matrix T will not have full rank when the time between measurements T is very small. When the time between measurements T is small enough, the second column of (4.1-18) becomes rounded off to zero, and the second and third columns of

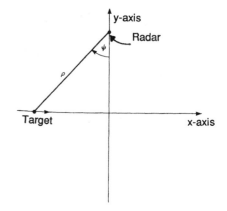

Figure 10.1-1 Geometry for example of target flying by radar. (From Morrison [5, p. 319].)

(4.1-28) likewise become rounded off to zero. Hence matrices T do not have full rank when time T between measurements is very small.

This singularity situation is improved sometimes if in addition to measuring range another parameter is measured, such as the Doppler velocity or the target angular position. Consider a target moving along the x axis as shown in Figure 10.1-1. Assume the radar is located as indicated and that it is only making slant range measurements of the target's position. At the time when the target passes through the origin, the tracker will have difficulty estimating the target's velocity and acceleration. This is because the target range only changes slightly during this time so that the target behaves essentially like a stationary target even though it could be moving rapidly. If, in addition, the radar measured the target aspect angle ψ, it would be able to provide good estimates of the velocity and acceleration as it passed through the origin. In contrast, if the target were being tracked far from the origin, way off to the right on the x axis in Figure 10.1-1, range only measurements would then provide a good estimate of the target's velocity and acceleration.

If the radar only measured target azimuth, then the radar measurements would convey more information when the target passed through the origin than when it was far from the origin. Thus it is desirable to make two essentially independent parameter measurements on the target, with these being essentially orthogonal to each other. Doing this would prevent the matrix inversion from tending toward singularity, or equivalently, prevent T from not having full rank.

Methods are available to help minimize the sensitivity to the computer round-off error problem discussed above. They are called square-root filtering [79] or voltage-processing filtering. This type of technique was introduced in Section 4.3. Specifically the Gram–Schmidt method was used to introduce this type of technique. In this chapter we will first give further general details on the technique followed by detailed discussions of the Givens, Householder, and Gram-Schmidt methods in Chapters 11 to 13. For completeness, clarity, convenience and in order that this chapter stand on its own, some of the results given in Section 4.3 will be repeated. However, it is highly recommended that if Sections 4.2 and 4.3 are not fresh in the reader's mind that he or she reread them before reading the rest of this chapter.

10.2 ORTHOGONAL TRANSFORMATION OF LEAST-SQUARES ESTIMATE ERROR

We proceed initially by applying an orthonormal transformation to $e(X_{n,n}^*)$ of (4.1-31) [79]. Let F be an orthonormal transformation matrix. It then follows from (4.3-9) to (4.3-11), and also from (4.3-16) and (4.3-17) and reference 79 (p. 57), that

$$F^T F = I = F F^T \tag{10.2-1}$$

and

$$F^{-1} = F^T \tag{10.2-2}$$

Also

$$\|FY\| = \|Y\| \tag{10.2-3}$$

where $\| \bullet \|$ is the Euclidean norm defined by (4.2-40) and repeated here [79, 101]:

$$\|y\| = (y^T y)^{1/2} \tag{10.2-4}$$

Thus $e(X_{n,n}^*)$ of (4.1-31) is the square of the Euclidean norm of

$$E = T X_{n,n}^* - Y_{(n)} \tag{10.2-5}$$

or

$$e(X_{n,n}^*) = \|E\|^2 = e_T \tag{10.2-6}$$

where e_T was first used in (1.2-33).

Applying an $s \times s$ orthonormal transformation F to E, it follows from (4.3-21), and also reference 79 (p. 57), that

$$e(X_{n,n}^*) = \|FE\|^2 = \|F T X_{n,n}^* - F Y_{(n)}\|^2$$
$$= \|(FT)X_{n,n}^* - (F Y_{(n)})\|^2 \tag{10.2-7}$$

Assume here that $X_{n,n}^*$ is an $m' \times 1$ matrix, that T is $s \times m'$, and that $Y_{(n)}$ is $s \times 1$. As indicated in Section 4.3 [see, e.g., (4.3-31) and (4.3-59)] and to be further indicated in the next section, F can be chosen so that the transformed matrix $T' = FT$ is given by

$$T' = FT = \begin{bmatrix} U \\ \overline{} \\ 0 \end{bmatrix} \begin{matrix} \}m' \\ \\ \}s - m' \end{matrix} \qquad (10.2\text{-}8)$$

$$\underbrace{}_{m'}$$

where U is an upper triangular matrix. For example U is of the form

$$U = \begin{bmatrix} u_{11} & u_{12} & u_{13} & u_{14} \\ 0 & u_{22} & u_{23} & u_{24} \\ 0 & 0 & u_{33} & u_{34} \\ 0 & 0 & 0 & u_{44} \end{bmatrix} \qquad (10.2\text{-}9)$$

for $m' = 4$. In turn

$$FTX_{n,n}^* = \begin{bmatrix} UX_{n,n}^* \\ \text{-------} \\ 0 \end{bmatrix} \begin{matrix} \}m' \\ \\ \}s - m' \end{matrix} \qquad (10.2\text{-}10)$$

and

$$FY_{(n)} = \begin{bmatrix} Y_1' \\ \text{---} \\ Y_2' \end{bmatrix} \begin{matrix} \}m' \\ \\ \}s - m' \end{matrix} \qquad (10.2\text{-}11)$$

On substituting (10.2-10) and (10.2-11) into (10.2-7) for $e(X_{n,n}^*)$, it is a straightforward matter to show that

$$e(X_{n,n}^*) = e(UX_{n,n}^* - Y_1') + e(Y_2') \qquad (10.2\text{-}12)$$

or equivalently

$$e(X_{n,n}^*) = \|UX_{n,n}^* - Y_1'\|^2 + \|Y_2'\|^2 \qquad (10.2\text{-}13)$$

This was shown in Section 4.3 for the special case where $s = 3$, $m' = 2$; see (4.3-49). We shall now show that it is true for arbitrary s and m'. Equations (10.2-12) and (10.3-13) follow directly from the fact that $FTX_{n,n}^*$ and $FY_{(n)}$ are column matrices so that $FTX_{n,n}^* - FY_{(n)}$ is a column matrix, E being given by (10.2-5). Let the elements of FE be designate as $\varepsilon_1' = 1, 2, \dots, s$. Hence from

(10.2-5), (10.2-10), and (10.2-11)

$$
FE = E' = \begin{bmatrix} \varepsilon'_1 \\ \varepsilon'_2 \\ \vdots \\ \varepsilon'_{m'} \\ \hline \varepsilon'_{m'+1} \\ \vdots \\ \varepsilon'_s \end{bmatrix} = \left[\begin{array}{c} UX^*_{n,n} - Y'_1 \\ \hline -Y'_2 \end{array} \right] \begin{array}{l} \} \, m' \\ \\ \} \, s - m' \end{array} \tag{10.2-14}
$$

From (10.2-3), (10.2-4), (10.2-6), and (10.2-14) it follows that

$$
e(X^*_{n,n}) = \|E\|^2 = \|FE\| = (FE)^T(FE) = \sum_{i=1}^{m'} \varepsilon_i^2 + \sum_{i=m'+1}^{s} \varepsilon_i^2 \tag{10.2-15}
$$

which yields (10.2-12) and (10.2-13) for arbitrary s and m', as we wished to show.

The least-squares estimate $X^*_{n,n}$ now becomes the $X^*_{n,n}$ that minimizes (10.2-13). Here, $X^*_{n,n}$ is not in the second term of the above equation so that this term is independent of $X^*_{n,n}$. Only the first term can be affected by varying $X^*_{n,n}$. The minimum $e(X^*_{n,n})$ is achieved by making the first term equal to zero by setting $\varepsilon'_1 = \varepsilon'_2 = \ldots = \varepsilon'_{m'} = 0$, as done in Section 4.3, to yield

$$
UX^*_{n,n} = Y'_1 \tag{10.2-16}
$$

The $X^*_{n,n}$ that satisfies (10.2-16) is the least-squares estimate being sought.

Because U is an upper triangular matrix, it is trivial to solve for $X^*_{n,n}$ using (10.2-16). To illustrate, assume that U is given by (10.2-9) and that

$$
X^*_{n,n} = \begin{bmatrix} x^*_1 \\ x^*_2 \\ x^*_3 \\ x^*_4 \end{bmatrix} \tag{10.2-17}
$$

and

$$
Y'_1 = \begin{bmatrix} y'_1 \\ y'_2 \\ y'_3 \\ y'_4 \end{bmatrix} \tag{10.2-18}
$$

We start with the bottom equation of (10.2-16) to solve for x_4^* first. This equation is

$$u_{44}x_4^* = y_4'$$ (10.2-19)

and trivially

$$x_4^* = \frac{y_4'}{u_{44}}$$ (10.2-20)

We next use the second equation from the bottom of (10.2-16), which is

$$u_{33}x_3^* + u_{34}x_4^* = y_3'$$ (10.2-21)

Because x_4^* is known, we can readily solve for the only unknown x_3^* to yield

$$x_3^* = \frac{y_3' - u_{34}x_4^*}{u_{33}}$$ (10.2-22)

In a similar manner the third equation from the bottom of (10.2-16) can be used to solve for x_2^*, and in turn the top equation then is used to solve for x_1^*.

The above technique for solving (10.2-16) when U is an upper triangular matrix is called the "back-substitution" method. This back-substitution method avoids the need to solve (10.2-16) for $X_{n,n}^*$ using

$$X_{n,n}^* = U^{-1}Y_1'$$ (10.2-23)

with the need to compute the inverse of U. The transformation of T to the upper triangular matrix T' followed by the use of the back-substitution method to solve (10.2-16) for $X_{n,n}^*$ is called voltage least-squares filtering or square-root processing. The use of voltage least-squares filtering is less sensitive to computer round-off errors than is the technique using (4.1-30) with W given by (4.1-32). (When an algorithm is less sensitive to round-off errors, it is said to be more accurate [79, p. 68].) The above algorithm is also more stable, that is, accumulated round-off errors will not cause it to diverge [79, p. 68].

In Section 4.3 we introduced the Gram–Schmidt method for performing the orthonormal transformation F. In the three ensuing sections, we shall detail this method and introduce two additional orthonormal transformations F that can make T have the upper triangular form of (10.2-8).

Before proceeding, we shall develop further the physical significance to the orthonormal transformation and the matrix U, something that we started in Section 4.3. We shall also give some feel for why the square-root method is more accurate, and then finally some additional physical feel for why and when inaccuracies occur. First, let us revisit the a physical interpretation of the orthonormal transformation.

10.2.1   Physical Interpretation of Orthogonal Transformation

Per our discussions in Sections 4.2 and 4.3 [see (4.2-2) and (4.3-54) and the discussion relating to these equations] we know that we can think of the transition–observation matrix T as consisting of m' column vectors $t_1, \ldots, t_{m'}$, with t_i being the ith column vector defined in an s-dimensional orthogonal hyperspace [101]. Thus T can be written as

$$T = [t_1 t_2 \ldots t_{m'}] \tag{10.2-24}$$

where

$$t_i = \begin{bmatrix} t_{1i} \\ t_{2i} \\ \vdots \\ t_{si} \end{bmatrix} \tag{10.2-24a}$$

whose entries represent the coordinates of t_i. As indicated in Section 4.3 and done again here in more detail for the case arbitrary s and m', the orthogonal transformation F puts these m' column vectors $t_1, \ldots, t_{m'}$ of the transition–observation matrix into a new orthogonal space. The coordinate directions in this new space are represented in the original orthogonal hyperspace by the s orthonormal unit row vectors of F. These row vectors are $f_i = q_i^T, i = 1, \ldots, s$; see (4.3-40a), (4.3-40b), and (4.3-58) to (4.3-58c). Thus

$$F = \begin{bmatrix} f_1 \\ f_2 \\ \vdots \\ f_s \end{bmatrix} \tag{10.2-25}$$

The coordinates of the unit vector f_i are defined by the entries of the ith row of F, which is given by the s-dimensional row matrix

$$f_i = [f_{i1} \quad f_{i2} \quad \cdots \quad f_{is}] \tag{10.2-26}$$

From the discussion on projection matrices given in Section 4.2, we know that the magnitude of the projection of the vector t_1 onto the unit vector f_i is given by $f_i t_1$; specifically, see (4.2-36) and the discussion immediately following it. [Note that the transpose of f_i is not needed because f_i is a row matrix and not a column matrix as was the case in (4.2-35) for $\hat{t}_1$.] The direction of this projection is given by the vector f_i. Thus paralleling (4.2-35) the component of t_1 onto f_i is given by the vector

$$p_i = (f_i t_1) f_i \tag{10.2-27}$$

In the new space represented by the unit row vectors of F, the transformed vector t_1 is paralleling (4.2-33), represented by

$$t_{1F} = \sum_{i=1}^{s} p_i = (f_1 t_1)f_1 + (f_2 t_1)f_2 + \cdots + (f_s t_1)f_s \qquad (10.2\text{-}28)$$

If we represent this vector in the F row coordinate system by a column matrix whose entries designate the amplitudes of the respective row unit vectors f_i then, paralleling (4.3-4),

$$t_{1F} = \begin{pmatrix} f_1 t_1 \\ f_2 t_1 \\ \vdots \\ f_s t_1 \end{pmatrix} \qquad (10.2\text{-}29)$$

But this is nothing more than the product of F with t_1:

$$Ft_1 = \begin{bmatrix} f_1 t_1 \\ f_2 t_1 \\ \vdots \\ f_s t_1 \end{bmatrix} \qquad (10.2\text{-}30)$$

Thus the transformation of the vector t_1 by F gives us the coordinates of t_1 in the row unit vector space of F, as we wished to show.

From (10.2-8) and (10.2-9) it follows that we wish to pick F such that

$$Ft_1 = \begin{bmatrix} f_1 t_1 \\ f_2 t_1 \\ f_3 t_1 \\ \vdots \\ f_s t_1 \end{bmatrix} = \begin{bmatrix} u_{11} \\ 0 \\ 0 \\ \vdots \\ 0 \end{bmatrix} \qquad (10.2\text{-}31)$$

Physically, this mean that the vector t_1 lies along the vector f_1 and is orthogonal to all the other f_i. Hence f_1 of F is chosen to lie along t_1, as done in Section 4.3; see (4.3-27) and the discussion after (4.3-54). Since the matrix F is to be orthonormal, $f_2, \ldots, f_s$ are picked to be orthonormal to f_1 and in this way make $f_i t_1 = 0$ for $i = 2, 3, \ldots, s$, thus forcing all the $i = 2, 3, \ldots, s$ coordinates to be zero in (10.2-31).

Now consider the projection of the vector represented by the second column of T, that is, t_2, onto the row space of F. From (10.2-8) and (10.2-9), and paralleling (4.3-28),

$$Ft_2 = \begin{bmatrix} f_1t_2 \\ f_2t_2 \\ f_3t_2 \\ \vdots \\ f_st_2 \end{bmatrix} = \begin{bmatrix} u_{12} \\ u_{22} \\ 0 \\ \vdots \\ 0 \end{bmatrix} \qquad (10.2\text{-}32)$$

This tells us that t_2 lies in the two-dimensional plane formed by the two row vectors f_1 and f_2 and is to be in turn orthogonal to the remaining $(s-2)$-dimensional space defined by $f_3, f_4, \ldots, f_s$. Consequently f_2 is picked to form, in conjunction with f_1, the plane containing the space spanned by the vectors t_1 and t_2. The row vector f_i, for $i \leq m'$, is chosen so that in conjunction with the vectors $f_1, \ldots, f_{i-1}$ the vectors $f_1, \ldots, f_i$ span the i-dimensional space defined by the vectors $t_1, \ldots, t_i$ and is to be orthogonal to the space defined by the remaining vectors $f_{i+1}, \ldots, f_s$. Thus

$$Ft_i = \left.\begin{bmatrix} u_{1i} \\ \vdots \\ u_{ii} \\ 0 \\ \vdots \\ 0 \\ \hline 0 \\ \vdots \\ 0 \end{bmatrix}\right\} \begin{matrix} m' \\ \\ s-m' \end{matrix} \qquad (10.2\text{-}33)$$

Thus $f_1, \ldots, f_{m'}$, span the same space as defined by $t_1, \ldots, t_{m'}$ and

$$Ft_{m'} = \begin{bmatrix} f_1t_{m'} \\ \vdots \\ f_{m'}t_{m'} \\ \hline f_{m'+1}t_{m'} \\ \vdots \\ f_st_{m'} \end{bmatrix} = \left.\begin{bmatrix} u_{1m'} \\ \vdots \\ u_{m'm'} \\ \hline 0 \\ \vdots \\ 0 \end{bmatrix}\right\} \begin{matrix} m' \\ \\ s-m' \end{matrix} \qquad (10.2\text{-}34)$$

Define F_1 and F_2 as the matirces formed by, respectively, the first m' rows of F and remaining $s - m'$ bottom rows of F. Then

$$
F_1 = \begin{bmatrix} f_1 \\ \vdots \\ f_{m'} \end{bmatrix} \tag{10.2-35}
$$

and

$$
F_2 = \begin{bmatrix} f_{m'+1} \\ \vdots \\ f_s \end{bmatrix} \tag{10.2-36}
$$

and

$$
F = \begin{bmatrix} F_1 \\ \text{---} \\ F_2 \end{bmatrix} \begin{matrix} \}m' \\ \\ \}s - m' \end{matrix} \tag{10.2-37}
$$

and from (10.2-8) and (10.2-31) to (10.2-34)

$$
FT = \begin{bmatrix} F_1 T \\ \text{------} \\ F_2 T \end{bmatrix} = \begin{bmatrix} U \\ \text{---} \\ 0 \end{bmatrix} \begin{matrix} \}m' \\ \\ \}s - m' \end{matrix} \tag{10.2-38}
$$

From the discussion given above and (10.2-38), we know that the row vectors of F_1 spans the m'-dimensional space of the column vectors of T while the row vectors of F_2 spans the $(s - m')$-dimensional space orthogonal to the space of T. Furthermore F projects the column vectors of T only onto the space defined by the first m' row vectors of F, with T being orthogonal to the remaining $(s - m')$-dimensional space spanned by the remaining $s - m'$ row vectors of F.

Consider now the transformation given by (10.2-11) that projects the data vector $Y_{(n)}$ onto the row space of T. It can be rewritten as

$$
FY_{(n)} = \begin{bmatrix} F_1 Y_{(n)} \\ \text{--------} \\ F_2 Y_{(n)} \end{bmatrix} = \begin{bmatrix} Y'_1 \\ \text{----} \\ Y'_2 \end{bmatrix} \tag{10.2-39}
$$

From the above it follows that Y'_1 is physically the projection of $Y_{(n)}$ onto the space spanned by F_1, or equivalently, spanned by $t_1, \ldots, t_{m'}$, while Y'_2 is physically the projection of $Y_{(n)}$ onto the space spanned by F_2, the $(s - m')$-dimensional space orthogonal to the space spanned by $t_1, \ldots, t_{m'}$. This is reminiscent of our discussion in Section 4.2 relative to Figure 4.2-1. From that discussion it follows that Y_1' is the projection of $Y_{(n)}$ onto the space spanned by the columns of T, which here is identical to the row spanned by F_1, while Y_2' is orthogonal to this space, which here is identical to the space spanned by the rows of F_2. The Y_2' part of $Y_{(n)}$ is due to the measurement noise $N_{(n)}$. It corresponds to the $Y_{(2)} - TX_2$ part of Figure 4.2-1. The Y_2' part of $Y_{(n)}$ does not enter into the determination of the least-squares estimate $X_{n,n}^*$. Only the part of $Y_{(n)}$ projected into the column space of T, designated as Y_1', enters into the determination of $X_{n,n}^*$, as was the case for Figure 4.2-1. Because (10.2-16) is true, the least-squares estimate of $X_{n,n}$, is that $X_{n,n}^*$ that combines the columns of U to form the projection of $Y_{(n)}$ onto the space spanned by the columns of T, that is, to form Y_1'. Thus the orthonormal transformation F projects $Y_{(n)}$ onto the space spanned by the columns of the matrix T and then sets $Y_1' = UX_{n,n}^*$ to find the least-squares estimate $X_{n,n}^*$ per the discussion relative to Figure 4.2-1. This discussion gives us good physical insight into this powerful and beautiful orthonormal transformation.

10.2.2 Physical Interpretation of U

It is apparent from the discussion in Section 10.2.1 that Y_1' and Y_2' of (10.2-11) and (10.2-39) represent the original measurement set $Y_{(n)}$ of (4.1-11) and (4.1-11a) in a new orthonormal s-dimensional space. Furthermore it is only Y_1' that is needed to estimate X_n, it being in the m'-dimensional space that X_n is constrained to whereas Y_2' is orthogonal to it. We can think of the m'-dimensional column matrix Y_1' as the equivalent set of measurement to $Y_{(n)}$ made in this m'-dimensional space, which is the space spanned by the columns of the T matrix. When $s > m'$, the overdetermined case, Y_1' represents the sufficient m'-dimensional measurements replacing the original s-dimensional vector $Y_{(n)}$. [Recall that $Y_{(n)}$ originally consisted of $L+1$ measurements each of dimension $r+1$, see (4.1-1a), (4.1-5), and (4.1-11a), so that $s = (r+1)(L+1) > m'$ whereas the equivalent sufficient statistic measurement Y_1 has only dimension m']. For the equivalent space let us find the equivalent measurement equation to that (4.1-11). Doing this gives further physical insight into the transformation to the matrix U.

Let

$$X_{n,n}^* = X_n + N_{m'}'' \tag{10.2-40}$$

where $N_{m'}''$ is the error in the least-squares estimate $X_{n,n}^*$ and X_n is the true value of X. Substituting (10.2-40) in (10.2-16) yields

$$U(X_n + N_{m'}'') = Y_1' \tag{10.2-41}$$

which in turn can be written as

$$Y_1' = UX_n + N_{m'}' \tag{10.2-42}$$

where

$$N'_{m'} = UN''_{m'} \tag{10.2-42a}$$

Equation (10.2-42) represents our sought-after equivalent measurement equation to that of (4.1-11). We see that Y'_1, U, and $N'_{m'}$ replace, respectively, $Y_{(n)}$, T, and $N_{(n)}$. Thus, physically the U represents the transition–observation matrix for the transformed m-dimensional space.

Because the transformation F leads to (10.2-16) and because (10.2-16) consists of m' equations and m' unknowns, we know from the discussion in Section 10.2 that the least-squares solution for X_n is given by (10.2-23).

It would be comforting to confirm that we also get the least-squares solution (10.2-23) if we apply our general least-squares solution obtained in Section 4.1, that is (4.1-30) with W given by (4.1-32), to the equivalent measurement system represented by (10.2-42). Using the fact that now

$$T = U \tag{10.2-43}$$

we obtain from (4.1-30) and (4.1-32) that

$$X^*_{n,n} = (U^T U)^{-1} U^T Y'_1 \tag{10.2-44}$$

which becomes

$$X^*_{n,n} = U^{-1} U^{-T} U^T Y'_1 \tag{10.2-45}$$

and which in turn yields (10.2-23), as we intended to show. In the above equation we have taken the liberty to use U^{-T} to represent $(U^T)^{-1}$.

Now let us obtain the covariance of the least-squares estimate $X^*_{n,n}$ using (10.2-42). From (10.2-16) and (10.2-42) we have

$$U X^*_{n,n} = U X_n + N'_{m'} \tag{10.2-46}$$

which can be rewritten as

$$X^*_{n,n} - X_n = U^{-1} N'_{m'} \tag{10.2-47}$$

Thus the covariance of $X^*_{n,n}$ becomes

$$\mathrm{COV}(X^*_{n,n} - X_n) = S^*_{n,n} = E[(X^*_{n,n} - X_n)(X^*_{n,n} - X_n)^T] \tag{10.2-48}$$
$$= U^{-1} E[N'_{m'} N'^T_{m'}] U^{-T} \tag{10.2-48a}$$
$$= U^{-1} \mathrm{COV}(N'_{m'}) U^{-T} \tag{10.2-48b}$$

(In replacing $E[N'_{m'}N'^T_{m'}]$ by $\text{COV}(N'_{m'})$ it is assumed that the mean of $N'_{m'}$ is zero. We shall see shortly that this is indeed the case). It remains now to find $\text{COV}(N'_{m'})$.

From (4.1-11)

$$TX_n = Y_{(n)} - N_{(n)} \tag{10.2-49}$$

Applying the orthonormal transformation F of (10.2-8) yields

$$FTX_n = FY_{(n)} - FN_{(n)} \tag{10.2-50}$$

which from (10.2-37) to (10.2-39) can be rewritten as

$$\begin{bmatrix} UX_n \\ 0 \end{bmatrix} = \begin{bmatrix} Y'_1 \\ Y'_2 \end{bmatrix} - \begin{bmatrix} F_1 N_{(n)} \\ F_2 N_{(n)} \end{bmatrix} \tag{10.2-51}$$

Comparing (10.2-51) with (10.2-42) we see that

$$N'_{m'} = F_1 N_{(n)} \tag{10.2-52}$$

Thus

$$\begin{aligned} \text{COV}(N'_{m'}) &= E[N'_{m'}N'^T_{m'}] \\ &= F_1 E[N_{(n)}N^T_{(n)}]F^T_1 \end{aligned} \tag{10.2-53}$$

Assume that

$$E[N_{(n)}N^T_{(n)}] = \sigma^2 I_s \tag{10.2-54}$$

where I_s is the $s \times s$ identity matrix. Then

$$\begin{aligned} \text{COV}(N'_{m'}) &= F_1 \sigma^2 I_s F^T_1 \\ &= \sigma^2 I_{m'} \end{aligned} \tag{10.2-55}$$

where use was made of (10.2-1). Substituting (10.2-55) into (10.2-48b) yields

$$S^*_{n,n} = \text{COV}(X^*_{n,n} - X_n) = \sigma^2 U^{-1}U^{-T} \tag{10.2-56}$$

Dividing both sides by σ^2, we see that $U^{-1}U^{-T}$ is the normalized covariance matrix of the least-squares estimate. Its elements are the VRF for the least-squares estimate $X^*_{n,n}$; see Sections 1.2.4.4 and 5.8. The term U^{-1} is called the "square root" of $U^{-1}U^{-T}$ [79, pp. 17–18]. (The square root of a matrix is nonunique. These square roots are related to one another by an orthonormal transformation. Any matrix B that is positive definite has a square root, which we designate as S, with $B = SS^T$ [79, pp. 17–18]. When S is a complex square root of B, $B = SS^H$, where H is the transpose complex conjugate [79]). Thus U^{-1} is a square root of the VRF matrix of the least-squares estimate $X_{n,n}$. It is because U^{-1} is the square root of the VRF matrix of a least-squares estimate that the method being described in this section is called the square-root method. It is important to emphasize that to obtain the square-root matrix one does not obtain a direct square root in the usual sense but instead obtains it via an orthonormal transformation as indicated above and in the following sections.

We can also obtain $S^*_{n,n}$ instead of by (10.2-56), by applying (4.5-2) to the equivalent measurement system represented by (10.2-42). For the new measurement $W(h) = U^{-1}$, and the covariance of the measurement noise $R_{(n)}$ becomes $COV(N'_{m'})$ given (10.2-55). Hence (4.5-2) becomes

$$S^*_{n,n} = U^{-1} \sigma^2 I_{m'} U^{-T} = \sigma^2 U^{-1} U^{-T} \tag{10.2-57}$$

as we wished to show. We have assumed that the mean of $N_{(n)}$ is zero. From (10.2-52) it then follows that the mean of $N'_{m'}$ is zero as required for (10.2-48b) to follow from (10.2-48a).

The development of Section 4.3, which introduced the Gram–Schmidt procedure, gave us a physical feel for why U is upper triangular; see specifically (4.3-24) to (4.3-29). Further physical insight into the elements of the matrix U is given in Chapter 13 when the Gram–Schmidt orthonormal transformation F is again discussed.

10.2.3 Reasons the Square-Root Procedure Provides Better Accuracy

Loosely speaking, by using the square-root algorithms, we are replacing numbers that range from 10^{-N} to 10^N by numbers that range from $10^{-N/2}$ to $10^{N/2}$ [78, p. 126]. As a result, when using the square-root algorithm, the computer needs a numerical precision half that required when using the non-square-root algorithm given by the normal equation (4.1-30) with the weight given by (4.1-32) for the least-squares solution. There is, however, a price paid—more operations (adds and multiplies) are needed with square-root algorithms. This shall be elaborated on in Section 14.1.

A simple example is now given that further illustrates the advantage of using the square-root algorithm. Assume B is diagonal matrix given by

$$B = \text{Diag}\,[\,1, \varepsilon, \varepsilon, \varepsilon\,] \tag{10.2-58}$$

If $\varepsilon = 0.000001$ and the computations were carried out to only five-decimal-place accuracy, then the above matrix would be interpreted as

$$B = \text{Diag}\,[\,1, 0, 0, 0\,] \tag{10.2-59}$$

which is a singular matrix and hence noninvertible. If, on the other hand, the square root of B given by S were used in the computation, then

$$S = \text{Diag}[1, \varepsilon^{1/2}, \varepsilon^{1/2}, \varepsilon^{1/2}] \qquad (10.2\text{-}60)$$

where

$$\varepsilon^{1/2} = 0.001 \qquad (10.2\text{-}60a)$$

and the five-decimal-place accuracy of the computer no longer presents a problem, S being properly evaluated as a nonsingular matrix.

A measure of the accuracy needed for inverting a matrix B is the condition number. The condition number C is the ratio of the magnitude of the largest to the smallest eigenvalue of the matrix [81–83, 89, 102, 103], that is, the condition number of the matrix B is

$$C(B) = \left| \frac{\lambda_M}{\lambda_m} \right| \qquad (10.2\text{-}61)$$

where λ_M and λ_m are, respectively, the largest and smallest eigenvalues of B. The eigenvalues of a general matrix B are given by the roots of the characteristic equation:

$$\det[B - \lambda I] = 0 \qquad (10.2\text{-}62)$$

where det stands for "determinant of". For a diagonal matrix the eigenvalues are given by the diagonal elements of the matrix. Thus for the matrix B of (10.2-58) the largest eigenvalue is $\lambda_M = 1$ and the smallest eigenvalue is $\lambda_m = \varepsilon = 0.000001$ and

$$C(B) = \frac{1}{\varepsilon} = 10^6 \qquad (10.2\text{-}63)$$

On the other hand, the condition number for the square root of B given by (10.2-60) is

$$C(S) = \left| \frac{\lambda_M}{\lambda_m} \right|^{1/2} = \frac{1}{\varepsilon^{1/2}} = 10^3 \qquad (10.2\text{-}64)$$

The dynamic range of B is thus 60 dB, whereas that of S is 30 dB. The computer accuracy, or equivalently, word length needed to ensure that B is nonsingular and invertible is [103]

$$\text{Wordlength \{to invert } B\} \geq \log_2 \left| \frac{\lambda_M}{\lambda_m} \right| \qquad (10.2\text{-}65)$$

In contrast the word length needed to ensure that S is invertible is

$$\text{Wordlength (to invert } S = B^{1/2}) \geq \tfrac{1}{2}\left(\log_2 \left| \frac{\lambda_M}{\lambda_m} \right| \right) \qquad (10.2\text{-}66)$$

For $\varepsilon = 0.000001$ the word lengths for B and S become, respectively, 20 and 10 bits.

10.2.4   When and Why Inaccuracies Occur

Consider the constant-velocity target least-squares estimate problem given in Sections 1.2.6 and 1.2.10, that is, the least-squares fit of a straight-line trajectory to a set of data points. The least-squares estimate trajectory is a straight line defined by two parameters, the slope of the line v_0^* and the y intercept x_0^*: see Figure 1.2-10. The least-squares estimate fit is given by the line for which the error e_T of (1.2-33) is minimum. Two plots of e_T versus v_0 and x_0 are given in Figure 10.2-1 for two different cases. The case on the left is for when the measured data points fit a line with little error. A situation for which this is the case is illustrated in Figure 10.2-2a. Such a situation is called well-conditioned. For the second case on the right of Figure 10.2-1 the slope v_0 of the line fitting through the data points is not well defined, but the x_0 intercept is well defined. This situation is illustrated in Figure 10.2-2b. This is called a bad-conditioned or an ill-conditioned situation. For the ill-conditioned situation of Figure 10.2-2b the minimum of e_T in the v_0 dimension is not sharply defined. Big changes in v_0 result in small changes in e_T. Thus, it is difficult to estimate v_0. To find the minimum point great accuracy is needed in calculating e_T, and even then one is not ensured to obtaining a good estimate. Cases like this need the square-root procedure.

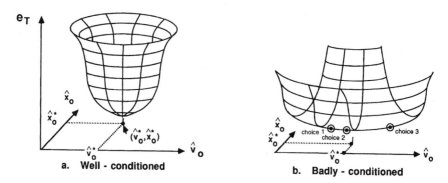

Figure 10.2-1 Surface of sum of squared differences e_T between trajectory range data and linear fit to trajectory as function of estimate for well-conditioned and badly conditioned cases. (After Scheffé [104].)

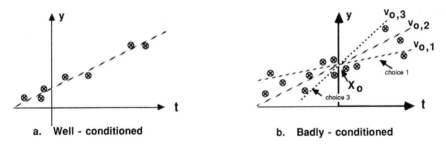

a. **Well - conditioned**   b. **Badly - conditioned**

Figure 10.2-2 Examples of trajectory range data y that lead to well-conditioned and badly conditioned cases. (After Scheffé [104].)

We now describe a way for telling when we have a well-conditioned or ill-conditioned situation. Consider the matrix

$$H = \begin{bmatrix} \dfrac{\partial^2 e_T}{\partial \hat{v}_0^2} & \dfrac{\partial e_T}{\partial \hat{v}_0 \partial \hat{x}_0} \\ \dfrac{\partial e_T}{\partial \hat{v}_0 \partial \hat{x}_0} & \dfrac{\partial^2 e_T}{\partial \hat{x}_0^2} \end{bmatrix} \tag{10.2-67}$$

which is called the curvature matrix or Hessian matrix [9]. The eigenvalues of this matrix give us the curvature along the $\hat{v}_0$ and $\hat{x}_0$ directions for the cases illustrated in Figure 10.2-1; see also Figure 10.2-3.

We now define the eigenvector of a matrix. The ith eigenvector of a matrix B is given by the column matrix X, which satisfies [105]

$$BX = \lambda_i X \tag{10.2-68}$$

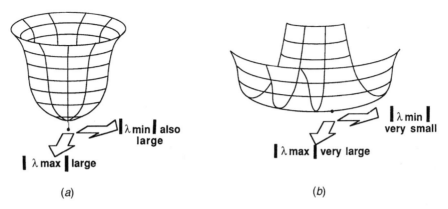

(a)   (b)

Figure 10.2-3 Eigenvalues of curvature matrix (Hessian matrix) for (a) well conditioned and (b) badly conditioned cases. (After Scheffé [104].)

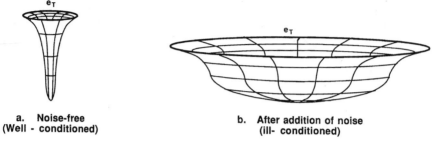

a. Noise-free
(Well - conditioned)   **b. After addition of noise**
                                            (ill- conditioned)

Figure 10.2-4   Effect of noise on conditioning. (After Scheffé [104].)

where λ_i is the ith eigenvalue of the matrix B. The $\hat{v}_0$ and $\hat{x}_0$ directions for the example of Figure 10.2-1 are the eigenvector directions; see Figure 10.2-3.

The addition of noise will cause one to go from the very well-conditioned case illustrated in Figure 10.2-4a to the ill-condition situation illustrated in Figure 10.2-4b.

There are three different orthonormal transformations F that can be used to transform the matrix T into the upper triangular form given by (10.2-8). One of these, the Gram–Schmidt, was introduced in Section 4.3, the other two are the Givens and Householder transformations. All three of these will be discussed in detail in the next three chapters. These three different transformations are mathematically equivalent in that they result in identical answers if the computations are carried out with perfect precision. However, they each have slightly different sensitivities to computer round-off errors. Because they are computationally different, they require different numbers of adds and multiples to arrive at the answers. Finally, the signal processor architectural implementation of the three algorithms are different. One of them (the Givens approach) lends itself to a particularly desirable parallel processor architecture, called the systolic architecture.

11

GIVENS ORTHONORMAL
TRANSFORMATION

11.1 THE TRANSFORMATION

The Givens orthonormal transformation for making a matrix upper triangular is made up of successive elementary Givens orthonormal transformations $G_1, G_2, \ldots$ to be defined shortly. Consider the matrix T expressed by

$$T = \begin{bmatrix} t_{11} & t_{12} & t_{13} \\ t_{21} & t_{22} & t_{23} \\ t_{31} & t_{32} & t_{33} \\ t_{41} & t_{42} & t_{43} \end{bmatrix} \tag{11.1-1}$$

First, using the simple Givens orthonormal transformation matrix G_1 the matrix T is transformed to

$$G_1 T = \begin{bmatrix} (t_{11})_1 & (t_{12})_1 & (t_{13})_1 \\ 0 & (t_{22})_1 & (t_{23})_1 \\ t_{31} & t_{32} & t_{33} \\ t_{41} & t_{42} & t_{43} \end{bmatrix} \tag{11.1-2}$$

The transformation G_1 forces the 2,1 term of the matrix T to be zero. Now applying another elementary Givens orthonormal transformation G_2 to the above matrix yields

$$G_2 G_1 T = \begin{bmatrix} (t_{11})_2 & (t_{12})_2 & (t_{13})_2 \\ 0 & (t_{22})_1 & (t_{23})_1 \\ 0 & (t_{32})_2 & (t_{33})_2 \\ t_{41} & t_{42} & t_{43} \end{bmatrix} \tag{11.1-3}$$

The second transformation G_2 forces the 3, 1 term to be zero. Applying in turn the third Givens orthonormal transformation G_3 to the above matrix now yields

$$G_3G_2G_1T = \begin{bmatrix} (t_{11})_3 & (t_{12})_3 & (t_{13})_3 \\ 0 & (t_{22})_1 & (t_{23})_1 \\ 0 & (t_{32})_2 & (t_{33})_2 \\ 0 & (t_{42})_3 & (t_{43})_3 \end{bmatrix} \tag{11.1-4}$$

Application of these successive elementary Givens orthonormal transformations has forced all the elements of the first column of T to zero below the first element. This process is now repeated for the second column of the above matrix with another set of elementary Givens orthonormal transformations so as to force all the elements below the diagonal of the second column to be zero. This process is next repeated for the third and last column of the matrix so as to force the elements below the diagonal to be zero, yielding the desired upper triangular matrix expressed by

$$\begin{bmatrix} U \\ -- \\ 0 \end{bmatrix} = \begin{bmatrix} u_{11} & u_{12} & u_{13} \\ 0 & u_{22} & u_{23} \\ 0 & 0 & u_{33} \\ 0 & 0 & 0 \end{bmatrix} \tag{11.1-5}$$

The first elementary Givens orthonormal transformation G_1 above is given as

$$G_1 = \left[\begin{array}{cc|cc} c_1 & s_1 & 0 & 0 \\ -s_1 & c_1 & 0 & 0 \\ \hline 0 & 0 & 1 & 0 \\ 0 & 0 & 0 & 1 \end{array} \right] \tag{11.1-6}$$

where

$$c_1 = \cos\theta_1 = \frac{t_{11}}{(t_{11}^2 + t_{21}^2)^{1/2}} \tag{11.1-6a}$$

and

$$s_1 = \sin\theta_1 = \frac{t_{21}}{(t_{11}^2 + t_{21}^2)^{1/2}} \tag{11.1-6b}$$

Examination of (11.1-6) indicates that each of its row matrices (vectors) have unit amplitude and are orthogonal with each other, as should be the case because G_1 is an orthonormal transformation. Physically the Givens orthonormal transformation G_1 can be thought of as projecting the column vectors of the matrix T onto the row space of G_1, as discussed in Section 10.2.1; see also Section 4.3. The new space to which G_1 transforms T is such that for the first column matrix of T, designated as t_1, the new second coordinate t_{21} is zero while the first coordinate, t_{11}, is not; see (11.1-2). Alternately, we can view the transformation G_1 as rotating the vector t_1 so as to achieve the same result. Specifically, we can think of G_1 as rotating the two-dimensional vector formed by the first two coordinates of t_1 such that it lines up with the x axis. Equivalently, we can view G_1 as physically performing a rotation of the first two rows of the matrix T so as to make the term t_{21} equal zero. This is the viewpoint that we shall first take. This rotation (called a Givens rotation) is achieved by the upper left-hand corner 2×2 matrix of (11.1-6), that is, by the matrix

$$G_1' = \begin{bmatrix} c_1 & s_1 \\ -s_1 & c_1 \end{bmatrix} \tag{11.1-7}$$

which acts on the first two rows of the matrix T as follows:

$$\begin{bmatrix} c_1 & s_1 \\ -s_1 & c_1 \end{bmatrix} \begin{bmatrix} t_{11} & t_{12} & t_{13} \\ t_{21} & t_{22} & t_{23} \end{bmatrix} = \begin{bmatrix} (t_{11})_1 & (t_{12})_1 & (t_{13})_1 \\ 0 & (t_{22})_1 & (t_{23})_1 \end{bmatrix} \tag{11.1-8}$$

Using (11.1-6a) and (11.1-6b), it is easy to verify that

$$\begin{aligned} (t_{11})_1 &= c_1 t_{11} + s_1 t_{21} \\ &= (t_{11}^2 + t_{21}^2)^{1/2} \end{aligned} \tag{11.1-9}$$

and that the lower left-hand component of the matrix on the right-hand side of (11.1-8) above is zero, that is,

$$\begin{aligned} (t_{21})_1 &= -s_1 t_{11} + c_1 t_{21} \\ &= 0 \end{aligned} \tag{11.1-10}$$

The rotation performed by the matrix G_1' of (11.1-7) is illustrated in Figure 11.1-1. Consider the vector

$$\bar{t}_0 = t_{11} i + t_{21} j \tag{11.1-11}$$

formed by the first two coordinates of the first column matrix t_1 of T. Here i and j are unit direction vectors for respectively the first and second coordinates of t_1, the x and y coordinates respectively. Figure 11.1-1a shows these vectors. The matrix G_1' rotates the vector $\bar{t}_0$ through an angle θ_1 defined by (11.1-6a) or (11.1-6b) to form a new vector $\bar{t}_1$ having the same magnitude but now aligned

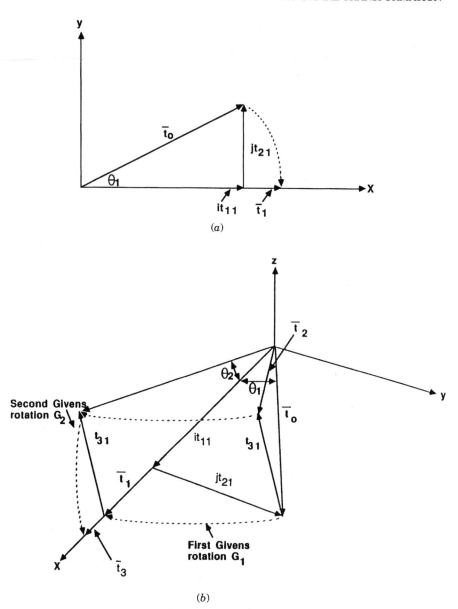

Figure 11.1-1   (*a*) Givens rotation of vector $\bar{t}_0$ in *x–y* plane onto *x* axis forming vector $\bar{t}_1$. (*b*) Two Givens rotations of three-dimensional vector $\bar{t}_2$ to form vector $\bar{t}_3$ aligned along *x* axis.

along the x axis so as to be given by

$$\bar{t}_1 = (t_{11}^2 + t_{21}^2)^{1/2} i + 0j \qquad (11.1\text{-}12)$$

The second Givens orthonormal transformation G_2 of (11.1-3) is given as

$$G_2 = \begin{bmatrix} c_2 & 0 & s_2 & 0 \\ 0 & 1 & 0 & 0 \\ -s_2 & 0 & c_2 & 0 \\ 0 & 0 & 0 & 1 \end{bmatrix} \qquad (11.1\text{-}13)$$

where

$$c_2 = \cos\theta_2 = \frac{(t_{11})_1}{[(t_{11})_1^2 + t_{31}^2]^{1/2}} \qquad (11.1\text{-}13a)$$

and

$$s_2 = \sin\theta_2 = \frac{t_{31}}{[(t_{11})_1^2 + t_{31}^2]^{1/2}} \qquad (11.1\text{-}13b)$$

This Givens rotation yields, for the top left-hand corner element of the transformed matrix,

$$\begin{aligned} (t_{11})_2 &= [(t_{11})_1^2 + t_{31}^2]^{1/2} \\ &= [t_{11}^2 + t_{21}^2 + t_{31}^2]^{1/2} \end{aligned} \qquad (11.1\text{-}14)$$

and, for the third element in the first column,

$$(t_{31})_2 = 0 \qquad (11.1\text{-}15)$$

Figure 11.1-1b illustrates the first rotation by θ_1 and the second rotation by the amount θ_2. Here the three-dimensional vector $\bar{t}_2$ is given by

$$\begin{aligned} \bar{t}_2 &= \bar{t}_0 + kt_{31} \\ &= it_{11} + jt_{21} + kt_{31} \end{aligned} \qquad (11.1\text{-}16)$$

where k is the unit vector along the z axis. This vector is first rotated by G_1 onto the x–z plane to form $\bar{t}_1 + kt_{13}$ and then rotated by G_2 onto the x axis to yield the new vector $\bar{t}_3$:

$$\bar{t}_3 = [t_{11}^2 + t_{21}^2 + t_{31}^2]^{1/2} i \qquad (11.1\text{-}17)$$

To get further physical insight into the Givens transformation G_1, we will now view it as transforming the column space of the matrix T onto the new space defined by the row vectors of the matrix G_1. We first consider the projection of the first column vector t_1 of T onto the row space of G_1. (A word is in order relative to our notation. Note that whereas t_1 is the whole s-dimensional first column of T, the vector $\bar{t}_0$ consists of only the first two coordinates of t_1. The vector $\bar{t}_1$ is just $\bar{t}_0$ rotated onto the x axis while the vector $\bar{t}_2$ is a vector formed by just the first three coordinates of t_1.) We will consider the projection of t_1 onto the successive rows of G_1, starting with the first, in order to obtain the coordinates of t_1 after projecting it into the row space of G_1.

Designate the first row of G_1 as the vector g_1 whose coordinates (in the original $x, y, z, \ldots$ coordinate system; see Section 4.3) are given by the entries of the first row of G_1. This vector g_1 is a unit vector lying in the plane defined by the first two coordinates of the column vector t_1 of T. This plane is the x, y space of Figure 11.1-2a. Moreover, g_1 is chosen to lie along the vector $\bar{t}_0$ formed by the first two coordinates of t_1 with the remaining coordinates being zero. Hence, from Figure 11.1-2a and (11.1-6)

$$g_1 = i \cos \theta_1 + j \sin \theta_1 + k \cdot 0 + 0 + \cdots \qquad (11.1\text{-}18)$$

where i, j, and k are the unit vectors in the x, y, and z directions of the rectangular coordinate system (see Figure 11.1-2c). They are also the unit vector directions for the vector formed by the first three coordinates of the column vectors of T, that is, of the vector t_1.

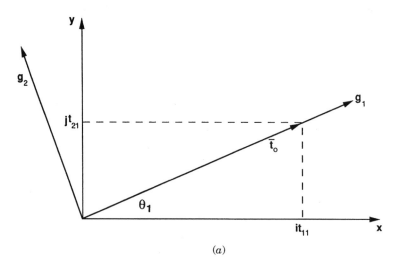

(a)

Figure 11.1-2 (a) Transformation of vector formed by first two coordinates of t_1, designated $\bar{T}_0$, into g_1, g_2 two-dimensional space from x, y space. (b) Transformation of vector formed by $\bar{t}_0 = g_1 \| \bar{t}_0 \|$ and $k t_{31}$(effectively first three coordinates of $\bar{t}_1$), designated as $\bar{t}_2$, into $(g_1)_2$, $(g_3)_2$ space. (c) Givens unit vectors.

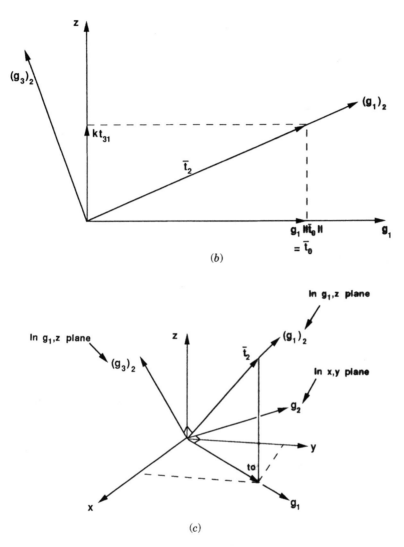

(b)

(c)

Figure 11.1-2   (*Continued*)

The projection of t_1 onto g_1 is given by the dot product of t_1 and g_1 [see (4.2-36)], or equivalently by the product of the first row of G_1 by the first column of T as given by (4.2-37) for the three-dimensional case. Because only the x, y coordinates of g_1 are nonzero, only the first two coordinates of t_1 enter into this product, or equivalent, projection. Thus we can think of g_1 as a two element vector to get

$$g_1 \cdot t_1 = g_1 \cdot \bar{t}_0 \qquad (11.1\text{-}19)$$

Applying (11.1-6a) and (11.1-6b) to $\bar{t}_0$ of (11.1-11) yields

$$\bar{t}_0 = i\|\bar{t}_0\| \cos \theta_1 + j\|\bar{t}_0\| \sin \theta_1 \qquad (11.1\text{-}20)$$

Substituting (11.1-18) and (11.1-20) into (11.1-19) and using (11.1-9) yield

$$g_1 \cdot t_1 = g_1 \cdot \bar{t}_0 = \|\bar{t}_0\| = (t_{11})_1 \qquad (11.1\text{-}21)$$

That is, the projection of $\bar{t}_0$ into g_1 is equal to the magnitude of $\bar{t}_0$, as we saw it should be from the above discussion relative to thinking of G_1 as doing a rotation of $\bar{t}_0$. This result also follows because, as indicated above, the unit vector g_1 was chosen to lie not only in the x, y plane but also along the vector $\bar{t}_0$; see (11.1-18) and (11.1-20).

Equation (11.1-21) gives us the first coordinate of the projection of t_1 onto the s-dimensional row space of G_1. Now let us determine the second coordinate obtained by projecting t_1 onto the second-row unit vector of G_1, which we designate as g_2. The vector g_2 is also in the x, y plane of Figure 11.1-2a but perpendicular to g_1. Specifically, from Figure 11.1-2a and (11.1-6) it follows that

$$g_2 = -i \sin \theta_1 + j \cos \theta_1 + k \cdot 0 + 0 + \cdots \qquad (11.1\text{-}22)$$

The projection of t_1 onto g_2 is given by

$$g_2 \cdot t_1 = g_2 \cdot \bar{t}_0 \qquad (11.1\text{-}23)$$

and substituting (11.1-20) and (11.1-22) into the above equation yields

$$g_2 t_1 = -\sin \theta_1 \cos \theta_1 + \cos \theta_1 \sin \theta_1 = 0 \qquad (11.1\text{-}24)$$

which, based on (11.1-2) and (11.1-8), is what it should be. Physically (11.1-24) should be zero because g_2 is orthogonal to the direction of g_1, which is lined up with the direction of t_1 and in turn $\bar{t}_0$.

Equation (11.1-24) gives us the second coordinate of the projection of t_1 onto the s-dimensional space of G_1. Now we determine the third coordinate by projecting t_1 onto the third-row unit vector of G_1, which we designate as g_3. Examining (11.1-6), we see that the third row consists of zero entries except for the third-column entry, which is a 1. Consequently, the projection of t_1 onto the third-row unit vector g_3 leaves the third coordinate of t_1 unchanged. Stated another way, g_3 is a unit vector aligned along the third coordinate of t_1. This follows because all the coordinates of g_3 are zero except the third, which is unity. Consequently, projecting t_1 onto g_3 leaves the third coordinate of t_1 unchanged. The same is true for the remaining coordinates of t_1; they are unchanged by the transformation G_1 because the ith-row vector g_i, $i > 2$, is a unit vector lined up with the ith coordinate of t_1; see (11.1-6). Although for (11.1-6) $s = 4$, this result is true when G_1 is an $s \times s$ matrix. In this case, the lower right-hand corner $(s - 2) \times (s - 2)$ matrix of G_1 is the identity matrix with the elements to its left being zero. Thus Figure 11.1-2a illustrates the projection of the vector $\bar{t}_0$, defined by the first two coordinates of t_1, onto the space defined by the first two unit row vectors g_1 and g_2 of G_1.

This completes the proof that G_1 projects T into the form given by (11.1-2). Designate this projected vector t_1 as $(t_1)_1$. In summary, G_1 transforms the columns of T to a new coordinate system for which the coordinates of T have the form given by (11.1-2) with $(t_{21})_1 = 0$. We will let $(t_1)_1$ represent the column vector t_1, the first column of T, in this new coordinate system.

Let us carry one step further our discussion of viewing the simple G_i Givens transformations as projecting the column space of the matrix T onto the space defined by the row vectors of G_i. This is now done by detailing this projection for G_2. On examining (11.1-13) it is seen that only the first and third coordinates are altered by this projection, the unit vectors of G_2 represented by the other rows of G_2 being unit vectors aligned along the other coordinates of the matrix T. Now consider the unit vectors $(g_1)_2$ and $(g_3)_2$ defined by the first and third rows of G_2. On examining (11.1-16), (11.1-13a), (11.1-13b), and (11.1-13), it can be verified that $(g_1)_2$ is chosen to line up with the vector defined by the first- and third-row coordinates of $(t_1)_1$, with all other coordinates set equal to zero, while $(g_3)_2$ is chosen to be orthogonal to $(g_1)_2$. This situation is illustrated in Figure 11.1-2b; see also Figure 11.1-2c. In this figure $\bar{t}_2$ is the vector formed by the first- and third-row coordinates of $(t_1)_1$; see (11.1-2) and (11.1-16). As a consequence of this choice for $(g_1)_2$, the projection of $\bar{t}_2$ onto $(g_1)_2$, which is the first coordinate of the vector $G_2(t_1)_1$, has a magnitude equal to the magnitude of $\bar{t}_2$, which is the desired $(t_{11})_2$ of (11.1-3); see also (11.1-14). In turn $\bar{t}_2$ equals the magnitude of the first three coordinates of t_1; see (11.1-14). Moreover, this last result follows because $(g_1)_2$ in the above is actually chosen to line up with the first three coordinates of the vector t_1.

Because $(g_3)_2$ is orthogonal to $(g_1)_2$ and in turn $\bar{t}_2$, the third coordinate of $G_2(t_1)_1$ given by $(g_3)_2(t_1)_1 = (g_3)_2\bar{t}_2$, equals zero. This results in the first column of G_2G_1T being in the form given in (11.1-3), as desired. Figure 11.1-2b illustrates the projection of the vector $\bar{t}_2$, defined by the vectors $\bar{t}_0$ and kt_{31}, onto the space defined by the first and third rows of G_2, or equivalently, the unit vector rows $(g_1)_2$ and $(g_3)_2$ of G_2. The physical significance of the succeeding Givens transformations follows in a similar manner. This completes our interpretation of the Givens transformations as an orthogonal coordinate transformation. In Chapter 13 we shall relate the Givens transformation to the Gram–Schmidt transformation introduced in Section 4.3. In Chapter 12 we shall relate the Givens transformation to the Householder transformation.

The transformations can be conveniently applied to T and $Y_{(n)}$ simultaneously by generating the augmented matrix T_0 given by

$$T_0 = [T|Y_{(n)}] \qquad (11.1\text{-}25)$$

and then applying the Givens orthonormal transformation to it. [This parallels the procedure followed in Section 4.3 where such an augmented matrix was generated to apply the Gram–Schmidt transformation to T and $Y_{(n)}$ simultaneously; see (4.3-54) and the discussions related to it]. To be concrete, assume that T is a 2×3 matrix given by

$$T = \begin{bmatrix} t_{11} & t_{12} \\ t_{21} & t_{22} \\ t_{31} & t_{32} \end{bmatrix} \qquad (11.1\text{-}26)$$

and

$$Y_{(n)} = \begin{bmatrix} y_1 \\ y_2 \\ y_3 \end{bmatrix} \qquad (11.1\text{-}27)$$

Also assume

$$X_{n,n}^* = \begin{bmatrix} x_1^* \\ x_2^* \end{bmatrix} \qquad (11.1\text{-}28)$$

Then (11.1-25) becomes

$$T_0 = \begin{bmatrix} t_{11} & t_{12} & y_1 \\ t_{21} & t_{22} & y_2 \\ t_{31} & t_{32} & y_3 \end{bmatrix} \qquad (11.1\text{-}29)$$

After three Givens rotations (11.1-29) becomes

$$T_0' = FT_0 = \begin{bmatrix} (t_{11})_3 & (t_{12})_3 & | & (y_1)_3 \\ 0 & (t_{22})_3 & | & (y_2)_3 \\ \hline 0 & 0 & | & (y_3)_3 \end{bmatrix} \qquad (11.1\text{-}30)$$

As shall be shown shortly, the square of the lower right element $(y_3)_3$ is the numerical value of the least-squares residue error given by (4.1-31), that is,

$$[(y_3)_3]^2 = (y_3)_3^2 = e(X_{n,n}^*) \qquad (11.1\text{-}31)$$

[This result was already shown in Section 4.3; specifically see (4.3-38).] The 2×2 upper left-hand corner matrix of t's corresponds to the upper triangular matrix of u's of (10.2-8) or (10.2-9). We have used the entries $(t_{ij})_3$ instead of u_{ij} in matrix T'_0 given by (11.1-30) in order to emphasize that three elementary Givens rotations are being used to transform the T matrix to the U matrix. The following gives step by step the Givens orthonormal rotation transformations for a 3×2 matrix T: Let

$$T = \begin{bmatrix} t_{11} & t_{12} \\ t_{21} & t_{22} \\ t_{31} & t_{32} \end{bmatrix}$$

(a) The first Givens rotation by matrix G_1 is given as

$$G_1 T = \begin{bmatrix} c_1 & s_1 & 0 \\ -s_1 & c_1 & 0 \\ 0 & 0 & 1 \end{bmatrix} \begin{bmatrix} t_{11} & t_{12} \\ t_{21} & t_{22} \\ t_{31} & t_{32} \end{bmatrix} = \begin{bmatrix} c_1 t_{11} + s_1 t_{21} & c_1 t_{12} + s_1 t_{22} \\ -s_1 t_{11} + c_1 t_{21} & -s_1 t_{12} + c_1 t_{22} \\ t_{31} & t_{32} \end{bmatrix}$$

$$= \begin{bmatrix} (t_{11})_1 & (t_{12})_1 \\ 0 & (t_{22})_1 \\ t_{31} & t_{32} \end{bmatrix}$$

where

$$c_1 = \frac{t_{11}}{\sqrt{t_{11}^2 + t_{21}^2}} \quad s_1 = \frac{t_{21}}{\sqrt{t_{11}^2 + t_{21}^2}}$$

$$(t_{11})_1 = \sqrt{t_{11}^2 + t_{21}^2}$$

Therefore

$$c_1 = \frac{t_{11}}{(t_{11})_1} \quad s_1 = \frac{t_{21}}{(t_{11})_1}$$

Thus

$$c_1 t_{11} + s_1 t_{21} = \frac{t_{11}}{(t_{11})_1} t_{11} + \frac{t_{21}}{(t_{11})_1} t_{21} = \frac{t_{11}^2 + t_{21}^2}{(t_{11})_1} = \frac{(t_{11})_1^2}{(t_{11})_1} = (t_{11})_1$$

$$-s_1 t_{11} + c_1 t_{21} = -\frac{t_{21}}{(t_{11})_1} t_{11} + \frac{t_{11}}{(t_{11})_1} t_{21} = 0 = (t_{21})_1$$

$$c_1 t_{12} + s_1 t_{22} = \frac{t_{11}}{(t_{11})_1} t_{12} + \frac{t_{21}}{(t_{11})_1} t_{22} = (t_{12})_1$$

$$-s_1 t_{12} + c_1 t_{22} = -\frac{t_{21}}{(t_{11})_1} t_{12} + \frac{t_{11}}{(t_{11})_1} t_{22} = (t_{22})_1$$

(b) The second Givens rotation by matrix G_2 yields

$$
G_2G_1T = \begin{bmatrix} c_2 & 0 & s_2 \\ 0 & 1 & 0 \\ -s_2 & 0 & c_2 \end{bmatrix} \begin{bmatrix} (t_{11})_1 & (t_{12})_1 \\ 0 & (t_{22})_1 \\ t_{31} & t_{32} \end{bmatrix}
$$

$$
= \begin{bmatrix} c_2(t_{11})_1 + s_2t_{31} & c_2(t_{12})_1 + s_2t_{32} \\ 0 & (t_{22})_2 = (t_{22})_1 \\ s_2(t_{11})_1 + c_2t_{31} & s_2(t_{12})_1 + c_2t_{32} \end{bmatrix}
$$

where

$$
c_2 = \frac{(t_{11})_1}{\sqrt{(t_{11})_1^2 + t_{31}^2}} \qquad s_2 = \frac{t_{31}}{\sqrt{(t_{11})_1^2 + t_{31}^2}}
$$

$$
(t_{11})_2 = \sqrt{(t_{11})_1^2 + (t_{31})^2}
$$

$$
c_2 = \frac{(t_{11})_1}{(t_{11})_2} \qquad s_2 = \frac{t_{31}}{(t_{11})_2}
$$

$$
c_2(t_{11})_1 + s_2t_{31} = \frac{(t_{11})_1}{(t_{11})_2}(t_{11})_1 + \frac{t_{31}}{(t_{11})_2}t_{31} = \frac{(t_{11})_2^2}{(t_{11})_2} = (t_{11})_2
$$

$$
-s_2(t_{11})_1 + c_2t_{31} = -\frac{t_{31}}{(t_{11})_2}(t_{11})_1 + \frac{(t_{11})_1}{(t_{11})_2}t_{31} = 0
$$

$$
c_2(t_{12})_1 + s_2t_{32} = \frac{(t_{11})_1}{(t_{11})_2}(t_{12})_1 + \frac{t_{31}}{(t_{11})_2}t_{32} = (t_{12})_2
$$

$$
-s_2(t_{12})_1 + c_2t_{32} = -\frac{t_{31}}{(t_{11})_2}(t_{12})_1 + \frac{(t_{11})_1}{(t_{11})_2}t_{32} = (t_{32})_2
$$

(c) The third Givens rotation by matrix G_3 yields

$$
G_3G_2G_1T = \begin{bmatrix} 1 & 0 & 0 \\ 0 & c_3 & s_3 \\ 0 & -s_3 & c_3 \end{bmatrix} \begin{bmatrix} (t_{11})_2 & (t_{12})_2 \\ 0 & (t_{22})_2 \\ 0 & (t_{32})_2 \end{bmatrix}
$$

$$
= \begin{bmatrix} (t_{11})_2 & (t_{12})_2 \\ 0 & \begin{array}{c} c_3(t_{22})_2 + s_3(t_{32})_2 \\ = (t_{22})_3 \end{array} \\ 0 & \begin{array}{c} -s_3(t_{22})_2 + c_3(t_{32})_2 \\ = (t_{32})_3 = 0 \end{array} \end{bmatrix} = \begin{bmatrix} (t_{11})_2 & (t_{12})_2 \\ 0 & (t_{22})_3 \\ 0 & 0 \end{bmatrix}
$$

EXAMPLE   295

where

$$c_3 = \frac{(t_{22})_2}{(t_{22})_3} \qquad s_3 = \frac{(t_{32})_2}{(t_{22})_3}$$

$$(t_{22})_3 = \sqrt{(t_{22})_2^2 + (t_{32})_2^2}$$

$$c_3(t_{22})_2 + s_3(t_{33})_2 = \frac{(t_{22})_2}{(t_{22})_3}(t_{22})_2 + \frac{(t_{32})_2}{(t_{22})_3}(t_{32})_2$$

$$= \frac{(t_{22})_2^2 + (t_{32})_2^2}{(t_{22})_3} \equiv (t_{22})_3$$

$$-s_3(t_{22})_2 + c_3(t_{32})_2 = -\frac{(t_{32})_2}{(t_{22})_3}(t_{22})_2 + \frac{(t_{22})_2}{(t_{22})_3}(t_{32})_2 = 0$$

As in Section 10.2, the Givens transformation yields a set of m' equations in the Gauss elimination form; see (10.2-16). It was indicated in Section 4.3 that the set of equations obtained using the voltage procedure will differ from those generally obtained if the normal Gauss elimination procedure is used. First note that the procedure for obtaining the Gauss elimination form is not unique. Usually, to eliminate the t_{21} term of (11.1-1), the factor t_{21}/t_{11} times the first row is subtracted from the second row. We could have just as well have subtracted $k(t_{21}/t_{11})$ times the first row from k times the second row, where k is an arbitrary constant. This is what we are doing when using the Givens transformation G_1 of (11.1-6), with c_1 and s_1 given by (11.1-6a) and (11.1-6b). In this case the constant k is c_1. This value of k was chosen over unity to make the rows of G_1 be unit vectors, which makes F an orthonormal transformation. Why do we want to make F an orthonormal matrix? It turns out that the sensitivity to computer round-off errors of our least-squares solution to (10.2-16) is minimized if the Gauss elimination form is obtained by the use of an orthonormal transformation, that is, if the matrices U and Y_1' are obtained from T and $Y_{(n)}$ using an orthonormal transformation matrix F [129].

11.2 EXAMPLE

By way of numerical example let

$$T = \begin{bmatrix} 1 & 1-\varepsilon \\ 1-\varepsilon & 1 \\ 1 & 1 \end{bmatrix} \tag{11.2-1}$$

We chose this T because it is one for which round-off errors can lead to a problem when ε is very small. Specifically, when ε is small enough, the computer round-off errors will cause the columns of T to be considered essentially identical so as to lead to a degenerate situation of the type discussed in Chapter 10. This is precisely the type of situation for which the Givens orthonormal transformation procedure is designed to work. The T example given by (11.2-1) was originally suggested by Golub [102; see also 79, p. 91].

We shall pick a fairly large ε for our example so as to be able to calculate the orthonormalized T on a hand calculator (such as the HP15C or HP32S) without running into computational round-off problems. Toward this end we pick $\varepsilon = 0.1$. Applying three Givens rotations yields, for the transformed T matrix,

$$G_1 G_2 G_3 T = T' = \begin{bmatrix} 1.676305461 & 1.670339961 \\ 0 & 0.141295480 \\ \hline 0 & 0 \end{bmatrix} \tag{11.2-2}$$

The following are step-by-step hand calculations for Givens orthonormal rotation transformations of the 3×2 matrix T of (11.2-1) for $\varepsilon = 0.1$:

$$T = \begin{bmatrix} 1 & 0.9 \\ 0.9 & 1 \\ 1 & 1 \end{bmatrix} = \begin{bmatrix} t_{11} & t_{12} \\ t_{21} & t_{22} \\ t_{31} & t_{32} \end{bmatrix}$$

(a) For the first Givens rotation

$$\sqrt{t_{11}^2 + t_{21}^2} = 1.345362405 = (t_{11})_1 \qquad c_1 = 0.743294146$$
$$s_1 = 0.668964732$$

so that

$$G_1 T = \begin{bmatrix} (t_{11})_1 & (t_{12})_1 \\ (t_{21})_1 & (t_{22})_1 \\ (t_{31})_1 & (t_{32})_1 \end{bmatrix} = \begin{bmatrix} 1.345362405 & 1.337929463 \\ 0 & 0.141225888 \\ 1 & 1 \end{bmatrix}$$

(b) For the second Givens rotation

$$\sqrt{(t_{11})_1^2 + (t_{31})_1^2} = 1.676305461 = (t_{11})_2 \qquad c_2 = 0.802575924$$
$$s_2 = 0.596549986$$

so that

$$G_2 G_1 T = \begin{bmatrix} (t_{11})_2 & (t_{12})_2 \\ (t_{21})_2 & (t_{22})_2 \\ (t_{31})_2 & (t_{32})_2 \end{bmatrix} = \begin{bmatrix} 1.676305461 & 1.670339961 \\ 0 & 0.141225888 \\ 0 & 0.004434121 \end{bmatrix}$$

EXAMPLE   **297**

(c) For the third Givens rotation

$$\sqrt{(t_{22})_2^2 + (t_{32})_2^2} = 0.141295480 = (t_{22})_3$$

so that

$$G_3 G_2 G_1 T = \begin{bmatrix} (t_{11})_3 & (t_{12})_3 \\ (t_{21})_3 & (t_{22})_3 \\ (t_{31})_3 & (t_{32})_3 \end{bmatrix} = \begin{bmatrix} 1.676305461 & 1.670339961 \\ 0 & 0.141295480 \\ 0 & 0 \end{bmatrix}$$

which agrees with (11.2-2).

It would be desirable to obtain an exact solution in terms of ε for the matrix FT and for the estimate $X_{n,n}^*$ obtained using this T and the measured data. In general, one would not have an exact solution. However, for the simple example given it is possible to get such an exact solution. This is convenient because it allows examination of the relative accuracies of various algorithms being described in these sections for obtaining the least-squares solution when T is given by (11.2-1). The effect of different computer precision could also be examined. This is exactly what is done in reference 79 (pp. 91–100). We will next provide the exact solution to FT.

It is possible to obtain the sought-after exact solution in terms of ε using three Givens rotations of (11.2-1). After much straightforward algebra one obtains [79, pp. 91–96]

$$(t_{11})_3 = p = [2 + (1 - \varepsilon)^2]^{1/2} \tag{11.2-3}$$

$$(t_{12})_3 = \frac{3 - 2\varepsilon}{p} \tag{11.2-4}$$

$$(t_{22})_3 = \frac{\varepsilon(\varsigma)^{1/2}}{p} \tag{11.2-5}$$

$$(y_1)_3 = \frac{y_1 + (1 - \varepsilon)y_2 + y_3}{p} \tag{11.2-6}$$

$$(y_2)_3 = \frac{(1 - \varepsilon)y_1 + y_2 + y_3 - (t_{12})_3(y_1)_3}{(t_{22})_3} \tag{11.2-7}$$

$$(y_3)_3^2 = y_1^2 + y_2^2 + y_3^2 - (y_1)_3^2 + (y_2)_3^2 \tag{11.2-8}$$

where

$$p = [2 + (1 - \varepsilon)^2]^{1/2} \tag{11.2-9}$$

and

$$\varsigma = 6 - 4\varepsilon + \varepsilon^2 \tag{11.2-10}$$

Using the above results for (11.1-30) it is a straightforward matter to now solve (10.2-16) by the back-substitution method. From the above (10.2-16) becomes

$$\begin{bmatrix} (t_{11})_3 & (t_{12})_3 \\ 0 & (t_{22})_3 \end{bmatrix} \begin{bmatrix} x_1^* \\ x_2^* \end{bmatrix} = \begin{bmatrix} (y_1)_3 \\ (y_2)_3 \end{bmatrix} \tag{11.2-11}$$

Solving yields, for the exact least-squares estimate solution,

$$x_1^* = \frac{1}{\varsigma} \left[\frac{(3-\varepsilon)y_1 - [3(1-\varepsilon)+\varepsilon^2]y_2}{\varepsilon} + y_3 \right] \tag{11.2-12}$$

and

$$x_2^* = \frac{1}{\varsigma} \left[\frac{(3-\varepsilon)y_2 - [3(1-\varepsilon)+\varepsilon^2]y_1}{\varepsilon} + y_3 \right] \tag{11.2-13}$$

[Reference 79 (pp. 91–95) actually uses a different procedure for obtaining the above exact solutions that we now outline. Let C_i be the ith column vector of the matrix (11.1-29) and let U_i be the ith column vector of the orthonormalized transformed version of this matrix given by (11.1-30). It turns out that after an orthonormal transformation the inner products of corresponding columns of these matrices are preserved, that is,

$$C_i^T C_j = U_i^T U_j \quad \text{for all } i,j \tag{11.2-14}$$

By evaluating (11.2-14) for all i, j inner products, it is a straightforward matter to solve for all the terms of (11.1-30), giving the solutions of (11.2-3) through (11.2-10). Specifically one proceeds by first computing the inner products for $i = j = 1$ to solve for $(t_{11})_3$. Next the inner products for $i = 1$ and $j = 2,3$ are obtained to solve for respectively $(t_{12})_3$ and $(y_1)_3$. Then we obtain the inner product for $i = j = 2$ to solve for $(t_{22})_3$. The inner product for $i=2$ and $j=3$ then gives $(y_2)_3$. Finally the last inner product of $i = j = 3$ yields the $(y_3)_3$. This procedure is very much similar to the back-substitution procedure.]

11.3 SYSTOLIC ARRAY IMPLEMENTATION

11.3.1 Systolic Array

The Givens orthonormal transformation for triangularizing the augmented matrix T_0 described above can be implemented using a parallel-circuit architecture called a systolic array that shall be defined shortly. It should be cautioned that the systolic array described here would not be used in those cases where the T matrix is known in advance; see the discussion at the end of Section

4.3. For these cases some of the computations for triangularizing the T matrix can be done off-line. For the sidelobe canceler problem described in Section 4.4 the equivalent matrix to the T matrix, the matrix V, consists of random varibles (the random auxiliary antenna data) and hence the triangularization of V cannot be done in advance. Because of the important application of the systolic array implementation to the sidelobe canceler problem and for completeness, we include its description here. It may have possible application to some least-squares tracking problems.

To introduce the idea of a systolic array, we consider the problem of matrix multiplication as given by

$$\begin{bmatrix} x_{11} & x_{12} & x_{13} \\ x_{21} & x_{22} & x_{23} \\ x_{31} & x_{32} & x_{33} \end{bmatrix} \begin{bmatrix} a_1 \\ a_2 \\ a_3 \end{bmatrix} = \begin{bmatrix} y_1 \\ y_2 \\ y_3 \end{bmatrix} \qquad (11.3\text{-}1)$$

a. Matrix - Vector Multiplication

$$\begin{bmatrix} X_{11} & X_{12} & X_{13} \\ X_{21} & X_{22} & X_{23} \\ X_{31} & X_{32} & X_{33} \end{bmatrix} \begin{bmatrix} a_1 \\ a_2 \\ a_3 \end{bmatrix} = \begin{bmatrix} Y_1 \\ Y_2 \\ Y_3 \end{bmatrix} \qquad Y_i = \sum_{j=1}^{3} Z_{ij}\, a_j$$
$$i = 1, 2, 3$$

b. **Systolic - Array Implementation**

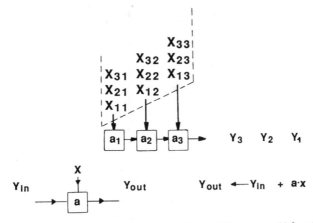

Figure 11.3-1 Matrix–vector multiplication with systolic array. (After McWhirter [140].)

The above matrix multiplication can be carried out with the systolic array shown in Figure 11.3-1. Each square of the systolic array does a multiplication and addition as indicated. The outputs y_i, $i = 1, 2, 3$, are generated as the outputs of the rightmost box. The inputs for a given column of the x_{ij} are shown staggered in the figure. This is necessary because the output of the box labeled a_{i-1} is needed before the calculations for the box labeled a_i can be carried out. To better understand what is happening, let us follow the computation of y_1. First $a_1 x_{11}$ is computed at the first clock cycle in the box labeled a_1. Then $a_1 x_{11} + a_2 x_{12}$ is calculated on the next clock cycle in the box labeled a_2. Finally, the desired output $a_1 x_{11} + a_2 x_{12} + a_3 x_{13} = y_1$ is computed on the third clock cycle in the box labeled a_3. While y_1 is being calculated, y_2 and y_3 are being calculated behind it.

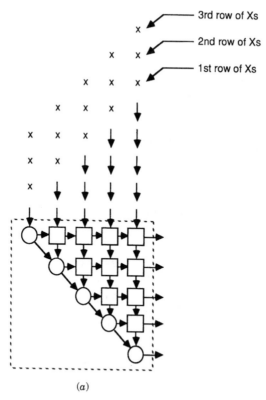

(a)

Figure 11.3-2 (a) Systolic array implementation of Givens orthonormal transformation. (b) Computations performed by circular boundary and rectangular internal cells of systolic array implementing Givens orthonormal transformation. (After Kung, H. T. and W. M. Gentleman, "Matrix Triangularization by Systolic Arrays," Proceedings of the SPIE, 1981, Vol. 298, Real Time Signal Processing IV.)

Now we will define a systolic array. It is a collection of individual computer boxes or elements, many of which do identical computations. These elements do calculations in parallel and most importantly have only a connection to adjacent computer boxes, or elements, as is the case for the systolic array of Figure 11.3-2. A key feature of the systolic array is that the computations are carried out in parallel by many computer processors, thus providing a high throughput. The computer elements making up the systolic array could be arranged as a one-dimensional array, as in Figure 11.3-1, a two-dimensional array (as is the case for the systolic array needed to do the Givens orthonormalization), or a three-dimensional array. We now give the two-dimensional systolic array needed to do the Givens orthonormal transformation.

The systolic array required to do the Givens orthonormal transformation is given in Figure 11.3-2 [106]. It consists of two types of computer elements, those in the square boxes and those in the circular boxes. The circular elements compute the c_i's and s_i's of (11.1-6a) and (11.1-6b) needed for the Givens rotation, the ith-row circular box calculating the c_i's and s_i's for the ith row of T_0. Assume T_0 is a $(m' + 1) \times s$ matrix. Then, except for the bottom circular box, the circular boxes calculate elements $(t_{ii})_k$. The bottom circular box calculates $(y_{m+1})_k$, where $k = 1, 2, \ldots, s'$ and s' is the number of Givens

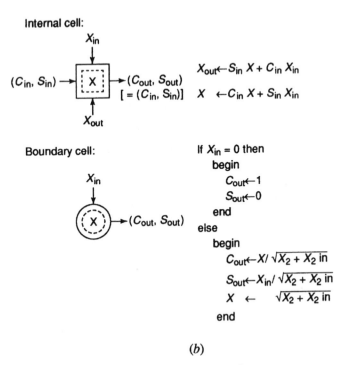

Internal cell:

$$X_{out} \leftarrow -S_{in} X + C_{in} X_{in}$$
$$X \leftarrow C_{in} X + S_{in} X_{in}$$

Boundary cell:

If $X_{in} = 0$ then
  begin
    $C_{out} \leftarrow 1$
    $S_{out} \leftarrow 0$
  end
else
  begin
    $C_{out} \leftarrow X / \sqrt{X^2 + X^2 \text{ in}}$
    $S_{out} \leftarrow X_{in} / \sqrt{X^2 + X^2 \text{ in}}$
    $X \leftarrow \sqrt{X^2 + X^2 \text{ in}}$
  end

(*b*)

Figure 11.3-2   (*Continued*)

rotations needed to complete the orthonormal transformation to obtain $T_0' = FT_0$; see, for example, (11.1-30). At the end of the computations the circular boxes contain

$$(t_{ii})_{s'} = u_{ii} \qquad i = 1, \ldots, m' \tag{11.3-2}$$

and the bottom circular box contains

$$(y_{m+1})_{s'} = [e_n(X_{n,n}^*)]^{1/2} \tag{11.3-3}$$

The inner square boxes calculate $(t_{ij})_k$ for $i \neq j$. At the completion of the triangularization the inner square boxes contain

$$(t_{ij})_{s'} = u_{ij} \qquad \text{for } i \neq j \tag{11.3-4}$$

The right-hand boundary square boxes calculate $(y_i)_k$ and at the end of the Givens orthonormal transformation contain

$$(y_i)_{s'} \quad i = 1, \ldots, m' \tag{11.3-5}$$

Figure 11.3-3 shows the step-by-step computations involved in the Givens orthonormalization of T given by (11.2-1) for $\varepsilon = 0.1$. At the end of the orthonormalization the array contains the orthonormalized T given by (11.2-2).

Having performed the Givens orthonormal transformation of T_0 using the systolic array, it is now possible to calculate the desired least-squares estimate $X_{n,n}^*$, which is an $m' \times 1$ array containing the elements $x_1^*, \ldots, x_{m'}^*$. This is done using the back-substitution method to solve (10.2-16) as described in Sections 4.3 and 10.2. This back substitution can also be carried out using another systolic array [107; see also 106]. Cascading these two systolic arrays together yields the systolic array of Figure 11.3-4 [106].

In the above systolic array implementation the circular elements have more complex computations than do the square elements. The circular elements require a square-root operation [see (11.1-6a) and (11.1-6b)] whereas the square elements only require multiplication and add operations. A square-root free implementation of the Givens rotation algorithm has been developed [106, 108, 109]. It is only when this square-root free algorithm is implemented that the diagonal connections between the circular elements indicated in Figure 11.3-2 are needed. For the square-root algorithm given above they are not needed [106; see Figure 11.3-2b].

For the sidelobe canceler problem of Section 4.4 we are interested in calculating the least-squares residual and not the weight vector. McWhirter [89, 94] has developed an efficient triangular systolic array (like that of Figure 11.3-2) which computes the residue without the need for the back-substitution systolic array of Figure 11.3-4. The residual is obtained as the output of the bottom circular element of the systolic array. Both square-root and square-root free

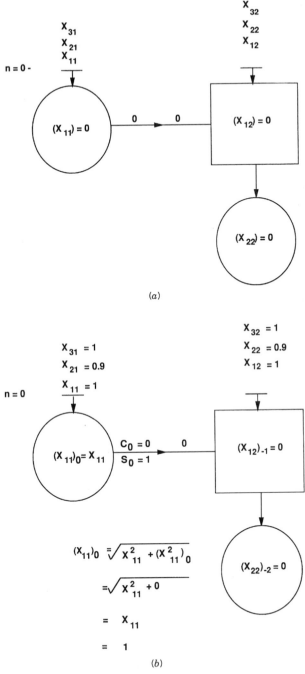

Figure 11.3-3 Step-by-step calculations involved in the Givens orthonormalization of T using a systolic array: (*a*) Initialization. (*b*) After first clock cycle at $n = 0$. (*c*) After second clock cycle at $n = 1$. (*d*) After third clock cycle at $n = 2$. (*e*) After fourth clock cycle at $n = 3$. (*f*) After fifth and final clock cycle at $n = 4$.

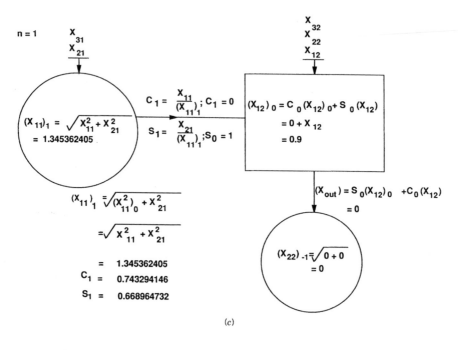

(c)

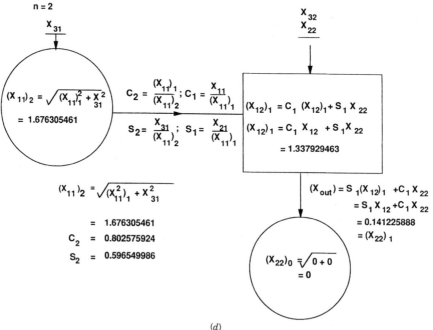

(d)

Figure 11.3-3 (*Continued*)

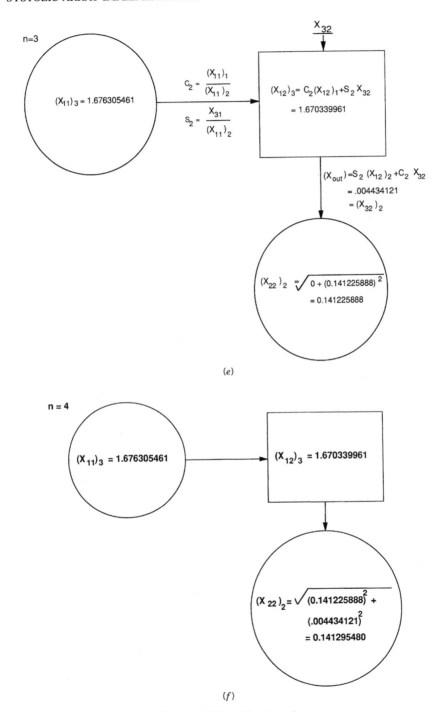

(e)

(f)

Figure 11.3-3 (*Continued*)

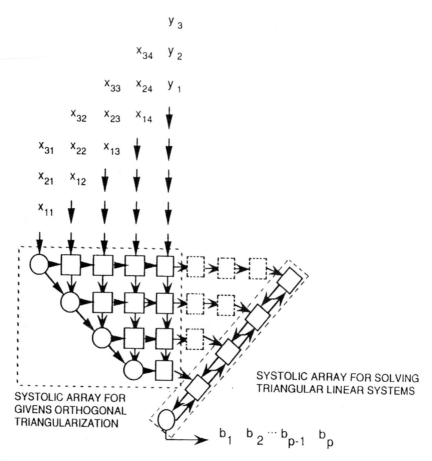

Figure 11.3-4 Systolic array implementation of least squares estimate. After Kung, H. T. and W. M. Gentleman, "Matrix Triangularization by Systolic Arrays," Proceedings of the SPIE, 1981, Vol. 298, Real Time Signal Processing IV.)

versions have been developed [89, 94]. These implementations offer significant reduction in hardware complexity and are less sensitive to round-off errors.

An alternative to using the above square-root algorithm or the square-root free algorithm is to use the Coordinate Rotation Digital Computer (CORDIC) algorithm [110] to perform the Givens transformations on the matrix T. This CORDIC algorithm will perform the Givens rotations just using additions and subtractions, no square-root operation being needed. The CORDIC algorithm provides a simple means for doing trigonometric calculations. It is the algorithm used on Hewlett-Packard calculators for trigonometric computations [111]. Given the two-dimensional vector (t_{11}, t_{21}) of Figure 11.1-1, the CORDIC algorithm will, in the first circular box, determine the rotated vector $\bar{t}_1$ of (11.1-12) and the angle θ_1 through which the vector was rotated, this angle being mathematically given by (11.1-6a) or (11.1-6b). It is necessary to know this angle θ_1 in order to rotate the remaining two-dimensional vector pairs (t_{1i}, t_{2i}) of the first two rows of the matrix T by θ_1 to complete the orthonormal transformation of the first two rows of the matrix T by G_1. The CORDIC algorithm will perform these latter rotations in the square boxes on the first row. The pair of numbers (t_{11}, t_{21}) forming the two-dimensional vector $\bar{t}_0$ of Figure 11.1-1 can be thought of as a complex number having a real part t_{11} and an imaginary part t_{21}. The CORDIC algorithm can just as well be applied to complex numbers to determine their magnitude and angle or to rotate them through some arbitrary angle θ.

11.3.2 CORDIC Algorithm

The process of using the CORDIC algorithm to determine the magnitude and angle of a complex number $x + jy$ is called *vectoring* [110]. The process of using the CORDIC algorithm to rotate a complex number $x + jy$ by some angle θ is called *rotation* [110]. Without loss of generality, in the following we will describe these two CORDIC algorithms as applied to complex numbers.

Assume that we are given a complex number $x_1 + jy$ having a magnitude R_1 and an angle θ_1, that is,

$$x_1 + jy_1 = R_1 \angle \theta_1 \tag{11.3-6}$$

The CORDIC *vectoring* algorithm rotates this complex number into the real axis to form the real number $R_1 \angle 0°$. In the process it determines the angle θ_1 of the complex number and its magnitude R_1. The rotation is carried out through a series of $n - 1$ small rotations of magnitude α_i for $i = 2, 3, \ldots, n - 1$, these rotations only requiring additions and subtractions. The basis for the CORDIC algorithm is that any angle θ can be approximated by the sum of such smaller clockwise or counterclockwise rotations α_i, specifically, by

$$\theta_1 \doteq \theta_n' = \sum_{i=2}^{n} r_i \alpha_i \qquad n \leq 2 \tag{11.3-7}$$

where

$$r_i = \pm 1 \tag{11.3-7a}$$

How r_i and α_i are determined will be given in the following paragraphs. Equation (11.3-7) applies as long as $|\theta_1| \leq 90°$. If

$$180° \geq |\theta_1| > 90° \tag{11.3-8}$$

it is easy to make $|\theta_1| \leq 90°$ by a simple rotation of α by $\pm 90°$ depending on whether respectively $\theta_1 < 0°$ or $\theta_1 > 0°$. This $90°$ rotation is illustrated in Figure 11.3-5. Mathematically, if $x_1 < 0$ and $y_1 > 0$, then $0 > 90°$ and the real

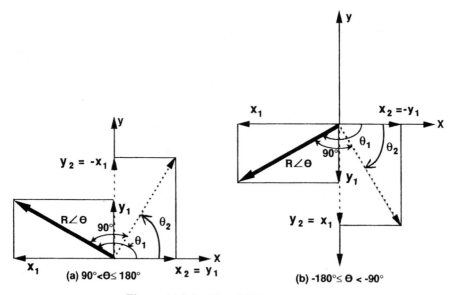

Figure 11.3-5 First CORDIC rotation.

and imaginary parts of the 90° rotated complex number are given by

$$x_2 = y_1 \qquad\qquad\qquad (11.3\text{-}9a)$$

$$y_2 = -x_1 \qquad\qquad\qquad (11.3\text{-}9b)$$

Thus the 90° rotated vector becomes

$$R_2 \angle \theta_2 = y_1 - jx_1 \qquad\qquad\qquad (11.3\text{-}10)$$

If $x_1 < 0$ and $y_1 < 0$, then $\theta_2 < -90°$ and the coordinates of the rotated vector complex number become

$$x_2 = -y_1 \qquad\qquad\qquad (11.3\text{-}11a)$$

$$y_2 = x_1 \qquad\qquad\qquad (11.3\text{-}11b)$$

Thus the 90° rotated complex number becomes

$$R_2 \angle \theta_2 = -y_1 + jx_1 \qquad\qquad\qquad (11.3\text{-}12)$$

If $|\theta_1| \leq 90°$, then no 90° rotation is necessary. For this case, for convenience, we determine the x_2 and y_2 coordinates by simply defining them to be identical

TABLE 11.3-1. CORDIC Algorithm Angle Rotations α_1 and Magnitude Changes K_i' That Result From Them

i	α_i	K_i'	K_i
1	(90°)		
2	45°		
3	26.56505118	1.414213562	1.414213562
4	14.03624347	1.118033989	1.581138830
5	7.125016349	1.030776406	1.629800601
6	3.576334375	1.007782219	1.642484066
7	1.789910608	1.001951221	1.645688915
8	0.895173710	1.000488162	1.646492278
9	0.447614171	1.000122063	1.646693254
10	0.223810500	1.000030517	1.646743506
11	0.111905677	1.000007629	1.646756069

to x_1 and y_1. Specifically,

$$x_2 = x_1 \tag{11.3-13a}$$

$$y_2 = y_1 \tag{11.3-13b}$$

Also then

$$R_2 \angle \theta_2 = R_1 \angle \theta_1 \tag{11.3-14}$$

The angle of rotation α_i is given by

$$\alpha_i = \tan^{-1} 2^{-(i-2)} \qquad i = 2, 3, \ldots, n \tag{11.3-15}$$

Table 11.3-1 gives a list of values of α_i for $i = 1, \ldots, 11$. The direction of the rotation of α_i is determined by the sign of r_i; see (11.3-7a). We shall show how to determine r_i shortly. The magnitude R of the complex number given by (11.3-6) is determined using the following.

CORDIC Vectoring Algorithm: For $i = 2, \ldots, n$, if

$$y_i \geq 0 \tag{11.3-16}$$

then

$$r_i = +1 \tag{11.3-17}$$

If

$$y_i < 0 \tag{11.3-18}$$

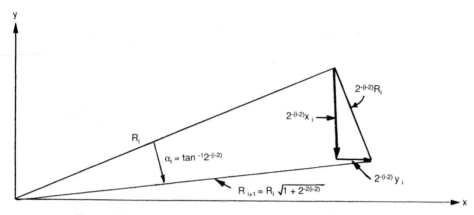

Figure 11.3-6   The ith CORDIC rotation by angle a_i for $r_i = 1$.

then

$$r_i = -1 \qquad (11.3\text{-}19)$$

Where

$$x_{i+1} = x_i + r_i 2^{-(i-2)} y_i \qquad (11.3\text{-}20a)$$

$$y_{i+1} = y_i - r_i 2^{-(i-2)} x_i \qquad (11.3\text{-}20b)$$

and

$$\theta'_{i+1} = \theta'_i + r_i \alpha_i \qquad (11.3\text{-}20c)$$

The above (11.3-20a) and (11.3-20b) rotation of $x_i + jy_i$ by α_i is illustrated in Figure 11.3-6. At $i = n$, if n is great enough, then

$$x_{n+1} \doteq K_{n+1} R_1 \qquad (11.3\text{-}21)$$

$$y_{n+1} \doteq 0 \qquad (11.3\text{-}22)$$

and

$$\theta_2 \doteq \theta'_{n+1} \qquad (11.3\text{-}23)$$

where

$$K_{n+1} = \prod_{i=2}^{n} K'_{i+1} \qquad (11.3\text{-}24)$$

and

$$K'_{i+1} = (1 + 2^{-2(i-2)})^{1/2} \qquad (11.3\text{-}25)$$

The above CORDIC vectoring algorithm given by (11.3-16) to (11.3-25) holds for θ_2 negative.

Table 11.3-1 tabulates K_1' for $i = 3, \ldots, 11$. We see from (11.3-21) that x_{n+1} gives R_1 to within a known factor K_{n+1} given by (11.3-24).

By way of examples let

$$x = x_1 = -0.788010754 \qquad (11.3\text{-}26)$$

$$y = y_1 = 0.615661475 \qquad (11.3\text{-}27)$$

It is apparent that $\theta_1 > 90°$, it being $142°$. Thus we first have to make $\theta_1 \leq 90°$ using (11.3-9a) and (11.3-9b) to yield

$$x_2 = 0.615661475 \qquad (11.3\text{-}28)$$

$$y_2 = 0.788010754 \qquad (11.3\text{-}29)$$

Now $\theta_2 = 142° - 90° = 52°$. Stepping through the CORDIC vectoring algorithm given above, we obtained the results of Table 11.3-2 for $i = 2, \ldots, 11$. This table indicates that for $i = 11$ we get $\theta_2 \doteq \theta_{11}' = 51.96646609$, which is accurate to 0.064% for this 11-bit representation of θ_1. Note that $\theta_2 \doteq \theta_{11}'$ can be represented by the sequence of ± 1's given by r_i for $i = 2, 3, \ldots, n$. For the example given above this sequence is given by $(+1, +1, -1, -1, +1, -1, -1, +1, +1, +1)$. The magnitude of $x_1 + jy_1$ equals 1.00000000. The magnitude of $x_{11} + jy_{11}$ equals 1.646756071. Dividing by K_{11} of Table 11.3-2 gives 1.000000001. Hence the magnitude $R_{11} = |x_{11} + jy_{11}|$ is accurate to 0.0000001%.

CORDIC Rotation Algorithm: For this algorithm the vector

$$R_1 \angle \phi_1 = x_1 + jy_1 \qquad (11.3\text{-}30)$$

is to be rotated by the angle θ_1. If $180° \geq |\theta_1| \geq 90°$, then we first have to rotate the complex number $R_1 \angle \phi_1$ given by (11.3-30) by $90°$ to produce the vector $R_2 \angle \phi_2$, which only has to be rotated by $\theta_2 = \theta_1 - 90°$. To rotate the complex number of (11.3-30) by $90°$, a procedure equivalent to that shown in Figure 11.3-5 or given by (11.3-9a) and (11.3-9b) or (11.3-11a) and (11.3-11b) is used. Next the rotation by θ_2 is performed by carrying out small rotations of α_i for $i = 2, 3, \ldots, n$ using the following:
   If

$$\theta_2 - \theta_i' \geq 0° \qquad (11.3\text{-}31)$$

then

$$r_i = +1 \qquad (11.3\text{-}32)$$

TABLE 11.3-2. CORDIC Vectoring Algorithm Example for $x_1 = -0.788010754$, $y_1 = 0.615661475$, $\theta_1 = 142°$, and $R_1 = 1$

i	X_i	Y_i	r_i	α_i	θ_i'	K_i'	K_i
1	−0.788010754	0.615661475	—	$(90°)^a$	$(90°)^a$	—	1.414213562
2	0.615661475	0.788010754	+1	45°	45°	1.414213562	1.581138830
3	1.403672229	0.172349278	+1	26.56505118	71.56505118	1.118033989	1.629800601
4	1.489846869	−0.529486836	−1	14.03624347	57.52880771	1.030776406	1.642484066
5	1.622218577	−0.157025118	−1	7.125016349	50.40379136	1.007782219	1.645688915
6	1.641846718	0.045752204	+1	3.576334375	53.98012574	1.001951221	1.646492278
7	1.644706231	−0.05863216	−1	1.789910608	52.19021513	1.000488162	1.646693254
8	1.64648206	−.005466146	−1	0.895173710	51.29504142	1.000122063	1.646743506
9	1.64656615	0.020260154	+1	0.447614171	51.74265559	1.000030517	1.646756069
10	1.646726897	0.007396337	+1	0.223810500	51.9646609	1.000007629	
11	1.646755789	0.00096381	+1	0.111905677			

^{a} Here $i = 1$ row needed because $\theta > 90°$ so that $90°$ rotation shown in parentheses is needed.

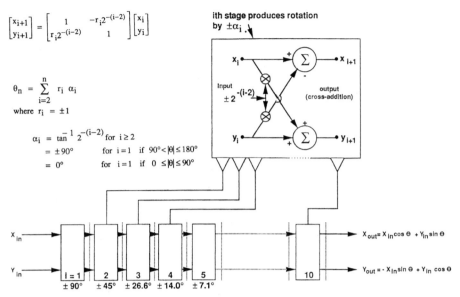

$$\begin{bmatrix} x_{i+1} \\ y_{i+1} \end{bmatrix} = \begin{bmatrix} 1 & -r_i 2^{-(i-2)} \\ r_i 2^{-(i-2)} & 1 \end{bmatrix} \begin{bmatrix} x_i \\ y_i \end{bmatrix}$$

ith stage produces rotation by $\pm \alpha_i$.

$$\theta_n = \sum_{i=2}^{n} r_i \alpha_i$$

where $r_i = \pm 1$

$$\alpha_i = \tan^{-1} 2^{-(i-2)} \text{ for } i \geq 2$$
$$= \pm 90° \quad \text{for } i = 1 \text{ if } 90° < |\theta| \leq 180°$$
$$= 0° \quad \text{for } i = 1 \text{ if } 0 \leq |\theta| \leq 90°$$

X_i → Σ → X_{i+1}

Input $\pm 2^{-(i-2)}$

output (cross-addition)

Y_i → Σ → Y_{i+1}

X_{in} → | i = 1 | 2 | 3 | 4 | 5 | ... | 10 | → $X_{out} = X_{in} \cos \theta + Y_{in} \sin \theta$

Y_{in} → → $Y_{out} = -X_{in} \sin \theta + Y_{in} \cos \theta$

$\pm 90°$ $\pm 45°$ $\pm 26.6°$ $\pm 14.0°$ $\pm 7.1°$

Figure 11.3-7 Circuit implementation of CORDIC rotation algorithm. (For simplicity multiplication of rotated output vector by factor K_{n+1} is omitted in writing rotated output vector (x_{out}, y_{out}).)

If

$$\theta_2 - \theta'_i < 0 \qquad (11.3\text{-}33)$$

then

$$r_i = -1 \qquad (11.3\text{-}34)$$
$$x_{i+1} = x_i - r_i 2^{-(i-2)} y_i \qquad (11.3\text{-}35)$$
$$y_{i+1} = y_i + r_i 2^{-(i-2)} x_i \qquad (11.3\text{-}36)$$
$$\theta'_{i+1} = \theta'_i + r_i \alpha_i \qquad i = 2, 3, \dots \qquad (11.3\text{-}37)$$

When n is large enough, $\theta_2 - \theta_{n+1} \doteq 0$ and x_{n+1} and y_{n+1} are the desired rotated vector real and imaginary coordinates multiplied by the known factor K_{n+1} of (11.3-24). Figure 11.3-7 gives a circuit for rotating $x_1 + jy_1$ by θ_1 from the sequence of α_i rotations given above. If the CORDIC r_i's that represent θ_1 are known in advance, the rotation can be performed just using (11.3-35) and (11.3-36) without having to use (11.3-37), which is just needed to determine the values of the r_i's from (11.3-31) to (11.3-34). This is exactly the situation one has for the Givens systolic array for Figure 11.3-2a. For this systolic array the CORDIC vectoring algorithm is used in the first circular element of each row to determine the angle θ_1 by which two-dimensional vectors in the square boxes of the same row have to be rotated. Because the CORDIC vectoring algorithm is used to determine θ_1, one has θ_1 in terms of the r_i's. Thus the rotation by θ_1 in the square boxes along the same row can be easily performed. Note that $\theta_2' = r_2\alpha_2$. The above CORDIC rotation algorithm given by (11.3-31) to (11.3-37) holds for θ_2 negative. The convention being used is a positive θ_2 results in a counterclockwise rotation.

The known factor K_{n+1} with which the rotated vectors are multiplied can be divided out by using a binary approximation for the necessary division so that just adds and subtracts are needed to perform this division. Alternately, it is possible to use a CORDIC algorithm that does not result in the rotated vector being multiplied by the known factor K_{n+1}; see [112]. This compensated CORDIC Rotation algorithm, which leaves the magnitude of the complex number unchanged after rotations, is considerably more complex than the uncompensated one given earlier. However, this may not be a significant disadvantage if serial arithmetic is used because usually the arithmetic part of serial computation is generally a small part of the total.

In the above discussion the entries of the matrix T_0 were assumed to be real and the Givens orthonormal transformation for real numbers was presented. When the t_{ij}'s or the y_i's (or both) are complex numbers as the latter would be for a radar data, the Givens orthonormal transformation of (11.1-7) is expressed as [83, 89, 113]

$$\begin{bmatrix} c_1 & s_1^* \\ -s_1 & c_1 \end{bmatrix} \tag{11.3-39}$$

where (11.1-6a) and (11.1-6b) become

$$c_1 = \frac{|t_{11}|}{(|t_{11}|^2 + |t_{21}|^2)^{1/2}} \tag{11.3-39a}$$

$$s_1 = \left(\frac{t_{21}}{t_{11}}\right)c_1 \tag{11.3-39b}$$

A CORDIC algorithm exists for rotating complex numbers [90, 91]. A wafer-scale implementation of the systolic array of Figure 11.3-2 was implemented by Rader at Lincoln Laboratories, MIT for the case where the data are complex [90–92]. This unit handled a matrix T having 64 columns. It was the size of a Walkman compact disc and consumed only 10 W.

12

HOUSEHOLDER ORTHONORMAL TRANSFORMATION '

In the preceding chapter we showed how the elementary Givens orthonormal transformation triangularized a matrix by successfully zeroing out one element at a time below the diagonal of each column. With the Householder orthonormal transformation all the elements below the diagonal of a given column are zeroed out simultaneously with one Householder transformation. Specifically, with the first Householder transformation H_1, all the elements below the first element in the first column are simultaneously zeroed out, resulting in the form given by (11.1-4). With the second Householder transformation H_2 all the elements below the second element of the second column are zeroed out, and so on.

The Householder orthonormal transformation requires fewer multiplies and adds than does the Givens transformation in order to obtain the transformed upper triangular matrix of (10.2-8) [103]. The Householder transformation, however, does not lend itself as easily to a systolic array parallel-processor type of implementation as did the Givens transformation. Hence the Householder may be preferred when a centralized processor is to be used but not if a custom systolic signal processor is used.

12.1 COMPARISON OF HOUSEHOLDER AND GIVENS TRANSFORMATIONS

Let us initially physically interpret the Householder orthonormal transformation as transforming the augmented matrix T_0 to a new coordinate system, as we did for the Givens transformations. For the first orthonormal Householder transformation H_1 the s rows are unit vectors, designated as h_i for the ith row, onto

315

which the columns of T_0 are projected. The first-row vector h_1 is chosen to line up with the vector formed by the first column of T_0, designated as t_1. Hence, the projection of t_1 onto this coordinate yields a value for the first coordinate in the transformed space equal to the full length of t_1, that is, equal to $\| t_1 \|$. The remaining $s - 1$ unit row vectors of H_i, designated as h_i, $i = 2, \ldots, s$, are chosen to be orthonormal to h_1. As a result they are orthonormal to t_1 since h_1 lines up with t_1. Hence the projections of t_1 onto the h_i, $i = 2, \ldots, s$, are all zero. As a result $H_1 T_0$ produce a transformed matrix of the form given by (11.1-4) as desired. This is in contrast to the simple Givens transformations $G_1, G_2, \ldots, G_{s-1}$ of the left-hand side of (11.1-4), which achieves the same outcome as H_1 does but in small steps. Here, G_1 first projects the column vector t_1 of T_0 onto a row space that is identical to that of the column space of T_0 with the exception of the first two coordinates. The first two coordinates, designated in Figure 11.1-2a as the x, y coordinates, are altered in the new coordinate system defined by G_1. The first of these two new coordinates, whose direction is defined by the first-row unit vector g_1, of G_1, is lined up with the vector formed by the first two coordinates of t_1, designated as $\bar{t}_0$; see (11.1-1) and Figure 11.1-2a. The second of these new coordinates, whose direction is defined by the second-row unit vector g_2 of G_1, is orthogonal to g_1 and in turn $\bar{t}_0$, but in the plane defined by the first two coordinates of t_1, that is, the x, y plane. As a result $G_1 T_0$ gives rise to the form given by (11.1-2).

The second Givens transformation G_2 projects the first-column vector $(t_1)_1$ of $G_1 T_0$ onto a row space identical to that of $(t_1)_1$ except for the first and third coordinates. The first-row unit vector $(g_1)_2$ of G_2 is lined up with the direction defined by the vector formed by the first and third coordinates of $G_1 T_0$. This vector was designated as $\bar{t}_2$; see (11.1-16) and Figures 11.1-2b,c. From the definition of $\bar{t}_2$ recall also that it is lined up with the first three coordinates of t_1. The third-row unit vector $(g_3)_2$ is lined up in a direction orthogonal to $(g_1)_2$ in the two-dimensional space formed by the first and third coordinates of the vector $(t_1)_2$; see Figures 11.1-2b,c. As a result the projection of $(t_1)_1$ onto the row space of G_2 gives rise to the form given by (11.1-3). Finally, applying the third Givens transformation G_3 yields the desired form of (11.1-4) obtained with one Householder transformations H_1. For this example T_0 is a 4×3 matrix, that is, $s = 4$. For arbitrary s, $s - 1$ simple Givens transformations $G_1, \ldots, G_{s-1}$ achieve the same form for the transformed matrix that one Householder transformation H_1 does, that is,

$$H_1 \equiv G_{s-1} \cdots G_2 G_1 \qquad (12.1\text{-}1)$$

Elements on the right and left hand sides of (12.1-1) will be identical except the sign of corresponding rows can be different if the unit row transform vectors of these transforms have opposite directions; see Section 13.1.

Let us recapitulate what the Givens transformations are doing. The first, G_1, projects T_0 onto a space whose first coordinate unit vector g_1 (defined by the first row of G_1; see Section 11.1) lines up with the direction of the two-dimensional vector formed by the first two coordinates of t_1. The second coordinate of the new space is along the unit vector g_2 (defined by the second row of G_1) orthogonal to the first coordinate but in the plane defined by the first two coordinates of t_1. The remaining coordinates of the space defined by G_1 are unchanged. In this transformed space the second coordinate of t_1 is zero; see (11.1-2). This in effect replaces two coordinates with one for the first two coordinates of t_1, the other coordinate being zero. The next Givens transformation now projects the transformed t_1, which is $(t_1)_1$, onto the row space of G_2. Here the first-row unit vector $(g_1)_2$ of G_2 is lined up with the vector formed by first and third coordinates of $(t_1)_1$, which is designated as $\bar{t}_2$ in (11.1-16) and Figures 11.1-2b,c. Again the second coordinate is orthogonal to the first but in the plane defined by the first and third coordinates of $(t_1)_1$. In this new coordinate system the second and the third coordinates of t_1 are zero; see (11.1-3). This simple Givens transformation again replaces two coordinates of $(t_1)_1$, the first and third, with one, the second being zero. The first and second simple Givens transformations together in effect line up the first row of $G_1 G_2$ with the vector formed by the first three coordinates of t_1. This continues until on the $s - 1$ Givens transformation the unit vector formed by the first row of $G_{s-1} \cdots G_2 G_1$ lines up with the vector formed by t_1 to produce the form given by (11.1-4). In a similar way, the first Householder transformation H_1 does in one transformation what $s - 1$ simple Givens transformations do. Similarly H_2 does in one transformation what $s - 2$ Givens transformations do; and so on.

12.2 FIRST HOUSEHOLDER TRANSFORMATION

We now develop in equation form the first Householder transformation H_1. A geometric development [80, 102, 109, 114] is used in order to give further insight into the Householder transformation. Consider the $s \times (m' + 1)$ dimensional augmented matrix T_0 whose columns are of dimension s. As done previously, we will talk of the columns as s-dimensional vectors in s-dimensional hyperspace. Designate the first column as the vector t_1. Form h' given by

$$h' = t_1 + \| t_1 \| i \tag{12.1-2}$$

where i is the unit vector along the x axis of this s-dimensional hyperspace, that is, the column space of T_0; see Figure 12.1-1. The vector i is given by (4.2-38) for $s = 3$. Physically let us view the first Householder transformation H_1 as a rotation of t_1. Then i in Figure 12.1-1 is the direction into which we want to rotate t_1 in order to have the transformed t_1, designated as $(t_1)_{1H}$, be

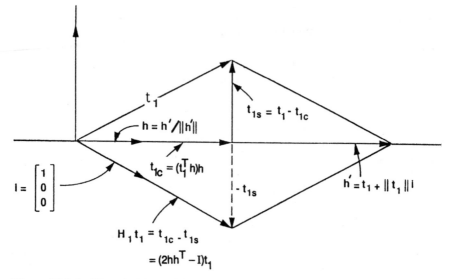

Figure 12.1-1   Householder reflection transformation H_1 for three-dimensional space.

given by

$$(t_1)_{1H} = H_1 t_1 = [\|\, t_1 \,\| \; 0 \quad 0 \quad \cdots \quad 0]^T \tag{12.1-3}$$

that is, in order for the first column of $H_1 T_0$ to have the form given by the first column of (11.1-4). In Figure 12.1-1, the vector h' forms the horizontal axis, which is not the x axis that is along the unit vector i. Because $H_1 t_1$ in Figure 12.1-1 is the mirror image of t_1 about this horizontal axis, the Householder transformation is called a "reflection."

Let h be the unit vector along h'; then

$$h = \frac{h'}{\|\, h' \,\|} \tag{12.1-4}$$

Let t_{1c} be the component of t_1 along the direction h. Then

$$t_{1c} = (t_1^T h) h \tag{12.1-5}$$

Designate t_{1s} as the component of t_1 along the direction perpendicular to h but in the plane formed by t_1 and i. Then

$$\begin{aligned} t_{1s} &= t_1 - t_{1c} \\ &= t_1 - (t_1^T h) h \end{aligned} \tag{12.1-6}$$

Then from geometry (see Figure 12.1-1)

$$
\begin{aligned}
H_1 t_1 &= t_{1c} - t_{1s} \\
&= (t_1^T h)h - [t_1 - (t_1^T h)h] \\
&= 2(t_1^T h)h - t_1 \\
&= 2h(t_1^T h) - t_1 \\
&= 2hh^T t_1 - t_1 \\
&= [2hh^T - I]t_1
\end{aligned}
\tag{12.1-7}
$$

Hence

$$
H_1 = 2hh^T - I
\tag{12.1-8}
$$

By picking h as given by (12.1-2) and (12.1-4), the vector t_1 is rotated (actually reflected) using (12.1-8) onto i as desired. After the reflection of t_1 the vector has the same magnitude as it had before the reflection, that is, its magnitude is given by $\| t_1 \|$. Moreover, as desired, the magnitude of the first element of the transformed vector $(t_1)_{1H}$ has the same magnitude value as the original vector t_1 with the value of the rest of the coordinate elements being zero. Specifically

$$
H_1 t_1 = H_1 \begin{bmatrix} t_{11} \\ t_{21} \\ \vdots \\ t_{s1} \end{bmatrix} = \begin{bmatrix} \| t_1 \| \\ 0 \\ 0 \\ \vdots \\ 0 \end{bmatrix} = (t_1)_{1H}
\tag{12.1-9}
$$

where

$$
\| t_1 \| = (|t_{11}|^2 + |t_{11}|^2 + \cdots + |t_{s1}|^2)^{1/2}
\tag{12.1-9a}
$$

The subscript 1 on H is used to indicate that this is the first Householder transformation on the matrix T_0, with more Householder transforms to follow in order to zero out the elements below the remaining diagonal elements.

In the above we have only talked about applying the $s \times s$ Householder transformation H_1 matrix to the first column of T_0. In actuality it is applied to all columns of T_0. This is achieved by forming the product $H_1 T_0$. Doing this yields a matrix of the form given by the right-hand side of (11.1-4) as desired. In this way we have physically reflected the first column of T_0 onto the first coordinate of the column space of T_0 (the x coordinate) by the use of the Householder transformation H_1.

12.3 SECOND AND HIGHER ORDER HOUSEHOLDER TRANSFORMATIONS

The second Householder transformation H_2 is applied to $H_1 T_0$ to do a collapsing of the vector formed by the lower $s-1$ elements of the second column of $H_1 T_0$ onto the second coordinate. Designate as $(t_2')_{1H}$ the $(s-1)$-dimensional vector formed by the lower $s-1$ elements of the second column of $H_1 T_0$, the elements starting with the diagonal element and including all the elements below it. It is to be reflected onto the $(s-1)$-dimensional vector $j' = [100 \cdots 0]^T$ that physically is the unit vector in the direction of the second coordinate of the second column of the matrix $H_1 T_0$. When this is done, all the elements below the diagonal of the seond column are made equal to zero. The top element is unchanged. The Householder transformation needed to do this is an $(s-1) \times (s-1)$ matrix that we designate as H_2'. This transformation H_2' operates on the $(s-1) \times (m')$ matrix in the lower right-hand corner of the transformed $s \times (m'+1)$ matrix $H_1 T_0$.

From H_2' we now form the Householder transformation matrix H_2 given by

$$H_2 = \begin{bmatrix} I_1 & 0 \\ 0 & H_2' \end{bmatrix} \begin{matrix} \}1 \\ \}s-1 \end{matrix} \qquad (12.2\text{-}1)$$

where I_1 is the 1×1 identity matrix and H_2 is now an $s \times s$ matrix. When the orthonormal transformation matrix H_2 is applied to $H_1 T_0$, it transforms all elements below the diagonal elements of the second column to zero and causes the transformed second diagonal element to have a magnitude equal to the magnitude of the $(s-1)$-dimensional column vector $(t_2')_{1H}$ before the transformation and leaves the top element of the second column unchanged. This procedure is now repeated for the third column of the transformed matrix $H_2 H_1 T_0$ and so on, the ith such transformation being given by

$$H_i = \begin{bmatrix} I_{i-1} & -0 \\ 0 & H_i' \end{bmatrix} \begin{matrix} \}i-1 \\ \}s-i+1 \end{matrix} \qquad (12.2\text{-}2)$$

The last transformation needed to triangularize the augmented matrix T_0 so as to have it in the form given by (11.1-30) is $i = m'+1$.

It is important to point out that it is not necessary to actually compute the Householder transformation H_i's given above using (12.1-8). The transformed columns are actually computed using the second form of the $H_1 t_1$ given by (12.1-7), specifically by

$$H_1 t_1 = -t_1 + 2(t_1^T h)h \qquad (12.2\text{-}3)$$

In turn the above can be written as

$$H_1 t_1 = -t_1 + \frac{2(t_1^T h')h'}{h'^T h'} \qquad (12.2\text{-}4)$$

for obtaining the transformed t_1 with h' given by (12.1-2). For detailed computer Householder transformation algorithms the reader is referred to references 79 to 81. Some of these references also discuss the accuracy of using the Householder transformation for solving the least-squares estimate relative to other methods discussed in this book.

Some final comments are worth making relative to the Householder transformation. In the above we applied $m' + 1$ Householder transformations so as to transform the augmented matrix T_0 (11.1-25) into an upper triangular matrix form. Specifically, applying the orthonormal transformation formed by

$$F = H_{m'+1}H_{m'}\cdots H_2H_1 \tag{12.2-5}$$

to (11.1-25), we obtain the transformed matrix T_0' given by

$$
\begin{aligned}
T_0' = FT_0 &= F[T\,|\,Y_{(n)}] \\
&= [FT\,|\,FY_{(n)}]
\end{aligned}
$$

$$
= \begin{bmatrix}
U & | & Y_1' \\
\hline
0 & | & Y_2' \\
\hline
0 & | & Y_3'
\end{bmatrix}
\begin{matrix}
\}\,m' \\
\\
\}\,1 \\
\\
\}\,s-m'-1
\end{matrix}
\tag{12.2-6}
$$

$$\underbrace{}_{m'}\quad\underbrace{}_{1}$$

with $Y_3' = 0$. [The above is the same result as obtained in Section 4.3 using the Gram–Schmidt onthogonalization procedure; see (4.3-60).] The vector Y_3' is forced to be zero by the last Householder transformation $H_{m'+1}$ of (12.2-5). To determine $X_{n,n}^*$ it is actually not necessary to carry out the last Householder transformation with the result that some computational savings are achieved. The last Householder transformation does not affect Y_1' and in turn $X_{n,n}^*$; see (10.2-16). Making $Y_3' = 0$ forces the residue error $e(X_{n,n}^*)$ of (4.1-31) to be given by $(Y_2')^2$, that is,

$$e(X_{n,n}^*) = (Y_2')^2 \tag{12.2-7}$$

This property was pointed out previously when (11.1-31) and (11.3-3) were given and before that in Section 4.3; see, for example, (4.3-56). If the last Householder transformation $H_{m'+1}$ is not carried out, then Y_3' is not zero and

$$e(X_{n,n}^*) = (Y_2')^2 + \|\,Y_3'\,\|^2 \tag{12.2-8}$$

A similar comment applies for the Givens transformation of Chapter 11.

13

GRAM–SCHMIDT ORTHONORMAL TRANSFORMATION

13.1 CLASSICAL GRAM–SCHMIDT ORTHONORMAL TRANSFORMATION

The Gram–Schmidt orthonormalization procedure was introduced in Section 4.3 in order to introduce the orthonormal transformation F applied to the matrix T. The Gram–Schmidt orthonormalization procedure described there is called the classical Gram–Schmidt (CGS) orthogonilization procedure. The CGS procedure was developed in detail for the case $s = 3$, $m' = 2$, and then these results were extrapolated to the general case of arbitrary s and m'. In this section we shall develop the CGS procedure in greater detail.

The CGS procedure is sensitive to computer round-off errors. There is a modified version of the Gram-Schmidt procedure that is not sensitive to computer round-off errors. This is referred to as the modified Gram–Schmidt (MGS). After developing the general CGS results, we shall develop the MGS procedure. One might ask why explain the CGS procedure at all. It is better to start with a description of the CGS procedure because, first, it is simpler to explain, second, it makes it easier to obtain a physical feel for the Gram–Schmidt orthogonalization, and third, it provides a physical feel for its relationship to the Householder and in turn Givens transformations. Hence, we shall first again start with a description of the CGS procedure.

As described in Section 4.3, starting with the $m' + 1$ vectors $t_1, t_2, \ldots, t_{m'+1}$, we transform these to $m' + 1$ orthogonal vectors, which we designate as $q'_1, q'_2, \ldots, q'_{m'+1}$. Having this orthogonal set, the desired orthonormal set of Section 4.3 (see Figure 4.3-1) can be obtained by dividing q'_i by its magnitude $\| q'_i \|$ to form q_i. We start by picking the first vector q'_1 equal to t_1, that is

$$q'_1 = t_1 \tag{13.1-1}$$

At this point the matrix

$$T_0 = [t_1 \quad t_2 \quad \cdots \quad t_{m'+1}] \tag{13.1-2}$$

can be thought of as being transformed to the matrix

$$T'_{(1)} = [q'_1 \quad t_2 \quad \cdots \quad t_{m'+1}] \tag{13.1-3}$$

Next q'_2 is formed by making it equal to t_2 less its component along the direction of q'_1. From (4.2-41) it follows that the vector component of t_2 along t_1, or equivalently q'_1, is given by

$$t_{2c} = \frac{(q'^T_1 t_2)q'_1}{q'^T_1 q'_1} \tag{13.1-4}$$

Let

$$r'_{12} = \frac{(q'^T_1 t_2)}{q'^T_i q'_1} \tag{13.1-5}$$

Then (13.1-4) becomes

$$t_{2c} = r'_{12}q'_1 \tag{13.1-6}$$

In turn

$$q'_2 = t_2 - t_{2c} \tag{13.1-7}$$

or

$$q'_2 = t_2 - r'_{12}q'_1 \tag{13.1-8}$$

At this point T_0 can be thought of as being transformed to

$$T'_{(2)} = [q'_1 \quad q'_2 \quad t_3 \quad \cdots \quad t_{m'+1}] \tag{13.1-9}$$

Next q'_3 is formed by making it equal to t_3, less the sum of its two components along the directions of the first two orthogonal vectors q'_1 and q'_2. Using (4.2-41) again yields that the component of t_3 along q'_1 and q'_2 is given by

$$t_{3c} = \frac{(q'^T_1 t_3)q'_1}{q'^T_1 q'_1} + \frac{(q'^T_2 t_3)q'_2}{q'^T_2 q'_2} \tag{13.1-10}$$

In general let

$$r'_{ij} = \frac{q'^T_i t_j}{q'^T_i q'_i} \qquad (13.1\text{-}11)$$

for $j > i > 1$. Then (13.1-10) can be written as

$$t_{3c} = r'_{13} q'_1 + r'_{23} q'_2 \qquad (13.1\text{-}12)$$

In turn q'_3 becomes

$$q'_3 = t_3 - t_{3c} \qquad (13.1\text{-}13)$$

or

$$q'_3 = t_3 - r'_{13} q'_1 - r'_{23} q'_2 \qquad (13.1\text{-}14)$$

At this point T_0 can be thought of as being transformed to

$$T_{(3)} = [q'_1 \quad q'_2 \quad q'_3 \quad t_4 \quad \cdots \quad t_{m'+1}] \qquad (13.1\text{-}15)$$

The formation of $q'_4, q'_5, \ldots, q'_{m'+1}$ follows the same procedure as used above to form $q'_1, q'_2,$ and q'_3. As a result it is easy to show that, for $j > 2$,

$$q'_j = t_j - \sum_{i=1}^{j-1} r'_{ij} q'_i \qquad (13.1\text{-}16)$$

and after the $(m' + 1)$st step T_0 is given by

$$T'_{m'+1} = [q'_1 \quad q'_2 \quad \cdots \quad q'_{m'+1}] = Q' \qquad (13.1\text{-}17)$$

For simplicity, for now let us assume $m' = 2$. Doing this makes it easier to develop the results we seek. The form of the results obtained using $m' = 2$ apply for the general case of arbitrary m' to which the results can be easily generalized. From (13.1-1), (13.1-8) and (13.1-14) it follows that

$$t_1 = q'_1 \qquad (13.1\text{-}18a)$$
$$t_2 = r'_{12} q'_1 + q'_2 \qquad (13.1\text{-}18b)$$
$$t_3 = r'_{13} q'_1 + r'_{23} q'_2 + q'_3 \qquad (13.1\text{-}18c)$$

We can rewrite the above equations in matrix form as

$$[t_1 \quad t_2 \quad t_3] = [q'_1 \quad q'_2 \quad q'_3] \begin{bmatrix} 1 & r'_{12} & r'_{13} \\ 0 & 1 & r'_{23} \\ 0 & 0 & 1 \end{bmatrix} \qquad (13.1\text{-}19)$$

where it should be emphasized that the matrix entries t_i and q'_1 are themselves column matrices. In turn, the above can be rewritten as

$$T_0 = Q'R' \qquad (13.1\text{-}20)$$

where T_0 is given by (13.1-2), Q' is given by (13.1-17), and for $m' = 2$

$$R' = \begin{bmatrix} 1 & r'_{12} & r'_{13} \\ 0 & 1 & r'_{23} \\ 0 & 0 & 1 \end{bmatrix} \qquad (13.1\text{-}21)$$

We can orthonormalize Q by dividing each orthogonal vector q'_i by its magnitude $\| q'_i \|$ to form the unitary vector q_i

$$q_i = \frac{q'_i}{\| q'_i \|} \qquad (13.1\text{-}22)$$

and in turn obtain

$$Q = [q_1 \quad q_2 \quad \cdots \quad q_{m'+1}] \qquad (13.1\text{-}23)$$

Let the magnitudes of the q'_i be given by the diagonal matrix

$$D_0 = \text{Diag} \, [\| q'_1 \|, \| q'_2 \|, \ldots, \| q'_{m'+1} \|] \qquad (13.1\text{-}24)$$

It is easy to see that Q is obtained by postmultiplying Q' by D_0^{-1}. Thus

$$Q = Q'D_0^{-1} \qquad (13.1\text{-}25)$$

Using (13.1-24) it follows that (13.1-20) can be written as

$$T_0 = Q'D_0^{-1}D_0R' \qquad (13.1\text{-}26)$$

which on using (13.1-25) becomes

$$T_0 = QR \qquad (13.1\text{-}27)$$

where

$$R = D_0R' \qquad (13.1\text{-}27a)$$

Substituting (13.1-24) and (13.1-21) into (13.1-27a) yields

$$R = \begin{bmatrix} \| q'_1 \| & r''_{12} & r''_{13} \\ 0 & \| q'_2 \| & r''_{23} \\ 0 & 0 & \| q'_3 \| \end{bmatrix} \qquad (13.1\text{-}28)$$

where

$$r''_{ij} = r'_{ij} \, \| q'_i \|$$

(13.1-28a)

Using (13.1-11) yields

$$r''_{ij} = \frac{q'^T_i t_j}{q'^T_i q'_i} \, \| q'_i \|$$

(13.1-29)

which becomes

$$r''_{ij} = q^T_i t_j$$

(13.1-30)

for $j > i > 1$.

Multiplying both sides of (13.1-27) by Q^T yields

$$Q^T T_0 = R$$

(13.1-31)

where use was made of the fact that

$$Q^T Q = I$$

(13.1-32)

which follows because the columns of Q are orthonormal. In the above Q is an $s \times (m' + 1)$ matrix and I is the $(m' + 1) \times (m' + 1)$ identity matrix. Because also the transformed matrix R is upper triangular, see (13.1-28), we obtain the very important result that Q^T is the desired orthonormal transformation matrix F of (10.2-8) or equivalently of (12.2-6) for the matrix T_0, that is,

$$F = Q^T$$

(13.1-33)

Strictly speaking Q^T should be a square matrix to obey all the properties of an orthonormal matrix F given by (10.2-1) to (10.2-3). In fact it is an $s \times (m' + 1)$ matrix. This problem can be readily remedied by augmenting Q^T to include the unit vectors $q_{m'+2}$ to q_s when $s > m' + 1$, where these are orthonormal vectors in the s-dimensional hyperspace of the matrix T_0. These vectors are orthogonal to the $(m' + 1)$-dimensional column space of T_0 spanned by the $m' + 1$ vectors q_1 to $q_{m'+1}$. Thus to form F, Q of (13.1-23) is augmented to become

$$Q = [q_1 \quad q_2 \quad \cdots \quad q_{m'+1} \quad q_{m'+2} \quad \cdots \quad q_s]$$

(13.1-34)

The matrix Q' is similarly augmented to include the vectors $q'_{m'+2}$ to q'_s. Also the matrix D_0 is augmented to include $s - m' - 1$ ones after the $(m' + 1)$st terms. It shall be apparent, shortly, if it is not already apparent, that the vectors $q_{m'+2}$ to q_s actually do not have to be determined in applying the Gram–Schmidt orthonormalization procedures. Moreover, the matrices Q, Q', and D_0 do not in fact have to be augmented.

For arbitrary m' and $s \geq m'$, it is now easy to show that R of (13.1-28) becomes (12.2-6) with $Y_3' = 0$. Hence, in general, (13.1-31) becomes the desired upper triangular form given by (12.2-6) with $Y_3' = 0$, that is,

$$R = Q^T T_0 = \left[\begin{array}{c|c} U & Y_1' \\ \hline 0 & Y_2' \\ \hline 0 & 0 \end{array}\right] \begin{array}{l} \}m' \\ \}1 \\ \}s - m' - 1 \end{array} \qquad (3.1\text{-}35)$$

$$\underbrace{}_{m'} \quad \underbrace{}_{1}$$

For our $m' = 2$ example of (13.1-28) and Section 4.3 [see (4.3-12) and (4.3-24) to (4.3-29a)]

$$R = \begin{bmatrix} \|q_1'\| & r_{12}'' & r_{13}'' \\ 0 & \|q_2'\| & r_{23}'' \\ 0 & 0 & \|q_3'\| \end{bmatrix} = \begin{bmatrix} u_{11} & u_{12} & y_1' \\ 0 & u_{22} & y_2' \\ 0 & 0 & y_3' \end{bmatrix} \qquad (13.1\text{-}36)$$

$$u_{11} = \|q_1'\| \qquad (13.1\text{-}37a)$$
$$u_{12} = r_{12}'' \qquad (13.1\text{-}37b)$$
$$u_{22} = \|q_2'\| \qquad (13.1\text{-}37c)$$
$$y_1' = r_{13}'' \qquad (13.1\text{-}37d)$$
$$y_2' = r_{23}'' \qquad (13.1\text{-}37e)$$
$$y_3' = \|q_3'\| \qquad (13.1\text{-}37f)$$

Because the bottom $s - m' - 1$ rows of R are zero, we can drop these rows in R above. This would be achieved if we did not augment Q. However, even though in carrying out the Gram–Schmidt procedure we do not have to augment Q, Q', and D_0, for pedagogic reasons and to be consistent with our presentations in Section 4.3 and Chapters 10 to 12, in the following we shall consider these matrices as augmented.

From (13.1-27a)

$$R' = D_0^{-1} R \qquad (13.1\text{-}38)$$

Thus, we have that, in general, for arbitrary $s > m'$,

$$R' = \left[\begin{array}{c|c} U' & Y_1'' \\ \hline 0 & 1 \\ \hline 0 & 0 \end{array}\right] \begin{array}{l} \}m' \\ \}1 \\ \}s - m' - 1 \end{array} \qquad (3.1\text{-}39)$$

$$\underbrace{}_{m'} \quad \underbrace{}_{1}$$

where

$$U' = D^{-1}U \qquad (13.1\text{-}39a)$$

$$Y_1'' = D^{-1}Y_1' \qquad (13.1\text{-}39b)$$

and D is the unaugmented D_0 of (13.1-24) without its $(m' + 1)$st entry, that is,

$$D = \text{Diag}[\, \|q_1'\|, \|q_2'\|, \ldots, \|q_{m'}'\| \,] \qquad (13.1\text{-}40)$$

On examining the above CGS procedure, we see that in the process of transforming the columns of T_0 into an orthonormal set Q, we simultaneously generate the desired upper triangular matrix given by (12.2-6) with $Y_3' = 0$, that is, (13.1-35) or equivalently (4.3-60). For instance, for our $m' = 2$ example, (13.1-28) is obtained using the r_{ij}'' and $\|\, q_1' \,\|$ terms needed to orthonormalize T_0. It now remains to give a physical explanation for why this happens. How is the orthonormal matrix Q related to the Householder transformations and in turn the simple Givens transformation? Finally, how are the elements of R given by (13.1-35) obtained using the CGS procedure related to those of (12.2-6) obtained using the Householder or Givens transformation?

The answer is that the orthonormal transform matrix F obtained using the CGS procedure is identical to those obtained using the Givens and Householder transformations. Thus all three methods will give rise to identical transformed augmented matrices $T_0' = FT_0$. This follows from the uniqueness of the orthonormal set of transformation vectors f_i of F needed to put T_0' in the upper triangular form. Putting T_0' in upper triangular form causes the solution to be in the Gauss elimination form. This form is not unique but is unique if the orthonormal transformation F is used. That is, except if some of the unit row vectors of F are chosen to have opposite directions for the transforms in which case the signs of the corresponding rows of the transformed matrices T_0' will have opposite sign. (If the entries of T_0' can be complex numbers then the unit vectors of F can differ by an arbitrary phase.) Also the identicalness of the F's for the three transforms applies only to the first $m' + 1$ rows. The remaining $s - m' - 1$ rows can be quite arbitrary as long as they are orthonormal to the first $m' + 1$ rows.

Let us now explain what is physically happening with the CGS orthogonalization procedure. To do this we will relate the CGS orthogonalization procedure with the Householder transformation. For the first Householder transformation H_1, the first row unit vector h_1 is chosen to line up in the direction of the vector t_1 of the first row of T_0; see the beginning of Chapter 12. This is exactly the way q_1 is chosen for the CGS procedure; see (13.1-1) and (4.3-1) and Figure 4.3-1. Next, for the second Householder transformation H_2 [see (12.2-1)], the second row unit vector $(h_2)_2$ is chosen to be orthonormal to h_1 of H_1 and in the plane formed by the vectors t_1 and t_2 or equivalently h_1 and t_2 Again, this is exactly how q_2', and equivalently q_2, is picked in the CGS method; see (13.1-8) and the related discussion immediately before it. That

$(h_2)_2$ is in the plane of h_1, and t_2 follows from the fact that the transformation H_2 leads to the second column of $H_2H_1T_0$ having only its top two elements. This means that t_2 has only components along h_1 and $(h_2)_2$, the unit vector directions for which the first two elements of the second column of $H_2H_1T_0$ gives the coordinates. The unit vectors h_1 and $(h_2)_2$ become the unit vectors for the first two rows of F formed from the Householder transformations as given by (12.2-5). This follows from the form of H_i as given in (12.2-2). We see that H_i for $i \geq 2$ has an identity matrix for its upper-left-hand corner matrix. Hence h_1 of H_1 and $(h_2)_2$ of H_2 are not effected in the product that forms F from (12.2-5). As a result, the projections of $t_1, t_2, \ldots, t_s$ onto h_1 are not affected by the ensuing Householder transformations $H_2, \ldots, H_{m+1}$. It still remains to verify that h_1 and $(h_2)_2$ are orthogonal. The unit vector h_1 is along the vector t_1 of T_0. The unit vector $(h_2)_2$ has to be orthogonal to h_1 because t_1 does not project the component along $(h_2)_2$, the 2,1 element of $H_2H_1T_0$ being zero as is the 2,1 element of FT_0 when F is formed by the Householder transformations; see (12.2-1) and (12.2-6). By picking the first coordinate of the row vector $(h_2)_2$ to be zero, we forced this to happen. As a result of this zero choice for the first entry of the row matrix $(h_2)_2$, the first column element of H_1T_0 does not project onto $(h_2)_2$, the first column of H_1T_0 only having a nonzero element in its first entry, the element for which $(h_2)_2$ is zero.

Next, for the Householder transform H_3 of (12.2-2), the third row unit vector $(h_3)_3$ is chosen to be orthonormal to h_1 and $(h_2)_2$ but in the space formed by h_1, $(h_2)_2$, and t_3 or equivalently t_1, t_2, and t_3. Again, this is exactly how q_3' is picked with the CGS procedure; see (13.1-14). In this way see that the unit row vectors of F obtained with Householder transformation are identical to the orthonormal column vectors of Q, and in turn row vectors of Q^T, obtained with the CGS procedure.

In Section 4.2 the Givens transformation was related to the Householder transformation; see (12.1-1) and the discussion just before and after it. In the above paragraphs we related the Householder transformation to the CGS procedure. We now can relate the Givens transformation directly to the CGS procedure. To do this we use the simple example of (4.2-1), which was the example used to introduce the CGS procedure in Section 4.3. For this case only three Givens transformations G_1, G_2, and G_3 are needed to form the upper triangular matrix T as given by (4.3-29). As indicated in Chapter 12, each of these Givens transformations represents a change from the immediately preceding orthogonal coordinate system to a new orthogonal coordinate system with the change being only in two of the coordinates of one of the unit vector directions making up these coordinate systems. Specifically, each new coordinate system is obtained by a single rotation in one of the planes of the s-dimensional orthogonal space making up the columns of the matrix T. This is illustrated in Figures 13.1-1 to 13.1-3.

We saw above that the CGS procedure forms during the orthogonalization process the upper triangular matrix R' of (13.1-21) and (13.1-39), which is related to the upper triangular matrix R [see (13.1-27a)], which becomes equal

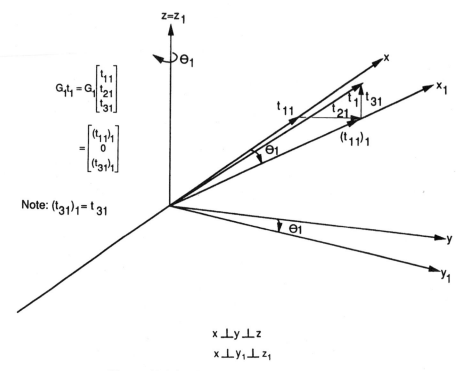

Figure 13.1-1 Givens transformation G_1 of t_1.

to FT_0 of (12.2-6) and (3.1-35) in general. As we shall see now, the CGS orthogonalization generation of R' gives us a physical significance for R' and U' and in turn for R and U.

Examination of (13.1-18a) to (13.1-18c) and (13.1-19) give us a physical explanation for why the CGS orthogonalization procedure produces an upper triangular R' and in turn upper triangular R. [The development given in Section 4.3, which introduced the CGS procedure, also gave us a physical feel for why R is upper triangular; see specifically (4.3-24) to (4.3-29).] Note that the orthogonal vectors $q'_1, \ldots, q'_i$ are chosen to form t_i; that is t_i is the weighted sum of $q'_1, \ldots, q'_i$ with the weight for q_i equaling 1; see (13.1-18a) to (13.1-18c). The ith column of R' gives the coefficients of the weightings for the q'_j's, $j = 1, 2, \ldots, m' + 1$ for forming t_i from the q'_i's; see (13.1-19). Because t_i is only formed by the weighted sum of $q'_1, \ldots, q'_i$, the coefficients of $q'_{i+1}, \ldots, q'_{m'+1}$ are zero, forcing the elements below the diagonal of the ith column to be zero, and in turn forcing R' to be upper triangular. Furthermore, physically, the coefficients of the ith column of R' give us the amplitude change that the orthogonal vectors $q'_1, \ldots, q'_i$ need to have to form t_i; see (13.1-18a) to (13.1-19). Worded another way, the i, j element r'_{ij} of R' times $\| q'_j \|$ gives the component of t_j along the direction q_j. Thus we now have a physical feel for the entries of R' and in turn U' [see (13.1-39)]. To get a physical interpretation of

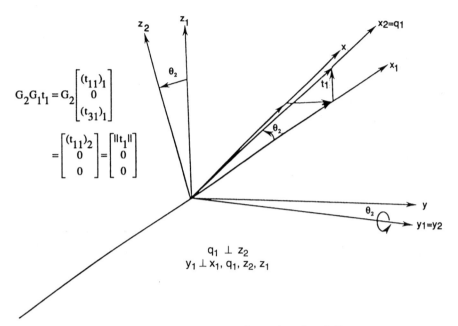

$$G_2G_1t_1 = G_2 \begin{bmatrix} (t_{11})_1 \\ 0 \\ (t_{31})_1 \end{bmatrix}$$

$$= \begin{bmatrix} (t_{11})_2 \\ 0 \\ 0 \end{bmatrix} = \begin{bmatrix} \|t_1\| \\ 0 \\ 0 \end{bmatrix}$$

$$q_1 \perp z_2$$
$$y_1 \perp x_1, q_1, z_2, z_1$$

Figure 13.1-2 Givens transformation G_2 of G_1t_1.

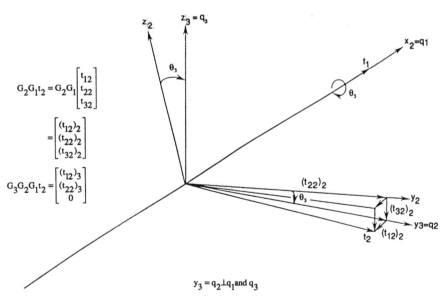

$$G_2G_1t_2 = G_2G_1 \begin{bmatrix} t_{12} \\ t_{22} \\ t_{32} \end{bmatrix}$$

$$= \begin{bmatrix} (t_{12})_2 \\ (t_{22})_2 \\ (t_{32})_2 \end{bmatrix}$$

$$G_3G_2G_1t_2 = \begin{bmatrix} (t_{12})_3 \\ (t_{22})_3 \\ 0 \end{bmatrix}$$

$$y_3 = q_2 \perp q_1 \text{ and } q_3$$

Figure 13.1-3 Givens transformation G_3 of $G_2G_1t_2$.

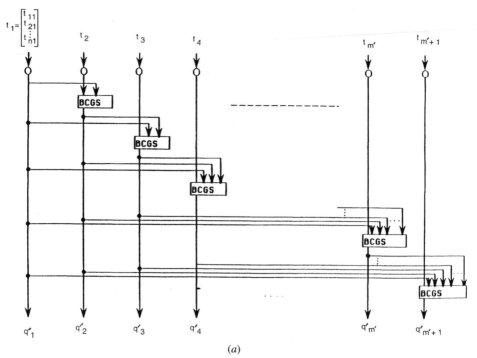

(a)

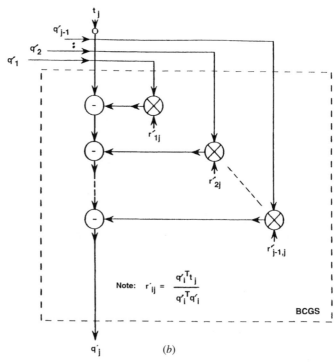

(b)

the elements of R we refer back to (4.3-24) to (4.3-26). We see that the entries of the ith column give us the magnitudes of the components of the ith column of T_0 along the unit vectors q_1, q_2, q_3. This physical interpretation of R also follows, after a little reflection, from the fact that the ith row of R is just $\| q_i' \|$ larger than that of R'; see (13.1-28a) and (13.1-28). Further insight can be gained by studying the circuit diagram for the CGS orthogonalization given in Figure 13.1-4.

Having obtained U and Y_1' from R, which is obtained using the CGS method, it is possible to obtain the least-square estimate $X_{n,n}^*$ using (10.2-16). It is worth noting that it is not necessary to determine R in order to determine U and Y_1' for (10.2-16). The least-squares estimate can be obtained directly from R'. Specificaly multiplying both sides of (10.2-16) by D^{-1} yields

$$D^{-1}UX_{n,n}^* = D^{-1}Y_1' \tag{13.1-41}$$

Applying (13.1-39a) and (13.1-39b) yields

$$U'X_{n,n}^* = Y_1'' \tag{13.1-42}$$

The matrices U' and Y_1'' are obtained directly from R'; see (13.1-39). The least-squares estimate $X_{n,n}^*$ is obtained directly from (13.1-42) using the back-substitution method. Hence R' obtained in the CGS orthogonalization of T_0 can be used directly to obtain the least-squares estimate $X_{n,n}^*$ without having to calculate R. The least-squares residue error is still given by

$$e(X_{n,n}^*) = (Y_2')^2 = \|q'(m' + 1)\| \tag{13.1-43}$$

where use was made of the generalized form of (13.1-28) with $\|q_3'\|$ replaced by $\|q_{m'+1}'\|$.

13.2 MODIFIED GRAM-SCHMIDT ORTHONORMAL TRANSFORMATION

As indicated previously, the Gram–Schmidt procedure described is referred to as the classical Gram–Schmidt procedure [76, 80–82, 101, 102, 115–118, 139]. There is another form called the modified Gram–Schmidt (MGS) procedure [76, 80–82, 101, 102, 115–118, 139]. The modified Gram–Schmidt, which we shall

◄ ───

Figure 13.1-4 (*a*) Circuit implementation of classical Gram–Schmidt (CGS) orthogonalization. Box BCGS generates from vector t_j an output vector q_j' orthogonal to q_1' to q_{j-1}'; see (13.1-14) and (13.1-16). (*b*) Basic classical Gram–Schmidt (BCGS) circuit. It generates from t_j the vector q_j' that is orthogonal to vectors $q_1', q_2', \ldots, q_{j-1}'$.

describe in this section, gives the same answers as the classical Gram–Schmidt procedure if there are no computer round-off errors. However, when computer round-off errors are present, the answers obtained are not the same. Answers obtained with the CGS method are much less accurate. This is illustrated very clearly in reference 101, using the following example for the augmented matrix T_0:

$$T_0 = \begin{bmatrix} 1 & 1 & 1 \\ 1.01 & 1 & 1 \\ 1 & 1.01 & 1 \\ 1 & 1 & 1.01 \end{bmatrix} \tag{13.2-1}$$

A computer round-off to four significant figures was assumed. Using the classical Gram–Schmidt procedure to obtain the orthonormal matrix Q yields [101]

$$Q = \begin{bmatrix} 0.4988 & -0.6705 \times 10^{-2} & -0.7765 \times 10^{-2} \\ 0.5037 & -0.7075 & -0.8193 \\ 0.4988 & 0.7066 & 0.4107 \\ 0.4988 & -0.6705 \times 10^{-2} & 0.4001 \end{bmatrix} \tag{13.2-2}$$

From (13.2-2), $q_2^T q_3 = 0.8672$, which theoretically should be zero since q_2 and q_3 should be orthonormal. On the other hand, using the modified Gram–Schmidt yields [101]

$$Q = \begin{bmatrix} 0.4988 & -0.6705 \times 10^{-2} & 0.3918 \times 10^{-2} \\ 0.5037 & -0.7075 & -0.4134 \\ 0.4988 & 0.7066 & -0.4061 \\ 0.4988 & -0.6705 \times 10^{-2} & 0.8151 \end{bmatrix} \tag{13.2-3}$$

with now $q_2^T q_3 = 0.00003872$, a result much closer to zero, the value it should have in the absence of errors.

The MGS does the same thing in principle as the CGS; that is, it obtains the orthogonal vector set Q' of (13.1-17) from T_0. The difference is that an algorithm less sensitive to computer round-off errors is used to calculate the q_i with the MGS procedure. As indicated, if there were no computer round-off errors, both the MGS and the CGS algorithms would give identical q_i' and r_{ij}'. However, because of computer round-off errors, the MGS procedure gives a more accurate result, as illustrated in the above example. The algorithm for the CGS for calculating q_j' is given by (13.1-16). It forms q_j' by subtracting from t_j the components of t_j parallel to $q_2', \ldots, q_{j-1}'$. The MGS does the same thing but uses a different algorithms for calculating the components of t_j parallel to $q_1', \ldots, q_{j-1}'$. Also, these components are subtracted sequentially as we shall see. We now develop the MGS method.

We start by developing q'_j for the MGS method. From (4.2-41) or (13.1-4) it follows that the vector component of t_j parallel to q'_1 is given by

$$(t_j)_1 = \frac{q_1'^T t_j}{q_1'^T q_1'} q'_1 = r_{1j} q'_1 \tag{13.2-4}$$

where r_{ij} without the prime is r'_{ij} for the MGS method given by

$$r_{1j} = \frac{q_1'^T t_j}{q_1'^T q_1'} \tag{13.2-4a}$$

Subtracting this component first from t_j yields

$$t_j^{(2)} = t_j - r_{1j} q'_1 \tag{13.2-5}$$

The above MGS calculation of the component of t_j parallel to q'_1, designated as $(t_j)_1$ in (13.2-4), is identical to that used for the CGS procedure. This is seen by examination of (13.1-16) and (13.1-11). Specifically, the first term on the right of (13.1-16) together with the first term (the $i = 1$ term) of the summation are identical to (13.2-5) when we note that r_{1j} of (13.2-4a) equals r'_{1j} of (13.1-11). We write r_{1i} here without a prime even though it is identical to r'_{1j} of (13.1-11) for the CGS algorithm. This because shortly we shall see a difference in the calculation of r_{ij} for the MGS algorithm when $i > 1$.

Next we want to calculate the component of t_j parallel to q'_2 so as to remove it also from t_j as done for the CGS algorighm in (13.1-16) by the $i = 2$ term of the summation. In (13.1-16) this component is calculated for the CGS algorithm by protecting t_j onto q'_2 to give

$$(t_j)_{2c} = \frac{q_2'^T t_j}{q_2'^T q_2'} q'_2 = r'_{2j} q'_2 \tag{13.2-6}$$

where

$$r'_{2j} = \frac{q_2'^T t_j}{q_2'^T q_2'} \tag{13.2-6a}$$

However, we could also have obtained this component of t_j parallel to q'_2 by projecting $t_j^{(2)}$ onto q'_2. The vector $t_j^{(2)}$ is the same as t_j except that it has had the component parallel to q'_1 removed. Hence $t_j^{(2)}$ and t_j have the same value for the component parallel to q'_2. The MGS algorithm uses $t_j^{(2)}$ instead of t_j for calculating the component parallel to q'_2 to yield

$$(t_j)_2 = \frac{q_2'^T t_j^{(2)}}{q_2'^T q_2'} q'_2 = r_{2j} q'_2 \tag{13.2-7}$$

where

$$r'_{2j} = \frac{q'^T_2 t^{(2)}_j}{q'^T_2 q'_2} \tag{13.2-7a}$$

Here, $(t_j)_2$ and $(t_j)_{2c}$ are identical if there are no computer round-off errors, that is,

$$(t_j)_2 = (t_j)_{2c} \tag{13.2-8}$$

and

$$r_{2j} = r'_{2j} \tag{13.2-8a}$$

However, if there are computer round-off errors, then different results are obtained for $(t_j)_2$ and $(t_j)_{2c}$ and for r_{2j} and r'_{2j}. It is because r_{2j} uses $t^{(2)}_j$ [see (13.2-7a)] while r'_{2j} uses t_j [see (13.2-6a) and (13.1-11)] that we use a prime to distinguish these two r's, even though physically they represent the same quantity, that is, the amplitude weighting on the vector q'_2 needed to make it equal to component of t_j along the vector q_2 direction.

It is because (13.2-7a) uses $t^{(2)}_j$, which does not contain the component parallel to q'_1, that the MGS procedure provides a more accurate calculation of r_{2j} and in turn the component of t_j parallel to q'_2 [81]. Because $t^{(2)}_j$ does not contain the component parallel to q'_1, it is smaller than t_j and, as a result, gives a more accurate r'_{2j} [81]. We require a smaller dynamic range when $t^{(2)}_j$ is used instead of t_j in the calculation of r_{2j} and in turn q'_j. This same type of computation improvement is carried forward in calculating r_{ij} for $i = 3, \ldots, j-1$ as we now see.

Next the MGS method subtracts $(t_j)_2$ from $t^{(2)}_j$ to yield

$$t^{(3)}_j = t^{(2)}_j - r_{2j} q'_2 \tag{13.2-9}$$

Then the MGS subtracts the component of t_j parallel to q'_3 from $t^{(3)}_j$ to form $t^{(4)}_j$, using $t^{(3)}_j$ instead of t_j (as done with the CGS algorithm) in the calculation of r_{3j} for better accuracy, specifically

$$t^{(4)}_j = t^{(3)}_j - r_{3j} q'_3 \tag{13.2-10}$$

where

$$r_{3j} = \frac{q'^T_3 t^{(3)}_j}{q'^T_3 q'_3} \tag{13.2-10a}$$

Continuing the MGS procedure gives for the ith step, where the component parallel to q_1' is removed,

$$t_j^{(i+1)} = t_j^{(i)} - r_{ij}q_i' \qquad (13.2\text{-}11)$$

$$r_{ij} = \frac{q_i'^T t_j^{(i)}}{q_i'^T q_i'} \qquad (13.2\text{-}12)$$

where r_{ij} applies for $j > i \geq 1$. Here $t_j^{(i)}$ is used with the MGS algorithm instead of t_j as was done for the CGS algorithm in (13.1-11). Finally, for the last step where the component parallel to q_{j-1}' is removed, we have

$$q_j' = t_j^{(j)} = t_j^{j-1} - r_{j-1j,j}q_{j-1}' \qquad (13.2\text{-}13)$$

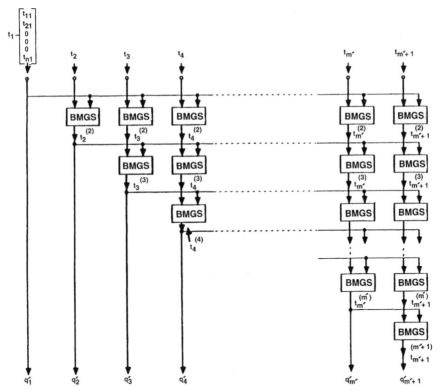

Figure 13.2-1 (*a*) Circuit implementation of modified Gram–Schmidt (MGS) orthogonalization. Box labeled BMGS generates from the two input vectors an output vector orthogonal to rightmost input vector. It is called a basic modified Gram–Schmidt (BMGS) orthogonalizer. Use of $(m' + 1)m'/2$ of these basic Gram–Schmidt orthogonalizers arranged as shown orthogonalizes $m + 1$ vectors $t_1, t_2, \ldots, t_{m'+1}$ into orthogonal set $(q_1', q_2', \ldots, q_{m'+1}')$. (*b*) The BMGS orthogonalizer circuits.

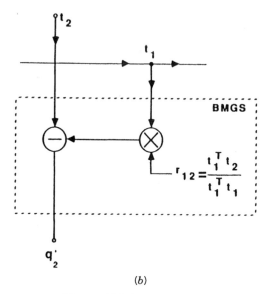

(b)

Figure 13.2-1 (*Continued*)

Again, it cannot be overemphasized that the classical and modified Gram–Schmidt algorithms give the same answer when there are no round-off errors.

In the literature the MGS algorithm is described usually slightly differently. The order of the computation is different but not as it affects the algorithm or its accuracy. When the component of t_j parallel to q_i' is removed, it is removed at the same time from t_k, $k = j + 1, \ldots, m' + 1$. Specifically, when q_1' and q_2' are formed,

$$T_{(2)}' = [q_1' \quad q_2' \quad t_3^{(2)} \quad \cdots \quad t_{m'+1}^{(2)}] \tag{13.2-14}$$

instead of (13.1-9). When q_3' is formed,

$$T_{(3)}' = [q_1' \quad q_2' \quad q_3' \quad t_4^{(3)} \quad \cdots \quad t_{m'+1}^{(3)}] \tag{13.2-15}$$

instead of (13.1-15). And so on.

A circuit implementation of the MGS algorithm is given in Figure 13.2-1. Comparing the circuit implementation for the MGS and CGS algorithms, respectively, in Figures 13.1-4 and 13.2-1 further emphasizes the difference between them. The reader should benefit further from discussions in reference 81 as well as the other references mentioned in this section relative to the Gram–Schmidt algorithm.

14

MORE ON VOLTAGE-PROCESSING TECHNIQUES

14.1 COMPARISON OF DIFFERENT VOLTAGE LEAST-SQUARES ALGORITHM TECHNIQUES

Table 14.1-1 gives a comparison for the computer requirements for the different voltage techniques discussed in the previous chapter. The comparison includes the computer requirements needed when using the normal equations given by (4.1-30) with the optimum least-squares weight W given by (4.1-32). Table 14.1-1 indicates that the normal equation requires the smallest number of computations (at least when $s > m$, the case of interest), followed by the Householder orthonormalization, then by the modified Gram–Schmidt, and finally the Givens orthogonalization. However, the Givens algorithm computation count does not assume the use of the efficient CORDIC algorithm.

The assumption is made that all the elements of the augmented matrix T_0 are real. When complex data is being dealt with, then the counts given will be somewhat higher, a complex multiply requiring four real multiplies and two real adds, a complex add requiring two real adds. The Householder algorithm has a slight advantage over the Givens and modified Gram–Schmidt algorithms relative to computer accuracy. Table 14.1-2 gives a summary of the comparison of the voltage least-squares estimation algorithms.

Before leaving this section, another useful least-squares estimate example is given showing a comparison of the poor results obtained using the normal equations and the excellent results obtained using the modified Gram–Schmidt algorithm. For this example (obtained from reference 82)

TABLE 14.1-1. Operation Counts for Various Least-Squares Computational Methods

Method	Asymptotic Number of Operations[a]
Normal equations (power method)	$\frac{1}{2}sm^2 + \frac{1}{6}m^3$
Householder orthogonalization	$sm^2 - \frac{1}{3}m^3$
Modified Gram–Schmidt	sm^2
Givens orthogonalization	$2\,sm^2 - \frac{2}{3}m^3$

Note: The matrix T is assumed to be an $s \times m$ matrix.

Source: From references 80 and 81.

[a] An operation is a multiply, or divide, plus and add.

$$
T = \begin{bmatrix} 1 & 1 & 1 \\ \varepsilon & 0 & 0 \\ 0 & \varepsilon & 0 \\ 0 & 0 & \varepsilon \end{bmatrix}
\tag{14.1-1}
$$

$$
Y_{(n)} = Y_{(4)} = \begin{bmatrix} 1 \\ 0 \\ 0 \\ 0 \end{bmatrix}
\tag{14.1-2}
$$

for (4.1-11a). We now slove for the least-squares estimate using the normal equations given by (4.1-30) and (4.1-32). We shall at first obtain the exact solution. Toward this end, we calculate

$$
T^T T = \begin{bmatrix} 1+\varepsilon^2 & 1 & 1 \\ 1 & 1+\varepsilon^2 & 1 \\ 1 & 1 & 1+\varepsilon^2 \end{bmatrix}
\tag{14.1-3}
$$

and

$$
T^T Y_{(4)} = \begin{bmatrix} 1 \\ 1 \\ 1 \end{bmatrix}
\tag{14.1-4}
$$

Then from (4.1-30) and (4.1-32) it follows that the exact solution (no round-off errors) is

$$
X_{n,n}^* = \frac{1}{3+\varepsilon^2} \begin{bmatrix} 1 \\ 1 \\ 1 \end{bmatrix}
\tag{14.1-5}
$$

TABLE 14.1-2. Voltage Least-Squares Estimate Algorithms (Orthonormal Transforms): Trade-offs

---

Householder
   Lowest cost on a serial (nonparallel, single-central-processor) machine
   Best numerical behavior (by a small margin)

Givens Rotations
   Introduces one zero at a time and as a result is more costly in number of
      computations required
   However, allows parallel implementations: linear and triangular systolic arrays
   Rotations can be efficiently implemented in hardware using CORDIC
      number representations
   Square-root free version requires computations equal to that of Householder but is
      no longer orthogonal and requires large dynamic range (generally alright if
      floating point used)

Modified Gram–Schmidt
   Like Givens MGS is more costly than Householder
   Like Givens is amenable to systolic implementation
   Provides joint order/time recursive updates

General Comment: Where accuracy is an issue, the orthonormal transforms
(voltage methods) are the algorithms of choice over the normal equation.
With the development of microcomputers having high precision, like 32 and 64 bits
floating point, accuracy is less of an issue. Where a high throughput is needed,
the systolic architecture offered by the Givens approach can provide the high
throughput of parallel processing.

Source: After Steinhardt [138].

Now we obtain the normal equation solution. Assume eight-digit floating-point
arithmetic. If $\varepsilon = 10^{-4}$, then $1 + \varepsilon^2 = 1.00000001$, which is rounded off to
1.0000000. The matrix $T^T T$ then is thought to contain all 1's for its entries and
becomes singular and noninvertable so that no least-squares estimate is
obtainable using the normal equations (4.1-30) and (4.1-32).

   Next let us apply the modified Gram–Schmidt algorithm to this same
example. The same eight-digit floating-point arithmetic is assumed. It follows
that Q' of (13.1-20) becomes

$$Q' = \begin{bmatrix} 1 & 0 & 0 & 0 \\ \varepsilon & -\varepsilon & -\frac{1}{2}\varepsilon & -\frac{1}{3}\varepsilon \\ 0 & \varepsilon & -\frac{1}{2}\varepsilon & -\frac{1}{3}\varepsilon \\ 0 & \varepsilon & \varepsilon & -\frac{1}{3}\varepsilon \end{bmatrix} \tag{14.1-6}$$

and U' and Y_1'' of (13.1-39) becomes

$$U' = \begin{bmatrix} 1 & 1 & 1 \\ 0 & 1 & \frac{1}{2} \\ 0 & 0 & 1 \end{bmatrix} \tag{14.1-7}$$

and

$$Y_1'' = \begin{bmatrix} 1 \\ \frac{1}{2} \\ \frac{1}{3} \end{bmatrix} \tag{14.1-8}$$

Finally substituting the above in (13.1-42) and solving using the back-substitution method yields

$$X_{n,n}^* = \frac{1}{3} \begin{bmatrix} 1 \\ 1 \\ 1 \end{bmatrix} \tag{14.1-9}$$

which is close to the exact solution of (14.1-5).

Those who desire to further pursue the voltage techniques described are urged to read references 76, 79, 81 to 83, 89, 91, 101 to 103, 115, 118 to 122, and 139. References 79, 115, 119, 121, and 122 apply the voltage techniques to the Kalman filter; see Section 14.5.

14.2   QR DECOMPOSITION

In the literature what is called the QR decomposition method is described for solving the least-squares estimation problem [81, 89, 102]. This involves the decomposition of the augmented matrix T_0 into a product of matrices designated as QR as done in (13.1-27) when carrying out the Gram–Schmidt algorithm in Chapter 13. Thus the Gram–Schmidt is a QR decomposition method. It follows that the same is true for the Householder and Givens methods. These give rise to the upper triangular R when the augmented T_0 is multiplied by an orthogonal transformation Q^T, that is, $Q^T T_0 = R$; see, for example (11.1-30) or (12.2-6), where $Q^T = F$ for these equations. Thus from (10.2-1), $QQ^T T_0 = T_0 = QR$, and the augmented matrix takes the form QR, as desired.

An additional physical interpretation of the matrices Q and R can be given that is worthwhile for obtaining further insight into this decomposition of the augmented matrix T_0. When Q is orthonormal, its magnitude is in effect unity, and it can be thought of as containing the phase information of the augmented matrix T_0. The matrix R then contains the amplitude information of T_0. The QR

can then be thought of as the polar decompostion of the augmented matrix T_0 into its amplitude R and phase Q components. That this is true can be rigorously proven by the use of Hilbert space [104].

14.3   SEQUENTIAL SQUARE-ROOT (RECURSIVE) PROCESSING

Consider the example augmented matrix T_0 given by (11.1-29). In this matrix time is represented by the first subscript i of two subscripts of the elements t_{ij}, and the only subscript i of y_i. Thus the first row represents the observation obtained first, at time $i = 1$, and the bottom row the measurement made last. For convenience we shall reverse the order of the rows so that the most recent measurement is contained in the top row. Then without any loss of information (11.1-29) becomes

$$T_0 = \begin{bmatrix} t_{31} & t_{32} & y_3 \\ t_{21} & t_{22} & y_2 \\ t_{11} & t_{12} & y_1 \end{bmatrix} \tag{14.3-1}$$

To solve for the least-squares estimate, we multiply the above augmented matrix T_0 by an orthonormal transformtion matrix to obtain the upper triangular matrix given by

$$T_0' = \begin{bmatrix} (t_{31})_3 & (t_{32})_3 & | & (y_3)_3 \\ 0 & (t_{22})_3 & | & (y_2)_3 \\ ----- & ----- & | & ----- \\ 0 & 0 & | & (y_1)_3 \end{bmatrix} \tag{14.3-2}$$

The entries in this matrix differ from those of (11.1-30) because of the reordering done. Using the back-substitution method, we can solve for the least-squares estimate by applying (14.3-2) to (10.2-16).

Assume now that we have obtained a new measurement at time $i = 4$. Then (14.3-1) becomes

$$T_0 = \begin{bmatrix} t_{41} & t_{42} & y_4 \\ t_{31} & t_{32} & y_3 \\ t_{21} & t_{22} & y_2 \\ t_{11} & t_{12} & y_1 \end{bmatrix} \tag{14.3-3}$$

We now want to solve for the least-squares estimate based on the use of all four measurements obtained at $i = 1, 2, 3, 4$. A straightforward method to obtain the least-squares would be to repeat the process carried out for (14.3-1), that is, triangularize (14.3-3) by applying an orthonormal transformation and then use the back-substitution procedure. This method has the disadvantage, however, of not making any use of the computations made previously at time $i = 3$.

To make use of the previous computations, one adds the new measurements obtained at time $i = 4$ to (14.3-2) instead of (14.3-1) to obtain the augmented matrix [79]

$$T_0' = \begin{bmatrix} t_{41} & t_{42} & y_4 \\ (t_{31})_3 & (t_{32})_3 & (y_3)_3 \\ 0 & (t_{22})_3 & (y_2)_3 \\ 0 & 0 & (y_1)_3 \end{bmatrix} \quad (14.3\text{-}4)$$

One can now apply an orthonormal transformation to (14.3-4) to upper triangularize it and use the back substitution to obtain the least-squares estimate. This estimate will be based on all the measurements at $i = 1, 2, 3, 4$. The computations required when starting with (14.3-4) would be less than when starting with (14.3-3) because of the larger number of zero entries in the former matrix. This is readily apparent when the Givens procedure is used. Furthermore the systolic array implementation for the Givens algorithm as represented by Figures 11.3-2 and 11.3-4 is well suited to implementing the sequential algorithm described above.

The sequential procedure outlined above would only be used if the least-squares estimate solutions are needed at the intermediate times $i = 1, 2, 3, 4, \ldots, j, \ldots$, etc. This is typically the case for the radar filtering problem. If we did not need these intermediate estimates, it is more efficient to only obtain the estimate at time $i = s$ at which the last measurement is made [79]. This would be done by only triangularizing the final augmented matrix T_0 containing all the measurements from $i = 1, \ldots, s$.

Sometimes one desires a discounted least-squares estimate: see Section 1.2.6 and Chapter 7. To obtain such an estimate using the above sequential procedure, one multiplies all the elements of the upper triangularized matrix T_0 by θ where $0 < \theta < 1$ before augmenting T_0 to include the new measurement as done in (14.3-4). Thus (14.3-4) becomes instead

$$T_0 = \begin{bmatrix} t_{41} & t_{42} & y_4 \\ \theta(t_{31})_3 & \theta(t_{32})_3 & \theta(y_3)_3 \\ 0 & \theta(t_{22})_3 & \theta(y_2)_3 \\ 0 & 0 & \theta(y_1)_3 \end{bmatrix} \quad (14.3\text{-}5)$$

The multiplication by θ is done at each update. Such a discounted (or equivalently weighted) least-squares estimate is obtained using the systolic arrays of Figures 11.3-2 and 11.3-4 by multiplying the elements of the systolic array by θ at each update; see references 83 and 89.

The sequential method described above is sometimes referred to in the literature as the sequential square-root information filter (SRIF) [79]. The reader is referred to reference 79 for the application of the sequential SRIF to the Kalman filter. The reader is also referred to references 78, 81 to 83, 89, 102, and 119.

14.4 EQUIVALENCE BETWEEN VOLTAGE-PROCESSING METHODS AND DISCRETE ORTHOGONAL LEGENDRE POLYNOMIAL APPROACH

Here we will show that the voltage-processing least-square approach of Section 4.3 and Chapters 10 to 13 becomes the DOLP approach of Section 5.3 when a polynomial of degree m is being fitted to the data, that is, when the target motion as a function of time is assumed to be described by a polynomial of degree m, and when the times between measurements are equal. Consider again the case where only range $x(r) = x_r$ is measured. Assume, [see (5.2-3)], that the range trajectory can be approximated by the polynomial of degree m

$$x(r) = x_r = p(t) = \sum_{j=0}^{m} a_j (rT)^j \qquad (14.4\text{-}1)$$

where for simplicity in notation we dropped the bar over the a_j. Alternately from (5.2-4)

$$x(r) = x_r = p(t) = \sum_{j=0}^{m} z_j r^j \qquad (14.4\text{-}2)$$

where

$$z_j = a_j T^j \qquad (14.4\text{-}2a)$$

where z_j is the scaled jth state derivative; see (5.2-8). The values of $x_r, r = 0, \ldots, s$, can be represented by the column matrix

$$X = \begin{bmatrix} x_0 \\ x_1 \\ \vdots \\ x_r \\ \vdots \\ x_s \end{bmatrix} \qquad (14.4\text{-}3)$$

Note that the column matrix X defined above is different from the column matrix used up until now. The X defined above is physically the matrix of the true ranges at times $r = 0, 1, 2, \ldots, s$. It is the range measurements y_r made without the measurement noise error, for example, N_n of (4.1-1) or $N_{(n)}$ of

(4.1-11a). The X defined up until now was the $m' \times 1$ column matrix of the process state vector. When this state vector X is multiplied by the observation matrix M or the transition–observation matrix T, it produces the range matrix X defined by (14.4-3). This is the only section where we will use X as defined by (14.4-3). Which X we are talking about will be clear from the text. Which of these two X's is being used will also be clear because the state vector X_n^* always has the lowercase subscript s or n or 3. The range matrix will usually not have a subscript or will have the capital subscript Z or B as we shall see shortly.

Applying (14.4-2) to (14.4-3) yields

$$
X = \begin{bmatrix} x_0 \\ x_1 \\ x_2 \\ \vdots \\ x_r \\ \vdots \\ x_s \end{bmatrix} = \begin{bmatrix} z_0 & + & 0 & + & \ldots & + & 0 \\ z_0 & + & z_1 & + & \ldots & + & z_m 1^m \\ z_0 & + & z_1 2 & + & \ldots & + & z_m 2^m \\ & & & \vdots & & & \vdots \\ z_0 & + & z_1 r & + & \ldots & + & z_m r^m \\ & & & \vdots & & & \vdots \\ z_0 & + & z_1 s & + & \ldots & + & z_m s^m \end{bmatrix}
\tag{14.4-4}
$$

or

$$
X = TZ_s
\tag{14.4-5}
$$

where

$$
Z_s = \begin{bmatrix} z_0 \\ z_1 \\ z_2 \\ \vdots \\ z_m \end{bmatrix}
\tag{14.4-5a}
$$

and

$$
T = \begin{bmatrix} 1 & 0 & 0 & \ldots & 0 \\ 1^0 & 1^1 & 1^2 & \ldots & 1^m \\ 2^0 & 2^1 & 2^2 & \ldots & 2^m \\ \vdots & \vdots & \vdots & & \vdots \\ r^0 & r^1 & r^2 & \ldots & r^m \\ \vdots & \vdots & \vdots & & \vdots \\ s^0 & s^1 & s^2 & \ldots & s^m \end{bmatrix} = \begin{bmatrix} 1 & 0 & 0 & \ldots & 0 \\ 1 & 1 & 1 & \ldots & 1 \\ 1 & 2^1 & 2^2 & \ldots & 2^m \\ \vdots & \vdots & \vdots & & \vdots \\ 1 & r & r^2 & \ldots & r^m \\ \vdots & \vdots & \vdots & & \vdots \\ 1 & s & s^2 & \ldots & s^m \end{bmatrix}
\tag{14.4-5b}
$$

Physically Z_s is the scaled state matrix Z_n of (5.4-12) for index time s instead of n; see also (5.2-8) and (14.4-2a). Here, T is the transition–observation matrix for the scaled state matrix Z_s.

Physically, (14.4-5) is the same as (4.1-11) when no noise measurements errors are present, that is, when $N_{(n)} = 0$. As indicated, Z_s is scaled version of the state matrix X_n of (4.1-11) or (4.1-2) and the matrix T of (14.4-5) is the transition–observation matrix of (4.1-11) or (4.1-11b) for the scaled state matrix Z_s, which is different from the transition–observation matrix T for X_s. The range $x(r)$ trajectory is modeled by a polynomial of degree m, hence X_s and in turn its scaled form Z_s have a dimension $m' \times 1$ where $m' = m + 1$; see (4.1-2) and discussion immediately following.

Now from (5.3-1) we know that $x(r)$ can be expressed in terms of DOLP $\phi_j(r), j = 0, \ldots, m$ as

$$x(r) = x_r = \sum_{j=0}^{m} \beta_j \phi_j(r) \tag{14.4-6}$$

where for simplicity we have dropped the subscript n on β_j. In turn, using (14.4-6), the column matrix of x_r, $r = 0, \ldots, s$, can be written as

$$X = \begin{bmatrix} x_0 \\ x_1 \\ x_2 \\ \vdots \\ x_r \\ \vdots \\ x_s \end{bmatrix} = \begin{bmatrix} \phi_0(0) & \phi_1(0) & \cdots & \phi_m(0) \\ \phi_0(1) & \phi_1(1) & \cdots & \phi_m(0) \\ \phi_0(2) & \phi_1(2) & \cdots & \phi_m(2) \\ \vdots & \vdots & & \vdots \\ \phi_0(r) & \phi_1(r) & \cdots & \phi_m(r) \\ \vdots & \vdots & & \vdots \\ \phi_0(s) & \phi_1(s) & \cdots & \phi_m(s) \end{bmatrix} \begin{bmatrix} \beta_0 \\ \beta_1 \\ \beta_2 \\ \vdots \\ \beta_r \\ \vdots \\ \beta_m \end{bmatrix} \tag{14.4-7}$$

or

$$X = PB \tag{14.4-8}$$

where

$$P = \begin{bmatrix} \phi_0(0) & \phi_1(0) & \cdots & \phi_m(0) \\ \phi_0(1) & \phi_1(1) & \cdots & \phi_m(1) \\ \phi_0(2) & \phi_1(2) & \cdots & \phi_m(2) \\ \vdots & \vdots & & \vdots \\ \phi_0(r) & \phi_1(r) & \cdots & \phi_m(r) \\ \vdots & \vdots & & \vdots \\ \phi_0(s) & \phi_1(s) & \cdots & \phi_m(s) \end{bmatrix} \tag{14.4-8a}$$

and

$$
B = \begin{bmatrix} \beta_0 \\ \beta_1 \\ \vdots \\ \beta_r \\ \vdots \\ \beta_m \end{bmatrix}
$$

(14.4-8b)

The DOLP $\phi_j(r)$ can be written as

$$\phi_0(r) = c_{00} \tag{14.4-9a}$$

$$\phi_1(r) = c_{10} + c_{11}r \tag{14.4-9b}$$

$$\phi_2(r) = c_{20} + c_{21}r + c_{22}r^2 \tag{14.4-9c}$$

$$\phi_m(r) = c_{m0} + c_{m1}r + c_{m2}r^2 + \ldots + c_{mm}r^m \tag{14.4-9d}$$

where the coefficients c_{ij} can be obtained from the equations of Section 5.3. In matrix form (14.4-9a) to (14.4-9d) is expressed by

$$P = TC \tag{14.4-10}$$

where T is defined by (14.4-5b) and C is given by the upper triangular matrix of DOLP coefficients:

$$
C = \begin{bmatrix}
c_{00} & c_{10} & c_{20} & \cdots & c_{m-1,0} & c_{m0} \\
0 & c_{11} & c_{21} & \cdots & c_{m-1,1} & c_{m1} \\
0 & 0 & c_{22} & \cdots & c_{m-1,2} & c_{m2} \\
0 & 0 & 0 & & \cdot & \cdot \\
\vdots & \vdots & \vdots & & \vdots & \vdots \\
0 & 0 & 0 & \cdots & c_{m-1,m-1} & c_{m,m-1} \\
0 & 0 & 0 & \cdots & 0 & c_{mm}
\end{bmatrix}
$$

(14.4-10a)

Substituting (14.4-10) into (14.4-8) yields

$$X = TCB \tag{14.4-11}$$

for the range matrix X in terms of the DOLP. Now x_r is the actual range. What we measure is the noise-corrupted range y_r, $r = 0, 1, 2, \ldots, s$, given by

$$y_r = x_r + v_r \tag{14.4-12}$$

as in (4.1-1) to (4.1-1c). In matrix form this becomes

$$Y = X + N \tag{14.4-13}$$

where

$$N = \begin{bmatrix} \nu_1 \\ \nu_2 \\ \vdots \\ \nu_s \end{bmatrix} \qquad (14.4\text{-}13a)$$

and Y is an $(s + 1) \times 1$ column matrix of y_r, $r = 0, 1, \ldots, s$. Thus here the numbers of measurements y_r equals $s + 1$; not s as in (4.1-10).

What we are looking for is an estimate of the ranges $x_0, \ldots, x_s$ designated as $x_0^*, \ldots, x_s^*$ or X^*. Specifically, we are looking for the least-squares estimate based on the range measurements $y_0, \ldots, y_s$ or Y of (14.4-12) and (14.4-13). If we use our polynomial expression for x_r, then we are looking for the least-squares estimate of the polynomial coefficients, that is, of the a_j of (14.4-1) or alternately the least-squares estimate of the scaled a_j, that is, z_i given by (14.4-2a) or (5.2-8) or the column matrix of (14.4-5). Designate the least-squares estimate of Z_s as Z_s^*. From (14.4-5)

$$X_Z^* = T Z_s^* \qquad (14.4\text{-}14)$$

where the subscript Z on the range matrix X^* is used to indicate that to obtain the estimate of the range matrix X of (14.4-3) we are using the polynomial fit for x_r given by (14.4-2) in terms of the scaled derivatives of x_i, that is, by the state coordinates z_i of (14.4-2a) and (14.4-5a). Similarly, if we use the DOLP representation for x_r, then the least-squares estimate of the range matrix X is obtained from the least-squares estimate of B, designated as B^*. Specifically, using (14.4-11) gives

$$X_B^* = T C B^* \qquad (14.4\text{-}15)$$

where the subscript B on X^* is used to indicate that the estimate is obtained using the DOLP representation for the range matrix X.

From (4.1-31) we know that the least-squares estimate of Z_s is given by the Z_s, which minimizes the magnitude of the column error matrix E_Z given by

$$E_Z = Y - X_Z = Y - T Z_s \qquad (14.4\text{-}16)$$

Similarly, the least-squares estimate of B is given by the B, which minimizes the magnitude squared of the column error matrix E_B given by

$$E_B = Y - X_B = Y - T C B \qquad (14.4\text{-}17)$$

To apply the voltage-processing procedure to (14.4-16) to obtain the least-squares estimate Z_s^* of Z_s, we want to apply an orthonormal transformation F to (14.4-16), which transforms it into m' equations of the m' unknowns of Z_s, with these equations being in the Gauss elimination form as done in Section 4.3 and Chapters 10 to 13. Note that because the $\phi_j(r)$ are the DOLP terms, the columns of P are orthonormal. This follows from (5.3-2). Hence the rows of the transpose of P, P^T, are orthonormal and P^T is an orthonormal transformation, like F of (13.1-33). Strictly speaking, P^T is not an orthonormal transformation because it is not square, it being of dimension $m' \times (s + 1)$. [This is because P only spans the m'-dimensional subspace of $(s+1)$-dimensional space of X. Specifically, it spans only the column space of T. Here, P could be augmented to span the whole $(s+1)$-dimensional, but this is not necessary. This is the same situation we had for F of (13.1-33)]. Let us try P^T as this orthonormal transformation.

Multiplying both sides of (14.4-16) by P^T and reversing the sign of E_Z yields

$$P^T E_Z = P^T T Z_s - P^T Y \tag{14.4-18}$$

or

$$E'_{1Z} = P^T T Z_s - Y'_1 \tag{14.4-19}$$

where

$$E'_{1Z} = P^T E_Z \tag{14.4-19a}$$
$$Y'_1 = P^T Y \tag{14.4-19b}$$

Now applying the transform P^T to (14.4-10) yields

$$P^T P = P^T T C \tag{14.4-20}$$

But

$$P^T P = I \tag{14.4-21}$$

where I is the $m' \times m'$ identity matrix. (Note that because P is not a square matrix, $P P^T \neq I$.) Thus

$$P^T T C = I \tag{14.4-22}$$

Postmultiplying both sides of (14.4-22) by C^{-1} yields

$$P^T T C C^{-1} = C^{-1} \tag{14.4-23}$$

or

$$C^{-1} = P^T T \tag{14.4-24}$$

Substituting (14.4-24) into (14.4-19) yields

$$E'_{1Z} = C^{-1}Z_s - Y'_1 \qquad (14.4\text{-}25)$$

Because C is upper triangular, it can be shown that its inverse is upper triangular (see problem 14.4-1). Thus C^{-1} is an upper triangular matrix and equivalent to U of the voltage-processing method. Thus

$$U = C^{-1} \qquad (14.4\text{-}26)$$

Then (14.4-25) becomes

$$E'_{1Z} = UZ_s - Y'_1 \qquad (14.4\text{-}27)$$

Thus the transformation P^T puts (14.4-16) in the form obtained when using the voltage-processing method, and hence P^T is a suitable voltage-processing method orthonomal transformation as we hoped it would be.

Equation (14.4-25) or (14.4-27) consists of $m' = m + 1$ equations with $m+1$ unknown z_i's to be solved for such that $\|E'_{1Z}\|$ is minimum. Hence, as was the case for (4.3-31) and (10.2-14), the minimum $\|E'_{1Z}\|$ is obtained by setting E'_{1Z} equal to zero to obtain for (14.4-25) and (14.4-27)

$$Y'_1 = C^{-1}Z_s^* \qquad (14.4\text{-}28)$$

which when using (14.4-26) becomes

$$Y'_1 = UZ_s^* \qquad (14.4\text{-}29)$$

Solving for Z_s^* in (14.4-28) or (14.4-29) gives the desired least-squares estimate Z_s^*. Equation (14.4-29) is equivalent to (10.2-16) of the voltage-processing method for estimating the unscaled $X_{n,n}^*$. Because C^{-1} and in turn U, are upper triangular, Z_s^* can be easily solved for using the back-substitution method, as discussed relative to solving (10.2-16) for $X_{n,n}^*$. However, because we know C when x_r is modeled to be a polynomial in time, it being the matrix of coefficients of the DOLP, we can solve (14.4-28) or (14.4-29) without using back substitution. Instead, we can solve for Z_s^* directly by writing (14.4-28) as

$$Z_s^* = CY'_1 = CP^T Y \qquad (14.4\text{-}30)$$

where use was made of (14.4-19b). [Note that Equation (14.4-27) for estimating Z is equivalent to (4.3-45) and (10.2-14) used in the process of estimating X_s^*, the unscaled Z_s, with Y'_2 not present. The orthonormal transformation P^T here, see (14.4-18), for example, is equivalent to that of F of (10.2-14) except there the rows of F were made to span the s-dimensional space of $Y_{(n)}$ while here, as indicated above, the rows of P^T only span the $m + 1 = m'$-dimensional space of the columns of T given by (14.4-5b). It is for this reason that Y_2' is not present in (14.4-25).]

Thus the voltage-processsing least-squares solution when fitting a polynomial fit of degree m to the data becomes a straightforward solution given by (14.4-30) with the orthonormal matrix P given by the matrix of DOLP given by (14.4-8a) and the upper triangular matrix C given by the matrix of DOLP coefficients given by (14.4-10a) with both of these matrices known beforehand. Thus the Gram-Schmidt, Givens, or Householder transformations do not have to be carried out for the voltage-processing method when a polynomial fit is made to data. In addition, the voltage-processing orthonormal transformation F equals P^T and hence is known beforehand. Also, the upper triangular matrix U is equal to C^{-1} and is known beforehand also. Moreover, U does not have to be calculated because (14.4-30) can be used to solve for Z_s^* directly.

It remains to relate the resultant voltage-processing method solution given by (14.4-30) with the DOLP solution. From (5.3-10) we know that the DOLP least-squares solution for β_j is given by

$$\beta_j^* = \sum_{r=0}^{s} \phi_j(r) y_r \qquad (14.4\text{-}31)$$

Using (14.4-8a), this can be written as

$$B^* = P^T Y \qquad (14.4\text{-}32)$$

On examining (14.4-30) we see that our Z_s^* obtained using the voltage-processing method with $F = P^T$ is given in terms of the DOLP solution for B^* given above. Specifically, applying (14.4-32) to (14.4-30) yields

$$Z_s^* = CB^* \qquad (14.4\text{-}33)$$

Computationally, we see that the voltage-processing solution is identical to the DOLP solution when the data is modeled by a polynomial of degree m and the times between measurements are equal.

Finally, the least-squares estimate of X of (14.4-3) can be obtained by applying (14.4-33) to (14.4-5) to yield

$$X^* = TCB^* \qquad (14.4\text{-}34)$$

which on applying (14.4-32) becomes

$$X^* = TCP^T Y \qquad (14.4\text{-}35)$$

Alternately, we could apply (14.4-32) directly to (14.4-8) to obtain

$$X^* = PP^T Y \qquad (14.4\text{-}36)$$

Both solutions, (14.4-35) and (14.4-36), are identical since, from (14.4-10), $P = TC$. It is just that the order of the calculations are different. In either case we do not need to obtain a matrix inverse or use the back-substitution method because here we know the matrices $C(= U^{-1})$ and $P^T(= F)$.

Let us recapitulate the important results obtained in this section. The main point is that the DOLP approach is equivalent to the voltage-processing approach. Moreover, when the voltage-processing method is used to provide a least-squares polynomial fit of degree $m, \ldots, s + 1$ consecutive data points equally spaced in time with $s + 1 > m' = m + 1$, the voltage-processing matrix U and the transformation matrix F of Chapters 10 to 13 are known in advance. Specifically, U is given by the inverse of the matrix C of the coefficient of the DOLP given by (14.4-10a). Moreover, C^{-1} does not need to be evaluated because we can obtain the scaled least-squares estimate of Z_s^* by using (14.4-30). Thus, for the case where a least-squares polynomial fit is being made to the data, the voltage-processing approach becomes the DOLP approach with U^{-1} given by C and $F = P^T$. This is indeed a beautiful result.

It is important to point out that U (and in turn U^{-1}) and F of the voltage-processing approach are known in advance, as discussed above, only if there are no missing measurements $y_j, j = 0, 1, 2, \ldots, s$. In the real world, some y_j will be missing due to weak signals. In this case T and in turn U are not known in advance, the drop out of data points being random events.

14.5 SQUARE-ROOT KALMAN FILTERS

The Kalman filters discussed up until now and in Chapter 18 use a power method of computation. They involve calculating the covariance matrix of the predicted and filtered state vectors; see, for example, (9.3-1a) to (9.3-1d) and (2.4-4a) to (2.4-4j). There are square-root Kalman filter algorithms that compute the square-root of these covariance matrices and hence are less sensitive to round-off errors [79, 115, 119, 121, 122]. A similar filter is the U-D covariance factorization filter discussed elsewhere [79, 119, 121, 122]. (In Section 10.2.2 we pointed out that U^{-1} is the square root of the covariance matrix $S_{n,n}^*$. Similarly $UD^{1/2}$ is another form of the square root of $S_{n,n}^*$ where U is an upper triangular matrix and D is a diagonal matrix. (Note that the square root of (13.1-39a) is of this form.) The question arises as to how to tell if one needs these square-root-type filters or can the simpler, conventional power method type filters be used. The answer is to simulate the power method filter on a general-purpose computer with double or triple precision and determine for what computation roundoff one runs into performance and stability problems. If for the accuracy to be used with the conventional filter, one does not run into a performance or stability problem, then the conventional filters described here can be used. If not, then the square-root filters should be considered.

15

LINEAR TIME-VARIANT SYSTEM

15.1 INTRODUCTION

In this chapter we extend the results of Chapters 4 and 8 to systems having time-variant dynamic models and observation schemes [5, pp. 99–104]. For a time-varying observation system, the observation matrix M of (4.1-1) and (4.1-5) could be different at different times, that is, for different n. Thus the observation equation becomes

$$Y_n = M_n X_n + N_n \qquad (15.1-1)$$

For a time-varying dynamics model the transition matrix Φ would be different at different times. In this case Φ of (8.1-7) is replaced by $\Phi(t_n, t_{n-1})$ to indicate a dependence of Φ on time. Thus the transition from time n to $n+1$ is now given by

$$X_{n+1} = \Phi(t_{n+1}, t_n) X_n \qquad (15.1-2)$$

The results of Section 4.1 now apply with M, Φ, and T replaced by M_n, $\Phi(t_n, t_{n-1})$, and T_n, respectively; see (4.1-5) through (4.1-31). Accordingly, the least-squares and minimum-variance weight estimates given by (4.1-32) and (4.5-4) apply for the time-variant model when the same appropriate changes are made [5]. It should be noted that with Φ replaced by $\Phi(t_n, t_{n-1})$, the results apply to the case of nonequal spacing between observations. We will now present the dynamic model differential equation and show how it can be numerically integrated to obtain $\Phi(t_n, t_{n-i})$.

15.2 DYNAMIC MODEL

For the linear, time-variant dynamic model, the differential equation (8.1-10) becomes the following linear, time-variant vector equation [5, p. 99]:

$$\frac{d}{dt}X(t) = A(t)X(t) \tag{15.2-1}$$

where the constant A matrix is replaced by the time-varying matrix $A(t)$, a matrix of parameters that change with time. For a process described by (15.2-1) there exists a transition matrix $\Phi(t_n + \zeta, t_n)$ that transforms the state vector at time t_n to $t_n + \zeta$, that is,

$$X(t_n + \zeta) = \Phi(t_n + \zeta, t_n)X(t_n) \tag{15.2-2}$$

This replaces (8.1-21) for the time-invariant case. It should be apparent that it is necessary that

$$\Phi(t_n, t_n) = I \tag{15.2-3}$$

15.3 TRANSITION MATRIX DIFFERENTIAL EQUATION

We now show that the transition matrix for the time-variant case satisfies the time-varying model differential equation given by (15.2-1), thus paralleling the situation for the time-invariant case; see (8.1-25) and (8.1-28). Specifically, we shall show that [5, p. 102]

$$\frac{d}{d\zeta}\Phi(t_n + \zeta, t_n) = A(t_n + \zeta)\Phi(t_n + \zeta, t_n) \tag{15.3-1}$$

The above equation can be numerically integrated to obtain Φ as shall be discussed shortly.

To prove (15.3-1), differentiate (15.2-2) with respect to ζ to obtain [5, p. 101]

$$\frac{d}{d\zeta}[\Phi(t_n + \zeta, t_n)X(t_n)] = \frac{d}{d\zeta}X(t_n + \zeta) \tag{15.3-2}$$

Applying (15.2-1) (15.2-2) yields

$$\frac{d}{d\zeta}[\Phi(t_n + \zeta, t_n)X(t_n)] = A(t_n + \zeta)X(t_n + \zeta)$$
$$= A(t_n + \zeta)\Phi(t_n + \zeta, t_n)X(t_n) \tag{15.3-3}$$

Because $X(t_n)$ can have any value, (15.3-1) follows, which is what we wanted to show.

One simple way to numerically integrate (15.3-1) to obtain $\Phi(t_n + \zeta, t_n)$ is to use the Taylor expansion. Let $\zeta = mh$, where m is an integer to be specified shortly. Starting with $k = 1$ and ending with $k = m$, we use the Taylor expansion to obtain [5, p. 102]

$$\Phi(t_n + kh, t_n) = \Phi[t_n + (k-1)h, t_n] + h\frac{d}{d\zeta}\Phi[t_n + (k-1)h, t_n] \quad (15.3\text{-}4)$$

which becomes [5, p. 102]

$$\Phi(t_n + kh, t_n) = \{I + hA[t_n + (k-1)h]\}\Phi[t_n + (k-1)h, t_n]$$
$$k = 1, 2, 3, \ldots, m \quad\quad (15.3\text{-}5)$$

At $k = m$ we obtain the desired $\Phi(t_n + \zeta, t_n)$. In (15.3-4) m is chosen large enough to make h small enough so that the second-order terms of the Taylor expansion can be neglected. The value of m can be determined by evaluating (15.3-5) with successively higher values of m until the change in the calculated value of $\Phi(t_n + \zeta, t_n)$ with increasing m is inconsequential.

Equation (15.2-2) is used to transition backward in time when rewritten as

$$X(t_n) = \Phi(t_n, t_n + \zeta)X(t_n + \zeta) \quad\quad (15.3\text{-}6)$$

The above is obtained by letting ζ be negative in (15.2-2). It thus follows that the inverse of $\Phi(t_n + \zeta, t_n)$ is

$$\Phi(t_n, t_n + \zeta) = [\Phi(t_n + \zeta, t_n)]^{-1} \quad\quad (15.3\text{-}7)$$

Thus interchanging the arguments of Φ gives us its inverse. In the literature the inverse of Φ is written as ψ and given by

$$\psi(t_n + \zeta, t_n) = [\Phi(t_n + \zeta, t_n)]^{-1} \quad\quad (15.3\text{-}8)$$

It is a straightforward matter to show that ψ satisfies the time-varying associated differential equation [5, p. 103]

$$\frac{d}{d\zeta}\psi(t_n + \zeta, t_n) = -\psi(t_n + \zeta, t_n)A(t_n + \zeta) \quad\quad (15.3\text{-}9)$$

thus paralleling the situation for the time-invariant case; see (8.1-30).

16

NONLINEAR OBSERVATION SCHEME AND DYNAMIC MODEL (EXTENDED KALMAN FILTER)

16.1 INTRODUCTION

In this section we extend the results for the linear time-invariant and time-variant cases to where the observations are nonlinearly related to the state vector and/or the target dynamics model is a nonlinear relationship [5, pp. 105–111, 166–171, 298–300]. The approachs involve the use of linearization procedures. This linearization allows us to apply the linear least-squares and minimum-variance theory results obtained so far. When these linearization procedures are used with the Kalman filter, we obtain what is called the extended Kalman filter [7, 122].

16.2 NONLINEAR OBSERVATION SCHEME

When the observation variables are nonlinearly related to the state vector coordinates, (15.1-1) becomes [5, pp. 166–171]

$$Y_n = G(X_n) + N_n \qquad (16.2\text{-}1)$$

where $G(X_n)$ is a vector of nonlinear functions of the state variables. Specifically,

$$G(X_n) = \begin{bmatrix} g_1(X_n) \\ g_2(X_n) \\ \vdots \\ g_n(X_n) \end{bmatrix} \qquad (16.2\text{-}2)$$

357

A common nonlinear observation situation for the radar is where the measurements are obtained in polar coordinates while the target is tracked in cartesian coordinates. Hence the state vector is given by

$$X(t) = X = \begin{bmatrix} x \\ y \\ z \end{bmatrix} \tag{16.2-3}$$

While the observation vector is

$$Y(t) = Y = \begin{bmatrix} R_s \\ \theta \\ \phi \end{bmatrix} \tag{16.2-4}$$

The nonlinear equation relating R_s, θ, and ϕ to x, y, and z are given by (1.5-3), that is $g_1(X)$, $g_2(X)$, and $g_3(X)$ are given by, respectively, (1.5-3a) to (1.5-3c). The inverse equations are given by (1.5-2). The least-squares and minimum-variance estimates developed in Chapters 4 and 9 require a linear observation scheme. It is possible to linearize a nonlinear observation scheme. Such a linearization can be achieved when an approximate estimate of the target trajectory has already been obtained from previous measurements.

One important class of applications where the linearization can be applied is when the target equations of motion are exactly known with only the specific parameters of the equations of motion not being known. Such is the case for unpowered targets whose equations of motion are controlled by gravity and possibly atmospheric drag. This occurs for a ballistic projectile passing through the atmosphere, an exoatmospheric ballistic missile, a satellite in orbit, and a planetary object. For these cases the past measurements on the target would provide an estimate of the target state vector $\bar{X}(t - \zeta)$ at some past time $t - \zeta$ (typically the last time measurements were made on the target). This nominal state vector estimate of $\bar{X}(t - \zeta)$ would be used to estimate the parameters in the known equations of motion for the target. In turn the equations of motion with these estimated parameters would be used to propagate the target ahead to the time t at which the next measurement is being made. This provides us with an estimate for the state vector $\bar{X}(t)$ that shall be used for linearizing the nonlinear observation measurements. Shortly we shall give the equations of motion for a ballistic projectile passing through the atmosphere in order to illustrate this method more concretely.

For those applications where the exact equations of motion are not known, such as when tracking an aircraft, the polynomial approximation of Chapters 5 to 7 can be used to estimate $\bar{X}(t - \zeta)$ and in turn $\bar{X}(t)$ with, the transition matrix Φ for a polynomial trajectory being used to determine $\bar{X}(t)$ from $\bar{X}(t - \zeta)$. The prediction from $t - \zeta$ to t cannot be made too far into the future because the predicted state vector would then have too large an error. In passing let us point out that the polynomial fit can be used also to obtain the initial state estimate

$\bar{X}(t - \zeta)$. For a satellite the insertion parameters can be used to obtain the initial trajectory parameters, and in turn the initial state vector $\bar{X}(t - \zeta)$. In the following paragraphs we will illustrate the linearization of the nonlinear observation equation of (16.2-1) by an example.

Assume a ballistic projectile for which a nominal state vector estimate $\bar{X}(t)$ has been obtained. If $\bar{X}(t)$ is reasonably accurate (as we shall assume it to be), it will differ from the true state vector $\bar{X}(t)$ by a small amount given by $\delta X(t)$, that is,

$$X(t) = \bar{X}(t) + \delta X(t) \tag{16.2-5}$$

Using $G(X_n)$ of (16.2-1) we can calculate the observation vector Y_n that one expects to see at time n (which corresponds to the time t). It is given by

$$\bar{Y}_n = G(\bar{X}_n) \tag{16.2-6}$$

This nominally expected value for the measurement vector $\bar{Y}_n$ will differ from the observed Y_n by a small amount δY_n given by

$$\delta Y_n = Y_n - \bar{Y}_n \tag{16.2-7}$$

Applying (16.2-1), (16.2-5), and (16.2-6) to the above equation yields

$$\delta Y_n = G(\bar{X}_n + \delta X_n) - G(\bar{X}_n) + N_n \tag{16.2-8}$$

Applying the Taylor series to the first term on the right-hand side of the above equation yields [5, p. 169]

$$\delta Y_n = M(\bar{X}_n) \, \delta X_n + N_n \tag{16.2-9}$$

where [5, p. 169]

$$[M(\bar{X}_n)]_{ij} = \left. \frac{dg_i(X)}{dx_j} \right|_{X=\bar{X}_n} \tag{16.2-10}$$

The second-order Taylor series terms have been dropped in the above equation.

By way of example, for the rectangular-to-spherical coordinate case $g_1(X)$, as indicated above, is given by the first equation of (1.5-3a) with $x_1 = x$, $x_2 = y$, $x_3 = z$, and

$$[M(\bar{X}_n)]_{11} = \frac{\bar{X}}{\bar{R}_s} \tag{16.2-11}$$

where

$$\bar{R}_s = (\bar{x}^2 + \bar{y}^2 + \bar{z}^2)^{1/2} \tag{16.2-11a}$$

Equation (16.2-9) is the sought after linearized observation equation, where δY_n replaces Y_n and δX_n replaces X_n. We shall shortly describe how to use the linearized observation equation given by (16.2-9) to obtain an improved estimate of the target trajectory. Briefly, what is done is the differential measurement vector δY_n is used to obtain an estimate $\delta X^*(t_n)$ of the differential state vector δX_n using the linear estimation theory developed up until now. This estimate $\delta X^*(t_n)$ is then added to the estimate $\bar{X}_n$ based on the past data to in turn obtain the new state vector estimate $X^*(t_n)$. This becomes clearer if we use the notation of Section 1.2 and let $X_{k,k}^*$ be the estimate at some past time k based on measurements made at time k and earlier. Using the target dynamics model, $X_{k,k}^*$ is used to obtain the prediction estimate at time n designated as $X_{n,k}^*$. From δY_n and $X_{n,k}^*$ an estimate for $\delta X(t_n)$, designated as $\delta X^*(t_n)$, is obtained. Adding the estimate $\delta X^*(t_n)$ to $X_{n,k}^*$ yields the desired updated estimate $X_{n,n}^*$. To obtain the estimate $\delta X^*(t_n)$, which for simplicity we write as δX^*, it is necessary to know the covariance matrices of $X_{n,k}^*$ and Y_n. The covariance of Y_n is assumed known. For the linear case the covariance of $X_{n,k}^*$ can be obtained from that of $X_{k,k}^*$ using target dynamics transition matrix Φ and (4.5-10), (9.2-1c), or (17.1-1) to be given shortly. If the target dynamics are nonlinear, then a linearization is needed. In Chapter 17 a detailed description is given of this linearization. Discussed in Section 16.3 is how this linearization is used to obtain the transition matrix for a target having a nonlinear dynamics model so that an equation equivalent to (4.5-10) or (9.2-1c) can be used to obtain the covariance of $X_{n,k}^*$.

16.3 NONLINEAR DYNAMIC MODEL

The linear time-invariant and time-variant differential equations given by (8.1-10) and (15.2-1), respectively, become, for a nonlinear target dynamics model [5, pp. 105–111],

$$\frac{d}{dt}X(t) = F[X(t), t] \tag{16.3-1}$$

where, as before, $X(t)$ is the state vector while F is a vector of nonlinear functions of the elements of X, and perhaps of time t if it is also time variant. To be more specific and by way of example let

$$X(t) = \begin{bmatrix} x_0(t) \\ x_1(t) \end{bmatrix} \tag{16.3-2}$$

and

$$F[X(t)] = \begin{bmatrix} f_0(x_0, x_1) \\ f_1(x_0, x_1) \end{bmatrix} \tag{16.3-3}$$

Then (16.3-1) becomes

$$\frac{d}{dt}x_0(t) = f_0[x_0(t), x_1(t)] \tag{16.3-4a}$$

$$\frac{d}{dt}x_1(t) = f_1[x_0(t), x_1(t)] \tag{16.3-4b}$$

As was done for the nonlinear observation equation given by (16.2-1), we would like to linearize (16.3-1) so that the linear estimation theory developed up until now can be applied. As discussed before, this is possible if we have an estimate of the target trajectory based on previous measurements and have in turn its state vector $\bar{X}(t)$ at time t. Differentiating (16.2-5) yields

$$\frac{d}{dt}X(t) = \frac{d}{dt}\bar{X}(t) + \frac{d}{dt}\delta X(t) \tag{16.3-5}$$

Using (16.3-1) and (16.3-5) yields

$$\frac{d}{dt}\bar{X}(t) + \frac{d}{dt}\delta X(t) = F[\bar{X}(t) + \delta X(t)] \tag{16.3-6}$$

For simplicity we have dropped the second variable t in F, the possible variation with time of F being implicitly understood.

Applying (16.3-6) to (16.3-4a) yields

$$\frac{d}{dt}\bar{x}_0(t) + \frac{d}{dt}\delta x_0(t) = f_0[\bar{x}_0(t) + \delta x_0(t), \bar{x}_1(t) + \delta x_1(t)] \tag{16.3-7}$$

Applying the Taylor expansion to the right-hand side of (16.3-7) yields [5, p. 109]

$$\frac{d}{dt}\bar{x}_0(t) + \frac{d}{dt}\delta x_0(t) = f_0(\bar{x}_0, \bar{x}_1) + \frac{df_0}{dx_0}\bigg|_{\substack{\bar{x}_0 \\ \bar{x}_1}} \cdot \delta x_0 + \frac{df_0}{dx_1}\bigg|_{\substack{\bar{x}_0 \\ \bar{x}_1}} \cdot \delta x_1 \tag{16.3-8}$$

where all second-order terms of the Taylor expansion have been dropped. By the same process we obtain for (16.3-4b) [5, p. 109]

$$\frac{d}{dt}\bar{x}_1(t) + \frac{d}{dt}\delta x_1(t) = f_1(\bar{x}_0, \bar{x}_1) + \frac{df_1}{dx_0}\bigg|_{\substack{\bar{x}_0 \\ \bar{x}_1}} \cdot \delta x_0 + \frac{df_1}{dx_1}\bigg|_{\substack{\bar{x}_0 \\ \bar{x}_1}} \cdot \delta x_1 \tag{16.3-9}$$

But from (16.3-1) [see also (16.3-4a) and (16.3-4b)]

$$\frac{d}{dt}\bar{X}(t) = F[\bar{X}(t)] \tag{16.3-10}$$

Hence (16.3-8) and (16.3-9) become [5, p. 109]

$$
\begin{pmatrix} \dfrac{d}{dt}\delta x_0(t) \\[2mm] \dfrac{d}{dt}\delta x_1(t) \end{pmatrix} = \begin{pmatrix} \dfrac{df_0}{dx_0} & \dfrac{df_0}{dx_1} \\[2mm] \dfrac{df_1}{dx_0} & \dfrac{df_1}{dx_1} \end{pmatrix}\Bigg|_{\substack{\bar x_0(t) \\ \bar x_1(t)}} \begin{pmatrix} \delta x_0(t) \\[2mm] \delta x_1(t) \end{pmatrix} \tag{16.3-11}
$$

The above can be rewritten as [5, p. 109]

$$
\frac{d}{dt}\delta X(t) = A[\bar X(t)]\delta X(t) \tag{16.3-12}
$$

where

$$
[A[\bar X(t)]]_{i,j} = \frac{df_i(X)}{dx_j}\Bigg|_{X=\bar X(t)} \tag{16.3-12a}
$$

Equation (16.3-12) is the desired linearized form of the nonlinear dynamics model given by (16.3-1). It is of the same form as (15.2-1). To achieve the linearization, the matrix $A(t)$ is replaced by $A[\bar X(t)]$ while the state vector X is replaced by the differential state vector δX.

We are now in a position to apply the linear estimation theory developed up until now to the differential state vector δX to obtain its estimate δX^*. Having this we can then form the new estimate X^* by adding δX^* to $\bar X$. We shall give the details of how this is done in the next section. Before doing this a few additional points will be made and an example of the linearization of the nonlinear dynamics model given.

Because (16.3-12) is linear and time variant, it follows from (15.2-2) that the transition equation for δX is [5, p. 111]

$$
\delta X(t_n + \zeta) = \Phi(t_n + \zeta, t_n; \bar X)\delta X(t_n) \tag{16.3-13}
$$

where Φ depends on $\bar X$ as well as on time. This transition matrix and its inverse satisfy the differential equations [5, p. 111]

$$
\frac{d}{d\zeta}\Phi(t_n + \zeta, t_n; \bar X) = A[\bar X(t_n + \zeta)]\Phi(t_n + \zeta, t_n; \bar X) \tag{16.3-14}
$$

$$
\frac{d}{d\zeta}\psi(t_n + \zeta, t_n; \bar X) = -\psi(t_n + \zeta, t_n; \bar X)A[\bar X(t_n + \zeta)] \tag{16.3-15}
$$

corresponding to the respective linear time-variant forms given by (15.3-1) and (15.3-9). The desired transition matrix $\Phi(t_n + \delta, t_n)$ or actually $\Phi(t_n, t_k)$ can be obtained by numerical integration of (16.3-14) using the dynamics model matrix $A[\bar X(t)]$ and the initial condition $\Phi(t_n, t_n; \bar X) = I$. This in turn lets us determine the covariance matrix of $X_{n,k}^*$ from that of $X_{k,k}^*$ as mentioned at the end of Section 16.2. The predicted estimate $X_{n,k}^*$ is determined from $X_{k,k}^*$ itself by numerically integrating the original nonlinear target dynamics equations given by (16.3-1); see also (16.3-4). Having $X_{n,k}^*$ and Y_n and their covariances, the minimum variance theory can be applied to obtain the combined estimate for $X_{n,n}^*$ as done in Section 4.5 and Chapter 9; see, for example (9.4-1). This is detailed in the next chapter. Before proceeding to that chapter an example linearization of the nonlinear differential dynamic equation shall be given. This example shall be used in the next chapter.

Assume we wish to track a ballistic projectile through the atmosphere. We will now develop its nonlinear differential equations of motion corresponding to (16.3-1). For simplicity, the assumption is made that the radar is located in the plane of the projectiles trajectory. As a result, it is necessary to consider only two coordinates, the horizontal coordinate x_1 and the vertical coordinate x_2. For further simplicity a flat earth is assumed. We define the target state vector as

$$X = \begin{bmatrix} x_1 \\ x_2 \\ \dot{x}_1 \\ \dot{x}_2 \end{bmatrix} \tag{16.3-16}$$

The derivative of the state vector becomes

$$\frac{dx}{dt} = \begin{bmatrix} \dot{x}_1 \\ \dot{x}_2 \\ \ddot{x}_1 \\ \ddot{x}_2 \end{bmatrix} \tag{16.3-17}$$

The acceleration components in the vector on the right-hand side of the above equation depend on the atmospheric drag force and the pull of gravity. Once we have replaced these acceleration components in (16.3-17) by their relationship in terms of the atmospheric drag and gravity, we have obtained the sought after form of the nonlinear dynamics model cooresponding to (16.3-1). The atmospheric drag equation for the projectile is approximated by [5, p. 105]

$$f_d = \tfrac{1}{2}\rho v^2 \alpha \tag{16.3-18}$$

where ρ is the atmospheric density, v is the projectile speed, and α is an atmospheric drag constant. Specifically,

$$\alpha = C_D A \tag{16.3-19}$$

where C_D is an atmospheric drag coefficient dependent on the body shape and A is the projection of the cross-sectional area of the target on a plane perpendicular to the direction of motion. The parameter α is related to the ballistic coefficient β of (2.4-6) in Section 2.4, by the relationship

$$\alpha = \frac{m}{\beta} \tag{16.3-20}$$

since β is given by (2.4-9). Physically, α represents the effective target drag area. The atmospheric density as a function of altitude is fairly well approximated by the exponential law given by [5, p. 105]

$$\rho = \rho_0 e^{-kx_2} \tag{16.3-21}$$

where ρ_0 and k are known constants.

To replace the acceleration components in (16.3-17) by their atmospheric drag and gravity relationships, we proceed as follows. First, the drag force is resolved into its x_1 and x_2 components by writing the velocity as a velocity vector given by

$$V = v\hat{V} \tag{16.3-22}$$

where $\hat{V}$ is the unit velocity vector along the ballistic target velocity direction. The atmospheric drag force can then be written as a vector F_d given by

$$F_d = -\tfrac{1}{2}\rho\alpha v^2 \hat{V} \tag{16.3-23}$$

Let $\hat{i}$ and $\hat{k}$ be the unit vectors along the x_1 and x_2 coordinates. Then

$$V = \dot{x}_1\hat{i} + \dot{x}_2\hat{k} \tag{16.3-24}$$

and

$$\hat{V} = \frac{\dot{x}_1\hat{i} + \dot{x}_2\hat{k}}{v} \tag{16.3-25}$$

Thus

$$F_d = -\tfrac{1}{2}\rho\alpha v(\dot{x}_1\hat{i} + \dot{x}_2\hat{k}) \tag{16.3-26}$$

and hence

$$m\ddot{x}_1 = -\tfrac{1}{2}\rho\alpha v\dot{x}_1 \tag{16.3-27}$$

and

$$m\ddot{x}_2 = -\tfrac{1}{2}\rho\alpha v\dot{x}_2 - mg \tag{16.3-28}$$

Substituting the above two equations in (16.3-1) and using (16.3-16) and (16.3-21) yield [5, p. 107]

$$
\begin{bmatrix} \dot{x}_1 \\ \dot{x}_2 \\ \ddot{x}_1 \\ \ddot{x}_2 \end{bmatrix} = \begin{bmatrix} \dot{x}_1 \\ \dot{x}_2 \\ -\dfrac{\rho_0 \alpha}{2m} e^{-kx_2} (\dot{x}_1^2 + \dot{x}_2^2)^{1/2} \dot{x}_1 \\ \dfrac{-\rho_0 \alpha}{2m} e^{-kx_2} (\dot{x}_1^2 + \dot{x}_2^2)^{1/2} \dot{x}_2 - g \end{bmatrix}
\tag{16.3-29}
$$

Applying (16.3-12a) yields [5, p. 110]

$A[\bar{X}(t)] =$

$$
\begin{bmatrix}
0 & | & 0 & | & 1 & | & 0 \\
0 & | & 0 & | & 0 & | & 1 \\
0 & | & ck\bar{v}\bar{\dot{x}}_1\exp(-k\bar{x}_2) & | & -c\left(\dfrac{\bar{v}^2 + \bar{\dot{x}}_1^2}{\bar{v}}\right)\exp(-k\bar{x}_2) & | & -\left(\dfrac{c\bar{\dot{x}}_1\bar{\dot{x}}_2}{v}\right)\exp(-k\bar{x}_2) \\
0 & | & ck\bar{v}\bar{\dot{x}}_2\exp(-k\bar{x}_2) & | & -\left(\dfrac{c\bar{\dot{x}}_1\bar{\dot{x}}_2}{\bar{v}}\right)\exp(-k\bar{x}_2) & | & -c\left(\dfrac{\bar{v}^2 + \bar{\dot{x}}_2^2}{\bar{v}}\right)\exp(-k\bar{x}_2)
\end{bmatrix}
$$

$$\tag{16.3-30}$$

where

$$c = \frac{\rho_0 \alpha}{2m} \tag{16.3-30a}$$

$$\bar{v} = (\bar{\dot{x}}_1^2 + \bar{\dot{x}}_2^2)^{1/2} \tag{16.3-30b}$$

$$\bar{x}_1 = \bar{x}_1(t) \quad \text{etc.} \tag{16.3-30c}$$

In the above α, the target effective drag area, was assumed to be known. More generally, it is unknown. In this case it must also be estimated based on the projectile trajectory measurements. The dependence of α and $\dot{\alpha}$ on the trajectory velocity and other state vector parameters provides a nonlinear differential equation of the form given by [5, p. 299]

$$\frac{d}{dt}\begin{pmatrix} \alpha(t) \\ \dot{\alpha}(t) \end{pmatrix} = F[x_1, x_2, \ddot{x}_1, \ddot{x}_2, \alpha(t), \ddot{\alpha}(t)] \tag{16.3-31}$$

The above equation is of the same form as the nonlinear differential target dynamics equation given by (16.3-1). The two-element state vector given on the left side of the above equation must be now estimated. This is done by adding this two-element state vector to the four-estimate state vector given by (16.3-16) to give a six-state vector instead of a four-state vector. [In Section 2.4 we gave an example where the target drag area β had to be estimated; see (2.4-6).]

17

BAYES ALGORITHM WITH ITERATIVE DIFFERENTIAL CORRECTION FOR NONLINEAR SYSTEMS

17.1 DETERMINATION OF UPDATED ESTIMATES

We are now in a position to obtain the updated estimate for the nonlinear observation and target dynamic model cases [5, pp. 424–443]. We shall use the example of the ballistic projectile traveling through the atmosphere for definiteness in our discussion. Assume that the past measurements have permitted us to obtain the state vector estimate $\bar{X}(t - \zeta)$ at the time $t - \zeta$, the last time observations were made on the target. As done at the end of Section 16.2, it is convenient to designate this past time $t - \zeta$ with the index k and write $\bar{X}(t - \zeta)$ as $X_{k,k}^*$. Assume that the measurements are being made in polar coordinates while the projectile is being tracked in rectangular coordinates, that is, the state-vector is given in rectangular coordinates, as done in Section 16.2. [Although previously the projectile trajectory plane was assumed to contain the radar, we will no longer make this assumption. We will implicitly assume that the radar is located outside the plane of the trajectory. It is left to the reader to extend (16.3-29) and (16.3-30) to this case. This is rather straightforward, and we shall refer to these equations in the following discussions as if this generalization has been made. This extension is given elsewhere [5, pp. 106–110].]

By numerically integrating the differential equation given by (16.3-29) starting with $\bar{X}(t - \zeta)$ at time $t - \zeta$ we can determine $\bar{X}(t)$. As before we now find it convenient to also refer to $\bar{X}(t) = \bar{X}_n$ as $X_{n,k}^*$, it being the estimate of the predicted state vector at time t (or n) based on the measurement at time $\xi < t$ (or $k < n$). We can compute the transition matrix $\Phi_{n,k}$ by numerical integration of the differential equation given by (16.3-14) with A given by (16.3-30). In turn $\Phi_{n,k}$ can be used to determine the covariance matrix of $\bar{X}(t)$

using [5, p. 431]

$$S_{n,k}^* = \Phi(t_n, t_k; \bar{X}) S_{k,k}^* \Phi(t_n, t_k; \bar{X})^T \qquad (17.1\text{-}1)$$

assuming we know $S_{k,k}$. Assume that at time t (or n) we get the measurement Y_n. Thus at time n we have Y_n and the estimate $X_{n,k}^*$, based on the past data, with their corresponding covariance matrices R_1 and $S_{n,k}^*$, respectively. What is desired is to combine these to obtain the updated estimate $X_{n,n}^*$. We would like to use the Bayes filter to do this. We can do this by using the linearized nonlinear observation equations. This is done by replacing the nonlinear observation equation (16.2-1) by its linearized version of (16.2-9). The linearized observation matrix $M(X_{n,n}^*)$ [which replaces the nonlinear G of (16.2-1)] is then used for M in (9.2-1) of the Bayes filter. Also Y_n and $X_{n,k}^*$ are replaced by their differentials in (9.2-1). They are designated as, respectively, δY_n and $\delta X_{n,k}^*$. These differentials are determined shortly.

One might think at first that the Bayes filter can be applied without the use of the linearization of the nonlinear observation equations. This is because in (9.2-1) for $\mathring{X}_{n,n}^*$, we can replace $(Y_n - M X_{n,k}^*)$ by $Y_n - G(\bar{X}_n)$ without using the linearization of the observations. However, as we see shortly, the calculation of H_n requires the use of $M(\bar{X}_n)$; see (17.1-5a) and (17.1-5b). Also $M(\bar{X})$ is needed to calculate $S_{n1,n}^*$ for use in the next update calculation, as done using (17.1-5b). Physically we need to calculate $M(\bar{X}_n)$ to find the minimum variance estimate $X_{n,n}^*$ as the weighted sum of Y_n and $X_{n,k}^*$ because to do this we need both of the variates to be combined in the same coordinate system and we need their variances in this same coordinate system. Calculating $M(\bar{X}_n)$ allows us to do the latter by using (16.2-9). The Bayes filter of (17.1-5) implicitly chooses the coordinates of $X_{n,s}^*$ for the common coordinate system.

The differentials are now determined. Using (16.2-6) with G given by (1.5-3), we can calculate $\bar{Y}_n$ in terms of $\bar{X}$ obtained as described above. Using in turn (16.2-7), we calculate δY_n to be given by

$$\delta Y_n = Y_n - \bar{Y}_n \qquad (17.1\text{-}2)$$

where

$$\bar{Y}_n = G(X_{n,k}^*) \qquad (17.1\text{-}2a)$$

from (16.2-6).

For convenience the differential $\delta X_{n,k}^*$ is referenced relative to $X_{n,k}^*$. If we knew X_n, the differential $\delta X_{n,k}^*$ could be given by

$$\delta X_{n,k}^* = X_{n,k}^* - X_n \qquad (17.1\text{-}3)$$

But we do not know X_n, it being what we are trying to estimate. As a result, we do the next best thing. The differential is referenced relative to our best estimate of X based on the past data, which is $X_{n,k}^*$. As a result

$$\delta X_{n,k}^* = X_{n,k}^* - X_{n,k}^* \qquad (17.1\text{-}4)$$
$$= 0$$

This may seem strange, but it is due to our choice of reference for $\delta X_{n,k}^*$. This will become clearer as we proceed.

We are now in a position to apply the Bayes filter. In place of Y_{n+1} and $\mathring{X}_{n+1,n}^*$ in (9.2-1) to (9.2-1d) we use respectively, δY_n and $\delta X_{n,k}^*$ [= 0, because of (17.1-4)]. Moreover, the Bayes filter of (9.2-1) to (9.2-1d) now becomes, for finding our updated differential estimate $\delta X_{n,n}^*$ [5, pp. 431–432],

$$\delta X_{n,n}^* = \delta X_{n,k}^* + \mathring{H}_n[\delta Y_n - M(X_{n,k}^*)\delta X_{n,k}^*] \qquad (17.1\text{-}5)$$

where $\delta X_{n,k}^* = 0$ in the above and

$$\mathring{H}_n = \mathring{S}_{n,n}^*[M(X_{n,k}^*)]^T R_1^{-1} \qquad (17.1\text{-}5a)$$

$$\mathring{S}_{n,n}^* = \{(\mathring{S}_{n,k}^*)^{-1} + [M(X_{n,k}^*)]^T R_1^{-1} M(X_{n,k}^*)\}^{-1} \qquad (17.1\text{-}5b)$$

But

$$\delta X_{n,n}^* = X_{n,n}^* - X_{n,k}^* \qquad (17.1\text{-}6)$$

all the diferential vectors being referred to $X_{n,k}^*$. The updated estimate $X_{n,n}^*$ is thus obtained from (17.1-6) to be

$$X_{n,n}^* = X_{n,k}^* + \delta X_{n,n}^* \qquad (17.1\text{-}7)$$

This is our desired updated estimate.

In obtaining the above update $X_{n,n}^*$, we have used a linearization of the nonlinear observation equations. The new updated estimate will have a bias, but it should be less than the bias in the original estimate $\bar{X}_n$. In fact, having the updated estimate $X_{n,n}^*$, it is now possible to iterate the whole process we just went through a second time to obtain a still better estimate for $X_{n,n}^*$. Let us designate the above updated estimate $X_{n,n}^*$ obtained on the first cycle as $(X_{n,n}^*)_1$ and correspondingly designate $\delta X_{n,n}^*$ as $(\delta X_{n,n}^*)_1$. Now $(X_{n,n}^*)_1$ can be used in place of $X_{n,k}^*$ to obtain a new δY_n and $\delta X_{n,k}^*$, which we shall call $(\delta Y_n)_2$ and $(\delta X_{n,k}^*)_2$, given by

$$(\delta Y_n)_2 = Y_n - (\bar{Y}_n)_2 \qquad (17.1\text{-}8)$$

where

$$(\bar{Y}_n)_2 = G[(X_{n,n}^*)_1] \qquad (17.1\text{-}8a)$$

and

$$(\delta X_{n,k}^*)_2 = X_{n,k}^* - (X_{n,n}^*)_1 \qquad (17.1\text{-}8b)$$

Note that $(\delta X_{n,k})_2$ is now not equal to zero. This is because $(X_{n,n}^*)_1$ is no longer $X_{n,k}^*$. The covariance of $(\delta X_{n,k})_2^2$ still is $S_{n,k}^*$ and that of $(\delta Y_n)_2$ still is R_1. Applying the Bayes filter, again using the new differential measurement and prediction estimate vector, yields $(\delta X_{n,n}^*)_2$ and in turn $(X_{n,n}^*)_2$ from

$$(X_{n,n}^*)_2 = (X_{n,n}^*)_1 + (\delta X_{n,n}^*)_2 \tag{17.1-9}$$

This procedure could be repeated with still better and better estimates obtained for $X_{n,n}^*$. The procedure would be terminated when [5, p. 433]

$$[(X_{n,n}^*)_{r+1} - X\binom{*}{n,n}_r]^T [(X_{n,n}^*)_{r+1} - X\binom{*}{n,n}_r] < \varepsilon \tag{17.1-10}$$

Generally, the first cycle estimate $(X_{n,n}^*)_1$ is sufficiently accurate. The above use of the Bayes algorithm with no iteration is basically the filter developed by Swerling [123] before Kalman; see the Appendix.

Once the final update $X_{n,n}^*$ has been obtained, the whole process would be repeated when a new observation Y_{n+m} is obtained at a later time $n + m$. The subscript $n + m$ is used here instead of $n + 1$ to emphasize that the time instant between measurements need not necessarily be equal. Then $X_{n+m,n}^*$ would be obtained by integrating forward the nonlinear equation of motion given by (16.3-1). This $X_{n+m,n}^*$ would be used to obtain $\delta X_{n+m,n}^*$, δY_{n+m}, and $M(X_{n+m,n}^*)$ using (17.1-2), (17.1-2a), (17.1-4), and (16.2-10). Integrating (16.3-14), $\Phi(t_{n+m}, t_n; \bar{X})$ would be obtained, from which in turn $S_{n+m,n}^*$ would be obtained using (17.1-1). Using the Bayes filter, specifically (17.1-5) to (17.1-5b), $\delta X_{n+m,n+m}^*$ would then be obtained and, in turn, the desired next update state vector $X_{n+m,n+m}^*$.

17.2   EXTENSION TO MULTIPLE MEASUREMENT CASE

We will now extend the results of Section 17.2 to the case where a number of measurements, let us say $L + 1$ measurements, are simultaneously used to update the target trajectory estimate as done in Section 9.5 for the Bayes and Kalman filters when the observation scheme and target dynamics model are linear. For concreteness we will still use the example consisting of a projectile passing through the atmosphere. Assume measurements are made at the $L + 1$ time instances $t_{n-L}, t_{n-L+1}, \ldots, t_{n-1}$, and t_n, where these times are not necessarily equally spaced. Let these $L + 1$ measurement be given by

$$Y_{n-L}, Y_{n-L+1}, \ldots, Y_{n-1}, Y_n \tag{17.2-1}$$

where Y_{n-i} is a measurement vector of the projectile position in polar coordinates; see (16.2-4). [This is in contrast to (5.2-1) where the measurement y_{n-i} was not a vector but just the measurement of one target parameter, e.g., range.] Let us put the $L + 1$ vector measurements in the form

$$
Y_{(n)} = \begin{pmatrix} Y_n \\ ---- \\ Y_{n-1} \\ ---- \\ \vdots \\ ---- \\ Y_{n-L} \end{pmatrix}
\tag{17.2-2}
$$

For the case of the projectile target being considered the observation scheme and the target dynamics model are nonlinear. To use the measurement vector (17.2-2) for updating the Bayes filter or Kalman filter, as done at the end of Chapter 9 through the use of T, requires the linearization of the $L + 1$ observations. Using the development given above for linearizing the nonlinear observation and dynamic model, this becomes a simple procedure [5] and will now be detailed.

As before, let $X_{k,k}^*$ be an estimate of the state vector of the projectile based on measurements prior to the $L + 1$ new measurements. Using the dynamic equations given by (16.3-1) we can bring $X_{k,k}^*$ forward to the times of the $L + 1$ new measurements to obtain $X_{n,k}^*, X_{n-1,k}^*, \ldots, X_{n-L,k}^*$ from which in turn we obtain $\bar{Y}_n, \bar{Y}_{n-1}, \ldots, \bar{Y}_{n-L}$ through the use of (17.1-2a) and then in turn obtain the differential measurement vector

$$
\delta Y_{(n)} = \begin{pmatrix} \delta Y_n \\ ----- \\ Y_{n-1} \\ ----- \\ \vdots \\ ----- \\ \delta Y_{n-L} \end{pmatrix}
\tag{17.2-3}
$$

by the use of (17.1-2).

Having the above $\delta Y_{(n)}$ we wish to obtain an update to the predicted X's given by $X_{n,k}^*, X_{n-1,k}^*, \ldots, X_{n-L,k}^*$. However, rather than update all $L + 1$ X's, it is better to update a representative X along the trajectory. This representative X would be updated using all the $L + 1$ measurements contained in $Y_{(n)}$ or equivalently $\delta Y_{(n)}$. First, using (16.2-9), we obtain

$$
\delta Y_{(n)} = \begin{bmatrix} M(\bar{X}_n)\, \delta X_n \\ ---------- \\ M(\bar{X}_{n-1})\, \delta X_{n-1} \\ ---------- \\ \vdots \\ ---------- \\ M(\bar{X}_{n-L})\, \delta X_{n-L} \end{bmatrix} + \begin{bmatrix} N_n \\ ---- \\ N_{n-1} \\ ---- \\ \vdots \\ ---- \\ N_{n-L} \end{bmatrix}
\tag{17.2-4}
$$

where M is defined by (16.2-10) and where, for simplicity of notation, $\bar{X}_{n-i}$ is used in place of $X^*_{n-i,k}$, which is $X^*_{k,k}$ brought forward to time $n - i$. It is the δX_n, $\delta X_{n-1}, \ldots$, δX_{n-L} that we want to update based on the differential measurement matrix $\delta Y_{(n)}$. Instead, as mentioned above, we will now reference all the differential state vectors δX_{n-i} at time $t_{n-i}, i = 0, \ldots, L$, to some reference time $t_{c,n} = t_{cn}$. It is generally best to choose the time t_{cn} to be at or near the center observation of the $L + 1$ observations. The transition matrix from the time t_{cn} to the time t_{n-i} of any measurement can be obtained by integrating (16.3-14). Using these transition matrices, (17.2-4) becomes

$$
\delta Y_{(n)} =
\begin{bmatrix}
M(\bar{X}_n)\Phi(t_n, t_{cn}; \bar{X}_{cn})\delta X_{cn} \\
\hdashline
M(\bar{X}_{n-1})\Phi(t_{n-1}, t_{cn}; \bar{X}_{cn})\delta X_{cn} \\
\hdashline
\vdots \\
\hdashline
M(\bar{X}_{n-L})\Phi(t_{n-L}, t_{cn}; \bar{X}_{cn})\delta X_{cn}
\end{bmatrix}
+
\begin{bmatrix}
N_n \\
\hdashline
N_{n-1} \\
\hdashline
\vdots \\
\hdashline
N_{n-L}
\end{bmatrix}
\tag{17.2-5}
$$

where $\bar{X}_{cn}$, also designated as $X^*_{cn,k}$, is the value of $X^*_{k,k}$ brought forward to time t_{cn}.

Equation (17.2-5) can now be written as

$$
\delta Y_{(n)} = T_{c,n}\, \delta X_{cn} + N_{(n)}
\tag{17.2-6}
$$

where

$$
T_{c,n} =
\begin{bmatrix}
M_n\Phi(t_n, t_{cn}; \bar{X}_{cn}) \\
\hdashline
M_{n-1}\Phi(t_{n-1}, t_{cn}; \bar{X}_{cn}) \\
\hdashline
\vdots \\
\hdashline
M_{n-L}\Phi(t_{n-L}, t_{cn}; \bar{X}_{cn})
\end{bmatrix}
\tag{17.2-6a}
$$

and

$$
N_{(n)} =
\begin{bmatrix}
N_n \\
\hdashline
N_{n-1} \\
\hdashline
\vdots \\
\hdashline
N_{n-L}
\end{bmatrix}
\tag{17.2-6b}
$$

and

$$M_{n-i} = M(\bar{X}_{n-i}) \tag{17.2-6c}$$

Referencing δX relative to $\bar{X}_{cn} = X^*_{cn,k}$ yields

$$\begin{aligned} \delta X^*_{cn,k} &= X^*_{cn,k} - X^*_{cn,k} \\ &= 0 \end{aligned} \tag{17.2-7}$$

The above parallels (17.1-4)

We can now apply the Bayes Filter of (17.1-5) to (17.1-5b) or Kalman filter of (9.3-1) to (9.3-1d) to update the projectile trajectory based on the $L+1$ measurements. This is done for the Bayes filter of (17.1-5) to (17.1-5b) by replacing M by T_{cn} and R_1 by $R_{(n)}$, which is the covariance matrix of $\delta Y_{(n)}$; specifically, (17.1-5) to (17.1-5b) become

$$\delta X^*_{cn,cn} = \delta X^*_{cn,k} + \overset{\circ}{H}_{cn}(\delta Y_{(n)} - T_{cn}\delta X^*_{cn,k}) \tag{17.2-8}$$

where

$$\overset{\circ}{H}_{cn} = \overset{\circ}{S}^*_{cn,cn} T^T_{cn} R^{-1}_{(n)} \tag{17.2-8a}$$

$$\overset{\circ}{S}^*_{cn,cn} = [(\overset{\circ}{S}^*_{cn,k})^{-1} + T^T_{cn} R^{-1}_{(n)} T_{cn}]^{-1} \tag{17.2-8b}$$

and from (17.1-1)

$$\overset{\circ}{S}^*_{cn,k} = \Phi(t_{cn}, t_k; \bar{X}_{cn}) \, \overset{\circ}{S}^*_{k,k} \Phi(t_{cn}, t_k; \bar{X}_{cn})^T \tag{17.2-8c}$$

Having $\delta X^*_{cn,cn}$, one can obtain the desired update

$$X^*_{cn,cn} = \bar{X}_{cn} + \delta X^*_{cn,cn} \tag{17.2-9}$$

Having this new first estimate $X^*_{cn,cn}$ for X at time t_{cn}, which we now designate as $(X^*_{cn,cn})_1$, we could obtain an improved estimate designated as $(X^*_{cn,cn})_2$. This is done by iterating the whole process described above with δX now referenced relative to $(X^*_{cn,cn})_1$, as done in (17.1-8b) when $Y_{(n)}$ consisted of one measurement.

If, in applying the above recursive Bayes filter, it was assumed that the variance of the estimate based on the past data was infinite or at least extremely large, then the recursive relation would actually degenerate into a nonrecursive minimum variance estimate based on the most recent $L+1$ measurements as given by

$$X^*_{cn,cn} = \bar{X}_{cn} + (T^T_{cn} R^{-1}_{(n)} T_{cn})^{-1} T^T_{cn} R^{-1}_{(n)} [Y_{(n)} - G(\bar{X}_{cn})] \tag{17.2-10}$$

The above equation follows from (4.1-30) with W given by (4.5-4) for the minimum variance estimate.

17.3 HISTORICAL BACKGROUND

The iterative differential correction procedure described in this chapter was first introduced by Gauss in 1795 [5]. There is an interesting story [5, 122, 124] relating to Gauss's development of his least-squares estimate and the iterative differential correction. At that time the astronomers of the world had been looking for a missing planet for about 30 years. There was Mercury, Venus, Earth, Mars, and then the missing planet or planetoid. It has been theorized that because the planet had fragmented into planetoids (asteroids) it was so difficult to locate. It was finally on January 1, 1801, that an Italian astronomer, Giuseppe Piazzi, spotted for the first time one of these planetoids. There was great rejoicing among the world's astronomers. However, the astronomers soon became concerned because the planetoid was out of view after 41 days, and they feared it would possibly not be found for another 30 years. At this time Gauss, who was then 23 years old, gathered the data that Piazzi had obtained on the planetoid Ceres. Over a period of a few months he applied his weighted least-squares estimate and the iterative differential correction techniques to determine the orbit of Ceres. In December of 1801, he sent his results to Piazzi who was then able to sight it again on the last day of 1801.

17.4 NONLINAR FILTERS

When the target dynamics or the observation equations are nonlinear, instead of using the extended Kalman filter (EKF) (which linearizes these equations) another option available is to use a nonlinear tracking filter [142–154].* These nonlinear filters can at times provide estimates whose rms estimates are two, three or more times better than the EKF.

---

*These references were provided by F. E. Daum.

18

KALMAN FILTER REVISITED

18.1 INTRODUCTION

In Section 2.6 we developed the Kalman filter as the minimization of a quadratic error function. In Chapter 9 we developed the Kalman filter from the minimum variance estimate for the case where there is no driving noise present in the target dynamics model. In this chapter we develop the Kalman filter for more general case [5, pp. 603–618]. The concept of the Kalman filter as a fading-memory filter shall be presented. Also its use for eliminating bias error buildup will be presented. Finally, the use of the Kalman filter driving noise to prevent instabilities in the filter is discussed.

18.2 KALMAN FILTER TARGET DYNAMIC MODEL

The target model considered by Kalman [19, 20] is given by [5, p. 604]

$$\frac{d}{dt}X(t) = A(t)X(t) + D(t)U(t) \tag{18.2-1}$$

where $A(t)$ is as defined for the time-varying target dynamic model given in (15.2-1), $D(t)$ is a time-varying matrix and $U(t)$ is a vector consisting of random variables to be defined shortly. The term $U(t)$ is known as the process-noise or forcing function. Its inclusion has beneficial properties to be indicated later. The matrix $D(t)$ need not be square and as a result $U(t)$ need not have the same dimension as $X(t)$. The solution to the above linear differential equation is

[5, p. 605]

$$X(t) = \Phi(t, t_{n-1})X(t_{n-1}) + \int_{t_{n-1}}^{t} \Phi(t, \lambda)D(\lambda)U(\lambda)d\lambda \qquad (18.2\text{-}2)$$

where Φ is the transition matrix obtained from the homogeneous part of (18.2-1), that is, the differential equation without the driving-noise term $D(t)U(t)$, which is the random part of the target dynamic model. Consequently, Φ satisfies (15.3-1).

The time-discrete form of (18.2-1) is given by [5, p. 606]

$$X(t_n) = \Phi(t_n, t_{n-1})X(t_{n-1}) + V(t_n, t_{n-1}) \qquad (18.2\text{-}3)$$

where

$$V(t, t_{n-1}) = \int_{t_{n-1}}^{t} \Phi(t, \lambda)D(\lambda)U(\lambda)d\lambda \qquad (18.2\text{-}4)$$

The model process noise $U(t)$ is white noise, that is,

$$E[U(t)] = 0 \qquad (18.2\text{-}5)$$

and

$$E[U(t)U(t')^T] = K(t)\delta(t - t') \qquad (18.2\text{-}6)$$

where $K(t)$ is a nonnegative definite matrix dependent on time and $\delta(t)$ is the Dirac delta function given by

$$\delta(t - t') = 0 \qquad t' \neq t \qquad (18.2\text{-}7)$$

with

$$\int_{a}^{b} \delta(t - t')\,dt = 1 \qquad a < t' < b \qquad (18.2\text{-}8)$$

18.3 KALMAN'S ORIGINAL RESULTS

By way of history as mentioned previously, the least-square and minimum-variance estimates developed in Sections 4.1 and 4.5 have their origins in the work done by Gauss in 1795. The least mean-square error estimate, which obtains the minimum of the ensemble expected value of the squared difference between the true and estimated values, was independently developed by

Kolmogorov [125] and Wiener [126] in 1941 and 1942, respectively. Next, the Kalman filter [19, 20] was developed, it providing an estimate of a random variable that satisfies a linear differential equation driven by white noise [see (18.2-1)]. In this section the Kalman filter as developed in [19] is summarized together with other results obtained in that study. The least mean-square error criteria was used by Kalman and when the driving noise is not present the results are consistent with those obtained using the least-squares error estimate, and minimum-variance estimate given previously.

Kalman [19] defines the optimal estimate as that which (if it exists) minimizes the expected value of a loss function $L(\varepsilon)$, that is, it minimizes $E[L(\varepsilon)]$, which is the expected loss, where

$$\varepsilon = x^*_{n,n} - x_n \tag{18.3-1}$$

where $x^*_{n,n}$ is an estimate of x_n, the parameter to be estimated based on the $n + 1$ observations given by

$$Y_{(n)} = (y_0, y_1, y_2, \ldots, y_n)^T \tag{18.3-2}$$

It is assumed that the above random variables have a joint probability density function given by $p(x_n, Y_{(n)})$. A scalar function $L(\varepsilon)$ is a loss function if it satisfies

$$\begin{align}
&\text{(i)} \quad L(0) = 0 \tag{18.3-3a}\\
&\text{(ii)} \quad L(\varepsilon') > L(\varepsilon'') > 0 \quad \text{if } \varepsilon' > \varepsilon'' > 0 \tag{18.3-3b}\\
&\text{(iii)} \quad L(\varepsilon) = L(-\varepsilon) \tag{18.3-3c}
\end{align}$$

Example loss functions are $L(\varepsilon) = \varepsilon^2$ and $L(\varepsilon) = |\varepsilon|$. Kalman [19] gives the following very powerful optimal estimate theorem

Theorem 1 [5, pp. 610–611] The optimal estimate $x^*_{n,n}$ of x_n based on the observation $Y_{(n)}$ is given by

$$x^*_{n,n} = E[x_n | Y_{(n)}] \tag{18.3-4}$$

If the conditional density function for x_n given $Y_{(n)}$ represented by $p(x_n | Y_{(n)})$ is (a) unimodel and (b) symmetric about its conditional expectation $E[x_n | Y_{(n)}]$.

The above theorem gives the amazing result that the optimum estimate (18.3-4) is independent of the loss function as long as (18.3-3a) to (18.3-3c) applies, it only depending on $p(x_n | Y_{(n)})$. An example of a conditional density function that satisfies conditions (a) and (b) is the Gaussian distribution.

In general, the conditional expectation $E[x_n|Y_{(n)}]$ is nonlinear and difficult to compute. If the loss function is assumed to be the quadratic loss function $L(\varepsilon) = \varepsilon^2$, then conditions (a) and (b) above can be relaxed, it now only being necessary for the conditional density function to have a finite second moment in order for (18.3-4) to be optimal. The estimate given by Theorem 1 is also called a Bayes estimate.

Before proceeding to Kalman's second powerful theorem, the concept of orthogonal projection for random variables must be introduced. Let λ_i and λ_j be two random variables. In vector terms these two random variables are independent of each other if λ_i is not just a constant multiple of λ_j. Furthermore, if [5, p. 611]

$$\lambda = \alpha_i \lambda_i + \alpha_j \lambda_j \qquad (18.3\text{-}5)$$

is a linear combination of λ_i and λ_j, then λ is said to lie in the two-dimensional space defined by λ_i and λ_j. A basis for this space can be formed using the Gram–Schmidt orthogonalization procedure. Specifically, let [5, p. 611]

$$e_i = \lambda_i \qquad (18.3\text{-}6)$$

and

$$e_j = \lambda_j - \frac{E\{\lambda_i\lambda_j\}}{E\{\lambda_i^2\}}\lambda_i \qquad (18.3\text{-}7)$$

It is seen that

$$E\{e_ie_j\} = 0 \qquad i \neq j \qquad (18.3\text{-}8)$$

The above equation represents the orthogonality condition. (The idea of orthogonal projection for random variables follows by virtue of the one-for-one analogy with the theory of linear vector space. Note that whereas in linear algebra an inner product is used, here the expected value of the product of the random variables is used.) If we normalize e_i and e_j by dividing by their respective standard deviations, then we have "unit length" random variables and form an orthonormal basis for the space defined by λ_i and λ_j. Let e_i and e_j now designate these orthonormal variables. Then

$$E\{e_ie_j\} = \delta_{ij} \qquad (18.3\text{-}9)$$

where δ_{ij} is the Kronecker δ function, which equals 1 when $i = j$ and equals 0 otherwise.

Let β be any random variable that is not necessarily a linear combination of λ_i and λ_j. Then the orthogonal projection of β onto the λ_i, λ_j space is defined by [5, p. 612]

$$\bar{\beta} = e_i E\{\beta e_i\} + e_j E\{\beta e_j\} \qquad (18.3\text{-}10)$$

Define

$$\tilde{\beta} = \beta - \bar{\beta} \qquad (18.3\text{-}11)$$

Then it is easy to see that [5, p. 612]

$$E\{\tilde{\beta}e_i\} = 0 = E\{\tilde{\beta}e_j\} \qquad (18.3\text{-}12)$$

which indicates that $\bar{\beta}$ is orthogonal to the space λ_i, λ_j. Thus β has been broken up into two parts, the $\bar{\beta}$ part in the space λ_i, λ_j, called the orthogonal projection of β onto the λ_i, λ_j space, and the $\tilde{\beta}$ part orthogonal to this space. The above concept of orthogonality for random variables can be generalized to an n-dimensional space. (A less confusing labeling than "orthogonal projection" would probably be just "projection.")

We are now ready to give Kalman's important Theorem 2.

Theorem 2 [5, pp. 612–613]   The optimum estimate $x_{n,n}^*$ of x_n based on the measurements $Y_{(n)}$ is equal to the orthogonal projection of x_n onto the space defined by $Y_{(n)}$ if
  1. The random variables $x_n, y_0, y_1, \ldots, y_n$ all have zero mean and either
  2. (a) x_n and $Y_{(n)}$ are just Gaussian or (b) the estimate is restricted to being a linear function of the measurement $Y_{(n)}$ and $L(\varepsilon) = \varepsilon^2$.

The above optimum estimate is linear for the Gaussian case. This is because the projection of x_n onto $Y_{(n)}$ is a linear combination of the element of $Y_{(n)}$. But in the class of linear estimates the orthogonal projection always minimizes the expected quadratic loss given by $E[\varepsilon^2]$. Note that the more general estimate given by Kalman's Theorem 1 will not be linear. However, if x_n and $Y_{(n)}$ are Gaussian then the optimal filter obtained using Theorem 1 is linear and identical to that of Theorem 2. We shall also shortly show that this filter is the Kalman filter given in Chapters 2 and 9.

Up till now the observations y_i and the variable x_n to be estimated were assumed to be scaler. Kalman actually gives his results for the case where they are vectors, and hence Kalman's Theorem 1 and Theorem 2 apply when these variables are vectors. We shall now apply Kalman's Theorem 2 to obtain the form of the Kalman filter given by him.

Let the target dynamics model be given by (18.2-1) and let the observation scheme be given by [5, p. 613]

$$Y(t) = M(t)X(t) \qquad (18.3\text{-}13)$$

Note that Kalman, in giving (18.3-13), does not include any measurement noise term $N(t)$. Because of this, the Kalman filter form he gives is different from that given previously in this book (see Section 2.4). We shall later show that his form can be transformed to be identical to the forms given earlier in this book. The measurement $Y(t)$ given in (18.3-13) is assumed to be a vector. Let us assume that observations are made at times $i = 0, 1, \ldots, n$ and can be represented by measurement vector given by

$$Y_{(n)} \equiv \begin{bmatrix} Y_{(n)} \\ --- \\ Y_{n-1} \\ --- \\ \vdots \\ --- \\ Y_0 \end{bmatrix} \qquad (18.3\text{-}14)$$

What is desired is the estimate $X^*_{n+1,n}$ of X_{n+1}, which minimizes $E[L(\varepsilon)]$. Applying Kalman's Theorem 2, we find that the optimum estimate is given by the projection of X_{n+1} onto $Y_{(n)}$ of (18.3-14). In reference 19 Kalman shows that this solution is given by the recursive relationships [5, p. 614]

$$\Delta^*_n = \Phi(n+1,n)P^*_n M^T_n (M_n P^*_n M^T_n)^{-1} \qquad (18.3\text{-}15a)$$

$$\Phi^*(n+1,n) = \Phi(n+1,n) - \Delta^*_n M_n \qquad (18.3\text{-}15b)$$

$$X^*_{n+1,n} = \Phi^*(n+1,n)X^*_{n,n-1} + \Delta^*_n Y_n \qquad (18.3\text{-}15c)$$

$$P^*_{n+1} = \Phi^*(n+1,n)P^*_n \Phi^*(n+1,n)^T + Q_{n+1,n} \qquad (18.3\text{-}15d)$$

The above form of the Kalman filter has essentially the notation used by Kalman in reference 19; see also reference 5. Physically, $\Phi(n+1,n)$ is the transition matrix of the unforced system as specified by (18.2-3). Defined earlier, M_n is the observation matrix, $Q_{n+1,n}$ is the covariance matrix of the vector $V(t_{n+1}, t_n)$, and the matrix P^*_{n+1} is the covariance matrix of the estimate $X^*_{n+1,n}$.

We will now put the Kalman filter given by (18.3-15a) to (18.3-15d) in the form of (2.4-4a) to (2.4-4j) or basically (9.3-1) to (9.3-1d). The discrete version of the target dynamics model of (18.2-3) can be written as [5, p. 614]

$$X_{n+1} = \Phi(n+1,n)X_n + V_{n+1,n} \qquad (18.3\text{-}16)$$

The observation equation with the measurement noise included can be written as

$$Y_n = M_n X_n + N_n \qquad (18.3\text{-}17)$$

instead of (18.3-13), which does not include the measurement noise. Define an augmented state vector [5, p. 614]

$$X'_n = \begin{bmatrix} X_n \\ ---- \\ N_n \end{bmatrix} \qquad (18.3\text{-}18)$$

and augmented driving noise vector [5, p. 615]

$$V'_{n+1,n} = \begin{bmatrix} V_{n+1,n} \\ ------- \\ N_{n+1} \end{bmatrix} \qquad (18.3\text{-}19)$$

Define also the augmented transition matrix [5, p. 615]

$$\Phi'(n+1,n) = \begin{bmatrix} \Phi(n+1,n) & | & 0 \\ -------------- & | & \\ 0 & | & 0 \end{bmatrix} \qquad (18.3\text{-}20)$$

and the augmented observation matrix

$$M'_n = (M_n \mid I) \qquad (18.3\text{-}21)$$

It then follows that (18.3-16) can be written as [5, p. 615]

$$X'_{n+1} = \Phi'(n+1,n)X'_n + V'_{n+1,n} \qquad (18.3\text{-}22)$$

and (18.3-17) as [5, p. 615]

$$Y_n = M'_n X'_n \qquad (18.3\text{-}23)$$

which have the same identical forms as (18.2-3) and (18.3-13), respectively, and to which Kalman's Theorem 2 was applied to obtain (18.3-15). Replacing the unprimed parameters of (8.3-15) with their above-primed parameters yields [5, p. 616]

$$X^*_{n,n} = X^*_{n,n-1} + H_n(Y_n - M_n X^*_{n,n-1}) \qquad (18.3\text{-}24a)$$

$$H_n = S^*_{n,n-1}M_n^T(R_n + M_n S^*_{n,n-1}M_n^T)^{-1} \qquad (18.3\text{-}24b)$$

$$S^*_{n,n} = (I - H_n M_n)S^*_{n,n-1} \qquad (18.3\text{-}24c)$$

$$S^*_{n,n-1} = \Phi(n,n-1)S^*_{n-1,n-1}\Phi(n,n-1)^T + Q_{n,n-1} \qquad (18.3\text{-}24d)$$

$$X^*_{n,n-1} = \Phi(n,n-1)X^*_{n-1,n-1} \qquad (18.3\text{-}24e)$$

where $Q_{n+1,n}$ is the covariance matrix of $V_{n+1,n}$ and R_{n+1} is the covariance matrix of N_{n+1}. The above form of the Kalman filter given by (18.3-24a) to (18.3-24e) is essentially exactly that given by (2.4-4a) to (2.4-4j) and (9.3-1) to (9.3-1d) when the latter two are extended to the case of a time-varying dynamics model.

Comparing (9.3-1) to (9.3-1d) developed using the minimum-variance estimate with (18.3-24a) to (18.3-24e) developed using the Kalman filter projection theorem for minimizing the loss function, we see that they differ by the presence of the Q term, the variance of the driving noise vector. It is gratifying to see that the two radically different aproaches led to essentially the same algorithms. Moreover, when the driving noise vector V goes to 0, then (18.3-24a) to (18.3-24e) is essentially the same as given by (9.3-1) to (9.3-1d), the Q term in (18.3-24d) dropping out. With V present X_n is no longer determined by X_{n-1} completely. The larger the variance of V, the lower the dependence of X_n on X_{n-1} and as a result the less the Kalman filter estimate $X_{n,n}^*$ should and will depend on the past measurements. Put in another way the larger V is the smaller the Kalman filter memory. The Kalman filter in effect thus has a fading memory built into it. Viewed from another point of view, the larger Q is in (18.3-24d) the larger $S_{n,n-1}^*$ becomes. The larger $S_{n,n-1}^*$ is the less weight is given to $X_{n,n-1}^*$ in forming $X_{n,n}^*$, which means that the filter memory is fading faster.

The matrix Q is often introduced for purely practical reasons even if the presence of a process noise term in the target dynamics model cannot be justified. It can be used to counter the buildup of a bias error. The shorter the filter memory the lower the bias error will be. The filter fading rate can be controlled adaptively to prevent bias error buildup or to respond to a target maneuver. This is done by observing the filter residual given by either

$$r_n = (Y_n - M_n X_{n,n}^*)^T (Y_n - M_n X_{n,n}^*) \qquad (18.3\text{-}25)$$

or

$$r_n = (Y_n - M_n X_{n,n}^*)^T (S_{n,n}^*)^{-1} (Y_n - M_n X_{n,n}^*) \qquad (18.3\text{-}26)$$

The quantity

$$s_n = Y_n - M_n X_{n,n}^* \qquad (18.3\text{-}27)$$

in the above two equations is often called the innovation process or just innovation in the literature [7, 127]. The innovation process is white noise when the optimum filter is being used.

Another benefit of the presence of Q in (18.3-24d) is that it prevents S^* from staying singular once it becomes singular for any reason at any given time. A matrix is singular when its determinent is equal to zero. The matrix S^* can become singular when the observations being made at one instant of time are perfect [5]. If this occurs, then the elements of H in (18.3-24a) becomes 0, and H becomes singular. When this occurs, the Kalman filter without process noise stops functioning — it no longer accepts new data, all new data being given a 0 weight by $H = 0$. This is prevented when Q is present because if, for example, $S_{n-1,n-1}^*$ is singular at time $n-1$, the presence of $Q_{n,n-1}$ in (18.3-24d) will make $S_{n,n-1}^*$ nonsingular.

APPENDIX

COMPARISON OF SWERLING'S AND KALMAN'S FORMULATIONS OF SWERLING–KALMAN FILTERS

PETER SWERLING

A.1 INTRODUCTION

In the late 1950s and early 1960s Swerling and Kalman independently developed what amounts to the same technique of recursive statistically optimum estimation [1–3]. It is of some interest to compare their formulations with respect to features that have some reasonably substantial relevance to actual applications of the last 35 years, particularly applications to tracking (include orbit prediction — as used henceforth, "tracking" includes but is not limited to orbit prediction).

Swerling was motivated by applications to estimating the orbits of earth satellites or other space vehicles. Though thus motivated, his actual formulation is presented in terms of an abstract n-component system vector with abstract dynamics, not specialized to orbit estimation. Kalman was motivated by the aim of deriving new ways to solve linear filtering and prediction problems. Swerling presented his results as recursive implementations of the Gauss method of least squares (somewhat genrealized to allow nondiagonal quadratic forms, which is necessary for statistical optimality); Kalman's development was presented as a recursive way to implement the solution of Wiener filtering and prediction. Specific features of the respective presentations were influenced by these motivations, but it is a simple matter to show either equivalence of the actual results or the straightforward extension of either author's results to the other's, as discussed below.

The defining essence of these statistically optimum recursive estimation methods is as follows:

(a) At any given time, all observational data available up to that time are employed to form a statistically optimum (minimum-variance) estimate of state parameters.

(b) However, at any given time one does not retain the whole record of all previous observations; rather, all observations up to that time are encapsulated in a current state vector estimate, that is, in a vector of n scalar estimates, n being the number of components of the state vector, together with the $n \times n$ error covariance matrix of the state vector estimate.

(c) When new observational data become available, they are optimally combined with the most recent state vector estimate based on previous observations to form a new ("updated") optimum estimate. The update equation involves the error covariance matrix of the estimate based on previous observations, which is also updated to a new error covariance matrix based on all observations including the latest.

The following is not an introduction to the subject. It is assumed that the reader is familiar with the basic concepts and nomenclature.

A.2   COMPARISON OF SWERLING'S AND KALMAN'S FORMULATIONS

A.2.1   Linear and Nonlinear Cases

Linear cases are those in which the relations between error-free observations and the state are linear and the state dynamics are described by linear equations. Nonlinear cases are those in which these relations are nonlinear. A large majority of applications, particularly to tracking, involve nonlinear relations.

In nonlinear cases, various matrices and vectors appearing in the update equations depend on the state vector or, more precisely, on an estimate of the state vector. In linear cases these dependences on the state vector disappear.

Swerling's initial formulation [1] included both linear and nonlinear cases (necessarily, since the motivation was orbit prediction). In fact, his initial formulation is directly for nonlinear cases; the linear case would be a specialization. Swerling actually gives two different formulations for nonlinear cases, which become identical in linear cases.

Kalman's initial formulation [2, 3] is for linear cases. However, it is a simple matter to extend Kalman's results to nonlinear cases. This has been done by later investigators in a variety of ways with respect to specific computational details. The resulting approaches have come to be known as the extended Kalman filter. Thus, the extended Kalman filter is the original Swerling filter.

A.2.2   System Noise

Kalman's initial formulation is in terms of estimating parameters of a system, which is considered to be a Markov random process.

Swerling's initial formulation considers the system to vary deterministically. In the first part of [1] he considers the system parameters to be constant. Then he extends the formulation to time-varying systems, the variation with time being deterministically described. (He actually gives two ways of extending the results to time-varying systems and two variations of the second way.)

It is a simple matter to extend Swerling's results to include Markov system statistics. The most straightforward way to do this is to use as a springboard his second way of treating time-varying systems in the section of reference 1 entitled "Modified Stagewise Procedure for Time-Varying Elements." Using somewhat simpler notation than in [1], the system dynamics are described by

$$X(t + \Delta t) = F[X(t), t, t + \Delta t] \qquad (A.2-1)$$

Then, absent a new observation, the estimate of X is extrapolated via Eq. (A.2-1) and its error covariance matrix is extrapolated in the manner dictated by the first-order expansion of F.

In order to treat the case where $X(t)$ is a Markov process, replace Eq. (A.2-1) by

$$X(t + \Delta t) = F[X(t), t, t + \Delta t] + \delta X(t, t + \Delta t) \qquad (A.2-2)$$

where δX is a zero-mean uncorrelated increment. Then, in the absence of a new observation,

(a) the estimate of X is extrapolated in the same way as before, that is, via Eq. (A.2-1) with $\delta X \equiv 0$.

(b) the error covariance matrix is extrapolated by adding the covariance matrix of δX to the extrapolation which would be obtained for $\delta X \equiv 0$.

The updating equations for incorporating new observational data are the same in either case.

In a later work [4] Swerling showed how to apply optimum recursive estimation to cases where both the system and the observation noise are random processes with essentially arbitrary covariance properties, that is, either or both can be non-Markov. This extension is by no means straightforward. The resulting update equations take the form of partial difference or partial differential equations.

A.2.3 Observation Noise

Swerling [1] considered observation noise to be correlated in blocks, with zero correlation between blocks and arbitrary intrablock correlation.

Kalman [2] does not explicitly include observation noise, but it is a simple matter to include Markov observation noise by augmenting the state vector to include observation noise components as well as system components. The same trick could be used to extend Swerling's formulation to include correlated Markov observation noise, once it has been extended to include Markov system noise as described in Section A.2.2. Kalman and Bucy in [3] explicitly assume white (uncorrelated) observation noise.

A.2.4 Matrix Inversion

The matrix inversion issue arises in relation to Swerling's formulation as follows. The method of updating the system estimate when incorporating a new observation depends on the error covariance matrix C, but the most "natural" way to update the error covariance matrix itself is to update its inverse C^{-1} and then invert. This imposes the need to invert an $n \times n$ matrix each time new observational data become available, n being the number of components of the system. Thirty-odd years ago this imposed an uncomfortable computing load; even with today's computing power, there is still a motive to reduce the order of matrices that need to be inverted.

Swerling [1, Eqs. (47) and (48)] stated a method of avoiding matrix inversion altogether by introducing new (scalar) observations one at a time and also assuming observation errors to be uncorrelated. Blackman in a 1964 review work [5] describes a method due to R. H. Battin of avoiding matrix inversion by introducing new observations one at a time [5, Section VI]; he also shows [5, Section VII] how to generalize these methods to allow new observations to be introduced k at a time with $k < n$, in which case matrices of order $k \times k$ need to be inverted.

A close scrutiny of Kalman's equations [2, Eqs. (28) to (32)] reveals that the order of matrices that need to be inverted is (in our present notation) $k \times k$ where new scalar observational data are introduced k at a time.

A.2.5 Imperfect Knowledge of System Dynamics

In many applications, including almost all tracking applications, the system dynamics are imperfectly known. Another way of saying this is that the function F in Eqs. (A.2-1) and (A.2-2) is not exactly known. Swerling [1] refers to this as "unknown perturbations;" much other literature refers to "plant model errors."

We will assume that such imperfect knowledge of F cannot validly be modeled as uncorrelated increments having the same status as δX in Eq. (A.2-2). This is the case in the vast majority of applications; in a sense it is true by definition, since if imperfect knowledge of F could be so modeled, it would be regarded as part of system noise.

An example arises in the tracking of maneuvering airborne targets when the higher derivatives of target motion are not described by known functions of the positions and velocities (in the six-component models) or of the positions, velocities, and accelerations (in the nine-component models).

In the presence of such plant model errors it is necessary to give the recursive estimation algorithm a fading memory to avoid unacceptable growth in the estimation errors. Swerling [1] suggests a method in which the elements of the extrapolated error covariance matrix are multiplied by a set of multiplicative factors. He does this in such a way that different components of the system state vector can be associated with different rates of fading memory, a feature that can be important in some applications. For example, suppose in orbit prediction that the orbit is described by the six osculating Keplerian elements. Certain orbital elements, for example, the inclination of the orbital plane, are much more stable in the presence of perturbations than others.

Swerling's multiplication method is ad hoc in that it does not purport to achieve statistical optimality. Indeed in order to attempt to achieve statistical optimality, it would be necessary to associate a statistical model with errors in the knowledge of F and in practice this usually is very difficult to do.

Kalman [2, 3] does not explicitly address this kind of imperfect knowledge of system dynamics. However, his algorithm contains a covariance matrix that is added to the extrapolated error covariance matrix, stemming from the presence of system noise (cf. the discussion in Section A.2.2). An additive covariance matrix of this type can also be used as an ad hoc way of providing a fading memory in the presence of plant model errors, even though the statistical properties of the latter differ from those assumed by Kalman for system noise. This is often the approach used to provide a fading memory (e.g., in the tracking of maneuvering targets). In fact, in tracking applications, when an additive deweighting matrix appears in the extrapolation of the error covariance matrix, in the majority of cases it is there to deal with the imperfect knowledge of system dynamics rather than with system noise of the type modeled by Kalman.

REFERENCES

1. Swerling, P., "First Order Error Propagation in a Stagewise Smoothing Procedure for Satellite Obvervations," *Journal of the Astronautical Sciences*, Vol. 6, No. 3, Autumn 1959.

2. Kalman, R. E., "A New Approach to Linear Filtering and Prediction Problems," *Transactions of the ASME, Journal of Basic Engineering*, Series 83D, March 1960, pp. 35–45.

3. Kalman, R. E., and R. S., Bucy., "New Results in Linear Filtering and Prediction Theory," *Transactions of the ASME, Journal of Basic Engineering*, Series 83D, March 1961, pp. 95–108.

4. Swerling, P., "Topics in Generalized Least Squares Signal Estimation," *SIAM Journal of Applied Mathematics*, Vol. 14, No. 5, September 1966.

5. Blackman, R. B., "Methods of Orbit Refinement," *Bell System Technical Journal*, Vol. 43, No. 3, May 1964.

PROBLEMS

1.2.1-1 Show that (1.2-11a) and (1.2-11b) for g and h independent of n can be put in the following standard feedback filter form [12]:

$$x^*_{n+1,n} = (g + h)y_n - g y_{n-1} + (2 - g - h)x^*_{n,n-1}$$
$$+ (g - 1)x^*_{n-1,n-2})$$

$$(\text{P1.2.1-1a})$$

In the recursion equation given by (P1.2.1-1a), $x^*_{n+1,n}$ is written in terms of the last two measurements y_n, y_{n-1} and the preceding two position predictions $x^*_{n,n-1}$, $x^*_{n-1,n-2}$. *Hint*: Substitute (1.2-11a) for $\dot{x}_{n+1,n}$ into (1.2-11b). Then use prediction equation (1.2-11b) for $x^*_{n,n-1}$ to solve for $\dot{x}^*_{n,n-1}$.

#1.2.1-2 Find recursion a equation similar to (P1.2.1-1a) for $x^*_{n,n}$.

#1.2.1-3 Repeat problem 1.2.1-2 for $\dot{x}^*_{n,n}$.

Notes: (1) Many of the problems derive important results: g–h recursive filter equation (problems 1.2.1-1 and 1.2.6-5), g–h filter bias error (1.2.4.3-1 and 1.2.6-4), g–h filter VRF for $x^*_{n+1,n}$ (1.2.4.4-1 and 1.2.6-2), relationship between g and h for critically damped filter (1.2.6-1) and for Benedict–Bordner filter (2.4-1), g–h filter transient error (1.2.6-3), stability conditions for g–h filter (1.2.9-1), relationship between g, h, T, σ_u^2, and σ_x^2 for steady-state Kalman filter (2.4-1), minimum-variance estimate (4.5-1), Bayes filter (9.2-1), Kalman filter without dynamic noise (9.3-1). These problems and their informative solutions form an integral part of the text. Other problems give a further feel for filter design parameters; see for instance the examples of Table P2.10-1 (in the solution section) and the problems associated with it.

(2) Problems marked with a dagger involve difficult algebraic manipulations.

(3) Problems marked with the pound symbol are those for which no solutions are provided.

(4) Equations and tables marked with double daggers are in the solution section. For example, Eq. (P1.2.6-1a)†† is found in the solution to Problem 1..2.6-1.

1.2.4.3-1   Using (P1.2.1-1a) derive the bias errors b^* given by (1.2-15). *Hint*: Apply constant-acceleration input

$$y_n = \tfrac{1}{2}\ddot{x}(nT)^2 \qquad n = 0, 1, \dots, n \qquad \text{(P1.2.4.3-1a)}$$

to filter. The output at time in steady state n will be the desired output plus the bias error b^*:

$$x^*_{n+1,n} = \tfrac{1}{2}\ddot{x}(n+1)^2 T^2 + b^* \qquad \text{(P1.2.4.3-1b)}$$

Similar expressions can be written for $x^*_{n,n-1}$ and $x^*_{n-1,n-2}$ that allow us to solve (P1.2.1-1a) for b^*.

#1.2.4.3-2   Using the results of problem 1.2.1-2 derive $b^*_{n,n}$ given by (1.2-16a).

#1.2.4.3-3   Using the results of problem 1.2.1-3 derive $\dot{b}^*_{n,n}$ given by (1.2-16b).

†1.2.4.4-1   Using (P1.2.1-1a) derive the VRF equation for $\sigma^2_{n+1,n}$ given by (1.2-19).

Hint: Square (P1.2.1-1a) and obtain its expected value making use of fact that y_n is independent of y_m for $m \neq n$ and that we are in steady-state conditions so that the expected values are independent of n. Multiply (P1.2.1-1a) rewritten for $x^*_{n,n-1}$ by y_{n-1} to obtain $E[x^*_{n,n-1}y_{n-1}]$ and by $x^*_{n-1,n-2}$ to obtain $E[x^*_{n,n-1}x^*_{n-1,n-2}]$.

#1.2.4.4-2   Using the results of problem 1.2.1-2, derive the VRF equation for $\sigma^2_{n,n}$ given by (1.2-20).

†#1.2.4.4-3   Derive (1.2-21).

1.2.5-1   (a) For a normalized acceleration $A_N = 3.0$, find the Benedict–Bordner filter for which $3\sigma_{n+1,n} = b^*$. Use Figure 1.2-7. Verify that these values for 3δ and b^*_N obtained from Figure 1.2-7 agree with the values obtained from (1.2-19) and (1.2-15).

(b) For $A_N = 3.0$ find the Benedict–Bordner filter for which E_{TN} of (1.2-31) is minimum. Use Figure 1.2-9.

(c) Compare the normalized total errors E_{TN} for these designs. How much lower is this total error when the minimum E_{TN} design is used? Compare the values of b^*/σ_x and $3\sigma_{n+1,n}/\sigma_x$ obtained for these designs.

1.2.5-2   Repeat problems 1.2.5-1(a) through (c) for $A_N = 0.001$.

#1.2.5-3   Repeat problems 1.2.5-1(a) through (c) for $A_N = 0.1$.

1.2.6-1 (a) Find the z-transform of (P1.2.1-1a) and in turn the transfer function $H_p(z) = X^*_{n+1,n}(z)/Y_n(z)$, where $X^*_{n+1,n}(z)$ and $Y_n(z)$ are the z-transforms of $x^*_{n+1,n}$ and y_n. For simplicity we will drop the arguments of the z-transform to give $X^*_{n+1,n}$ and Y_n for the z-transforms.

(b) Find the poles of the g–h filter transfer function H_p.

(c) Using the solution to (b), prove the relationship (1.2-36) between h and g for the critically damped filter.

†1.2.6-2 Obtain the g–h filter VRF for $\sigma^2_{n+1,n}$ given by (1.2-19) using the z-transform function $H_p(z)$ of Problem 1.2.6-1.

†1.2.6-3 Derive the expression for the transient error of a g–h filter given by (1.2-28).

Hint: Use z-transforms. Specifically apply the z-transform of the input ramp function to the g–h filter prediction z-transform obtained from problem 1.2.6-1, specifically given by (P1.2.6-1c). Subtract from this the ideal predicted output to generate the transient error sequence ε_n, $n = 0, 1, 2, \ldots, n$. Use (P1.2.6-2d)†† to evaluate the transient error with $H_p(z)$ replaced by $E(z)$, the z-transform of ε_n.

1.2.6-4 From the z-transform of the transient error ε_n for constant-acceleration input given by (P1.2.4.3-1a) and the final-value theorem [130]

$$\lim_{n \to \infty} \varepsilon_n = \lim_{z \to 1} (1 - z^{-1}) E(z) \qquad \text{(P1.2.6-4a)}$$

where $E(z)$ is the z-transform of ε_n, derive b^* given by (1.2-15).

†1.2.6-5 Obtain the g–h recursive feedback filter given by (P1.2.1-1a) using the z-transform of (1.2-8a), (1.2-8b), (1.2-10a) and (1.2-10b) for g and h constant.

Hint: Divide the z-transform of these equations by the z-transform of y_n, that is $Y_n(z)$, to obtain z-transforms of the filter transfer functions for $x^*_{n+1,n}$, $x^*_{n,n}$, $\dot{x}^*_{n,n}$, and $\dot{x}^*_{n+1,n}$. We will have four equations with four unknowns that can be used to solve for the four unknown transfer functions. Obtaining the appropriate inverse transfer function of the resulting $H_p(z)$ gives (P1.2.1-1a); see reference 131.

1.2.6-6 Repeat problem 1.2.5-1 for the critically damped filter using Figures 1.2-13 to 1.2-15.

1.2.6-7 Repeat problem 1.2.6-6 for $A_N = 0.001$.

1.2.6-8   Repeat problem 1.2.6-6 for $A_N = 0.003$.

#1.2.6-9   Using the procedure outlined in problem 1.2.6-3, derive the transient error given by (1.2-29)

#1.2.6-10   Using the procedure outlined in problem 1.2.6-4, derive $b_{n,n}^*$ of (1.2-16a).

#1.2.6-11   Using the procedure of problem 1.2.6-5, derive the $g–h$ recursive feedback filter for $x_{n,n}^*$.

#1.2.6-12   Repeat problem 1.2.6-6 for $A_N = 0.1$.

†**#1.2.6-13**   Using the procedure of problem 1.2.6-3 derive (1.2-30).

#1.2.7-1   Verify the results of examples 2 and 2a of Table 1.2-6 for the $g–h$ Bendict–Bordner filter.

1.2.7-2   (a) Verify results of examples 1a, 1b, and 1c of Table 1.2-7 for the critically dampled $g–h$ filter.
   (b) Repeat example 1c for $\ddot{x}_{max} = 1g$ and $\sigma_{n+1,n} = 150$ ft.
   (c) Repeat (b) with $\sigma_x = 167$ ft and $\sigma_{n+1,n}/\sigma_x = 3$.

1.2.7-3   Assume an air route surveillance radar with pulse width $= 1\,\mu s$ so that the range resolution is $500\,\text{ft}$; range accuracy $\sigma_x = 500\,\text{ft}$; scan period $T = 10\,\text{sec}$; $g_{max} = 0.2\,g = 6.4\,\text{ft/sec}^2$. Require $\sqrt{\text{VAR}(x_{n+1,n}^*)} = 608$ ft.
   (a) Design a $g–h$ Benedict-Bordner filter.
   (b) Determine b^*, $D_{x_{n+1,n}^*}/\Delta v^2$.

#1.2.7-4   (Continuation of problem 1.2-7-3)
   (a) Simulate the filter of problem 1.2.7-3 on the computer. See how well it tracks the turn maneuver of Figure P1.2.7-4. At the start of the turn maneuver it is assumed that the filter is in steady state with $x_{n,n-1}^* = y_n$.
   Assume that the range error is Gaussian with $500\,\text{ft}$ rms error independent from measurement to measurement; the range window for update is $\pm 3[(608)^2 + (500)^2]^{1/2} = 2,360\,\text{ft}$; and the range-to-radar distance is much greater than $16.46\,\text{nmi}$. Does the filter maintain track during the maneuver? Plot the tracking-filter error $(y_n - x_{n,n-1}^*)$ versus n.
   (b) Repeat for $R = 10\,\text{nmi}$ and $1\,\text{nmi}$

#1.2.7-5   Repeat problems 1.2.7-3 and 1.2.7-4 for a fading-memory filter.

#1.2.7-6   Assume $\sigma_x = 150\,\text{ft}$ and $\ddot{x}_{max} = 2g$. Find the Benedict $g–h$ filter for which $3\delta = b_N^* = 1$. What is T and b^*. Use Figure 1.2-7.

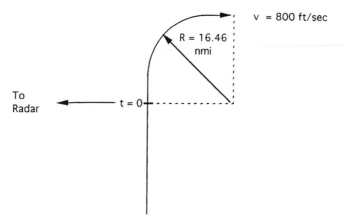

P1.2.7-4 Trajectory of maneuvering target.

#1.2.7-7 Repeat problem 1.2.7-6 for a critically damped g–h filter.

#1.2.7-8 Design a Benedict–Bordner g–h filter for $\sigma_x = 150\,\text{ft}$, $\ddot{x}_{max} = 2g = 64\,\text{ft/sec}^2$, and $(\sigma_{n+1,n}/\sigma_x) = 1$. Use (1.2-22) to solve for b^* and then (1.2-15) to solve for T. Find $\sigma_{n+1,n}$, $\sigma_{n,n}$, and $\dot{\sigma}_{n+1,n} = \sqrt{\text{VAR}(\dot{x}^*_{n+1,n})}$ using (1.2-19) to (1.2-21) and check that these same values are obtained using Figure 1.2-7.

1.2.9-1 Using the z-transform function $H_p(z)$ obtained in problem 1.2.6-1, which is

$$H_p(z) = \frac{z[(g+h)z - g]}{z^2 + (g+h-2)z + (1-g)} \qquad \text{(P.1.2.6-1c)}^{\dagger\dagger}$$

derive the stability conditions given by (1.2-37a) to (1.2-37c). *Hint*: Use the transform

$$u = \frac{z-1}{z+1} \qquad \text{(P.1.2.9-1a)}$$

which maps the interior of the unit circle of the z-plane into the left-hand plane [132].

1.2.10-1 (Continuation of problem 1.2.7-3)
Specify the track initiation filter for problem 1.2.7-3 and its weights g_n and h_n. Determine for which n a switch is to take place from a track initiation filter to the Benedict–Bordner filter designed in problem 1.2.7-3.

#1.2.10-2 (Continuation of problems 1.2.7-3 and 1.2.10-1)

(a) Simulate the filters of problems 1.2.7-4 and 1.2.10-1 going from detection at $t = 0$ $(n = 0)$ to steady state operation at $t = 100(n = 10)$ sec $= 1000$ sec. Assume the approaching constant-velocity target is detected at $t = 0$ $(n = 0)$; the range error is Gaussian with 500 ft rms error independent from measurement to measurement; and the range window for update is $\pm 2,360$ ft.

(b) Find the probability of $x^*_{n+1,n}$ not falling in the range window in steady state.

(c) Repeat (a) and (b) for a range window of $\pm(\frac{2}{3})2,360 = \pm 1,575$ ft.

1.2.10-3 Specify the track initiation filter for problem 1.2.7-5.

1.2.10-4 Verify (1.2-41) using (1.2-19), (1.2-35a), and (1.2-35b).

#1.2.10-5 For a critically damped g–h filter for what θ is $\mathrm{VRF}(x^*_{n+1,n}) = 3.0$, 1.0, 0.1, 0.03.

1.3-1 Show that (1.3-10a) to (1.3-10c) satisfy (1.3-9).

#2.3-1 Derive (2.3-4) from (2.3-1).

2.4-1 Show that g and h are related by (2.1-4) for the steady-state two-state Kalman filter having the dynamic model given by (2.4-10). Also show that σ^2_x, T, and σ^2_u are related to g and h by (2.1-5) for this case. *Hint:* Start with (2.4-4e), (2.4-4f), and (2.4-4j). Substitute (2.4-4j) into (2.4-4f). Expand these matrices out, equating corresponding terms on each side of the equation. Assume steady-state conditions, in which case the term of $S^*_{n,n-1}$ equal those of $S^*_{n-1,n-2}$, so that the subscript n can be dropped. This is true also for H_n and Q_n. Solving the equations that result leads to (2.1-4) and (2.1-5).

2.4-2 Using the results of problem 2.4-1 and extending these results by using (2.4-4j) to obtain $S^*_{n-1,n-1}$ from $S^*_{n,n-1}$, which we will for smplicity denote as respectively $\bar{S}$ and $\hat{S}$ in steady state, obtain expressions for the steady-state components of the matrices $\bar{S}$ and $\hat{S}$ for the g–h Kalman filter having the dynamics model gives by (2.1-1a) and (2.1-1b) with $\mathrm{VAR}(u_n) = \sigma^2_u$. Show that $\hat{s}_{12} = \hat{s}_{21}$ and $\bar{s}_{12} = \bar{s}_{21}$.

#2.4-3 (a) Using (2.1-5) find steady-state g–h Kalman filter parameters g and h for the case where the target dynamics model is given by (2.1-1a) and (2.1-1b) with u_n independent of u_{n+j} for $j \neq 0$ with the variance of u_n equal to σ^2_u independent of n for $\sigma_x = 150$ ft, $\sigma_u = 90.9$ ft/sec, and $T = 1.421$ sec. (Note that the T used here is that of the Benedict–Bordner filter of problem 1.2.7-8. Also the σ_u used here was obtained from (2.1-6) using $B = 1$ and $T = 1.421$ sec and $\ddot{x}_{max} = 2g = 64\,\text{ft/sec}^2$ of problem 1.2.7-8.) Using the results of problem 2.4-2, solve for $\sigma_{n+1,n}$, $\sigma_{n,n}$, $\dot{\sigma}_{n+1,n}$, and $\dot{\sigma}_{n,n}$. Using (2.4-4f), show how $\hat{s}_{ij}$ depends on $\bar{s}_{ij}$ and T and from the relationship obtain $\sigma_{n+1,n}$ and $\dot{\sigma}_{n+1,n}$ from $\sigma_{n,n}$ and $\dot{\sigma}_{n,n}$ and verify that the numerical values agree with those obtained above. Plot $\sigma_{n+\varepsilon,n}$ and $\dot{\sigma}_{n+\varepsilon,n}$ versus time, where $0 \le \varepsilon \le 1$, for $n = 0,1,2,3$, assuming track starts for $n \ll 0$.

(b) For the Benedict–Bordner filter having the same g, h, and T obtained for (a) above, find $\sigma_{n+1,n}$, $\sigma_{n,n}$, and $\dot{\sigma}_{n+1,n}$ for $\sigma_u^2 = 0$. Plot $\sigma_{n+\varepsilon,n}$ versus time, $0 \le \varepsilon \le 1$, $n = 0,1,2,3$. Compare these values with those obtained for the steady-state g–h Kalman filter above. Explain why values for $\sigma_{n+1,n}$, $\sigma_{n,n}$ and $\dot{\sigma}_{n+1,n}$ are smaller. Also find b^* for $\ddot{x}_{max} = 2g = 64\,\text{ft/sec}^2$, the value of problem 1.2.7-8. Note that $b^* \ne 3\sigma_{n+1,n}$. What do we have to do to have the equality hold.

#2.8-1 Following the procedure outlined in problem 2.4-1 derive (2.8-4) and (2.8-5) for the Asquith–Friedland filter.

#2.9-1 Prove that (2.9-8) leads to (2.9-9) for small T/τ.

2.10-1 Design the g–h–k filter using design curves of Figures 2.10-1 to 2.10-10. Assume $\sigma_x = 170\,\text{ft}$, $\sigma_{n+1,n}/\sigma_x = 2$, $\tau = 3\,\text{sec}$, $\ddot{x} = 1\,g$. Find (a) T, g, h, k, and $\sigma_{n,n}$; (b) $\dot{\sigma}_{n+1,n}$, $\dot{\sigma}_{n,n}$, $\ddot{\sigma}_{n+1,n}$, $\ddot{\sigma}_{n,n}$, and $\sigma_{m,n}$, $\dot{\sigma}_{m,n}$, $\ddot{\sigma}_{m,n}$, where m is time at the midpoint of the data interval; that is, find the midpoint smoothed position, velocity, and acceleration.

2.10-2 Repeat problem 2.10-1(a) for $\tau = 20\,\text{sec}$, all other assumptions remaining the same.

2.10-3 Repeat problem 2.10-1(a) for $\sigma_a = 5\,g$, $\sigma_x = 30\,\text{ft}$, and $\tau = 5\,\text{sec}$, all other assumptions remaining the same.

2.10-4 Repeat problem 2.10-1(a) for $\tau = 20\,\text{sec}$, $\sigma_{n+1,n}/\sigma_x = 3.33$, and σ_x and $\ddot{x}$ still 170 ft and 1g, respectively.

2.10-5 Design a Singer filter for $\tau = 20\,\text{sec}$, $T = 4\,\text{sec}$, $\sigma_x = 170\,\text{ft}$, and $\ddot{x} = 1\,g$. Find σ_{n+1} and σ_n.

2.10-6 For the example problem in Section 2.10 design a critically damped g–h–k filter with $b^* = 3\sigma_{n+1,n}$. To obtain the maximum jerk $\dddot{x}$, use (2.5-13) with $C = 1$ and use $\tau = 3\,\text{sec}$ in place of T and let $\sigma_w = \sigma_a$ in (2.4-13). Use Figure 1.3-1. Compare g, h, k, $\delta = \sigma_{n+1,n}/\sigma_x$, and $\sigma_{n,n}/\sigma_x$ values with those of the example of Section 2.10.

2.10-7 Repeat problem 2.10-6 so as to obtain a critically damped g–h–k filter that minimizes E_{TN}. Use Figure 1.3-3.

2.10-8 Repeat problem 2.10-6 for an optimum g–h–k filter. Use Figure 1.3-4.

2.10-9 Repeat problem 2.10-6 to obtain an optimum g–h–k filter that minimizes E_{TN}. Use Figure 1.3-6.

#2.10-10 Design a g–h–k filter using design curves of Figures 2.10-1 to 2.10-10. Assume $\sigma_x = 50$ ft, $\sigma_{n+1,n}/\sigma_x = 1$, $\tau = 10$ sec, and $\ddot{x} = 1$ g. Find (a) T, g, h, k, and $\sigma_{n,n}$; (b) $\dot{\sigma}_{n+1,n}$, $\dot{\sigma}_{n,n}$, $\ddot{\sigma}_{n+1,n}$, $\ddot{\sigma}_{n,n}$ and $\sigma_{m,n}$, $\dot{\sigma}_{m,n}$, $\ddot{\sigma}_{m,n}$, where m is time at the midpoint of the data interval; that is, find the midpoint smoothed position, velocity, and acceleration.

#2.10-11 Design a Singer filter for $\tau = 2$ sec, $T = 2$ sec, $\sigma_x = 30$ ft, and $\ddot{x} = 1$ g. Find g, h, k, σ_{n+1}, and σ_n.

#2.10-12 For problem 2.10-10 design a critically damped g–h–k filter with $b^* = 3\sigma_{n+1,n}$. To obtain the maximum jerk $\dddot{x}$, use (2.4-13) with $C = 1$ and use $\tau = 10$ sec in place of T and let $\sigma_w = \sigma_a$. Use Figure 1.3-1. Compare g, h, k, $\delta = \sigma_{n+1,n}/\sigma_x$, and $\sigma_{n,n}/\sigma_x$ values with those of problem 2.10.10.

#2.10-13 Repeat problem 2.10-12 so as to obtain a critically damped g–h–k filter that minimizes E_{TN}. Use Figure 1.3-3.

#2.10-14 Repeat problem 2.10-12 for an optimum g–h–k. Use Figure 1.3-4.

#2.10-15 Repeat problem 2.10-12 to obtain an optimum g–h–k filter that minimizes E_{TN}. Use Figure 1.3-6.

#2.10-16 Repeat problem 2.10-11 for $\tau = 10$ sec, everything else being the same.

#2.10-17 The steady-state Singer g–h–k Kalman filter should degenerate to the steady-state g–h Kalman filter having the target dynamics model given by (2.1-1a) and (2.1-1b) with $VAR(u_n) = \sigma_u^2$ when $\tau/T \doteq 0.5$. We will check this in this problem by comparing a steady-state Singer filter design equivalent to the steady-state g–h Kalman filte design of problem 2.4-3. For this purpose, for the equivalent Singer filter design use $T = 1.421$ sec, $\sigma_x = 150$ ft, and $\sigma_a = \ddot{x}_{max} = 64$ ft/sec^2, the value of $\ddot{x}_{max}$ used to obtain σ_u^2 in problem 2.4-3. Find g, h, k; $\sigma_{n+1,n}$, $\sigma_{n,n}$, $\dot{\sigma}_{n+1,n}$, and $\dot{\sigma}_{n,n}$ for the equivalent Singer filter and compare these values with those obtained for the steady-state g–h Kalman filter of problem 2.4-3.

3.5.1.4-1 (a) Assume a radar carrier frequency $f_c = 1232.5$ MHz, a chirp signal with a bandwidth $B_s = 1$ MHz, and uncompressed pulse width $T_U = 2000$ μsec (typical Cobra Dane radar track parameters). Calculate Δr, Δt, and ΔR for a target Doppler velocity of 15,000 ft/sec.

(b) Assume a radar carrier frequency of 1275 MHz, a chirp signal with a bandwidth $B_s = 200$ MHz, and uncompressed pulse width $T_U = 1000$ µsec (typical Cobra Dane radar wide-band parameters). Calculate Δr, Δt, and ΔR for a target Doppler velocity of 15,000 ft/sec.

(c) Repeat (a) for a chirp waveform with $T_U = 150$ µsec and $B_s = 5$ MHz, $f_c = 1232.5$ MHz, and target Doppler velocity of 15,000 ft/sec (possible Cobra Dane track parameters).

(d) Repeat (c) with $T_U = 1500$ µsec, everything else remaining same (another possible set of Cobra Dane track parameters).

3.5.1.4-2 Redo the first and second examples at the end of Section 3.5.1.4 for the calculation of ΔR of (3.5-14) when the acceleration a_d is 10 g, all other assumptions being the same.

#3.5.1.4-3 Assume a radar carrier frequency $f_c = 10$ GHz, a chirp signal with a bandwidth $B_s = 1$ MHz, and uncompressed pulse width $T_U = 200$ µsec. Calculate Δr, Δt, and ΔR for a target Doppler velocity of 3000 ft/sec.

#3.5.1.4-4 For problem 3.5.1.4-3 calculate the change in ΔR due to a target acceleration of 10 g.

3.5.2-1 (a) Use the same assumptions as for problem 2.10-5 except that an upchirp waveform is used having $B_s = 1$ MHz and $T_U = 4$ msec with the carrier frequency $f_c = 10$ GHz $= 10^{10}$ Hz. Find $\sigma_{n,n}$. Compare to the $\sigma_{n,n}$ obtained in problem 2.10-5 with a nonchirped waveform.

(b) Repeat (a) for a downchirped waveform. Compare to results for nonchirped and upchirped waveforms.

#3.5.2-2 (a) Use the same assumptions as for problem 2.10-16 except that an upchirp waveform is used having $B_s = 5$ MHz and $T_U = 200$ µsec with carrier frequency $f_c = 14$ GHz. Find $\sigma_{n,n}$. Compare to the $\sigma_{n,n}$ obtained in problem 2.10-16 with a nonchirped waveform.

(b) Repeat (a) for a downchirped waveform. Compare to results for nonchirped and upchirped waveforms.

4.1-1 (a) Derive (4.1-17) by substituting (4.1-15) and (2.4-3a) into (4.1-13).
  #(b) Derive (4.1-18).

#4.1-2 Verify that (4.1-15) is the inverse of (2.4-1b) by multiplying Φ by Φ^{-1}. Similarly verify that (4.1-26) is inverse of (4.1-4).

#4.1-3 Verify (4.1-28).

4.1-4 (a) For the constant-velocity target with $L = 1$, use T gives by (4.1-17) to obtain $\hat{W}$ from (4.1-32). Verify that this $\hat{W}$ is T^{-1}. Why is this the case? Note that the $\hat{W}$ obtained here is for the filtered estimate $X^*_{n,n}$; see (4.1-30).

    #(b) Repeat for $L = 2$.

4.5-1 Differentiate (4.5-13) with respect to $X^*_{n,n}$, as done in (4.1-46), to derive (4.5-4). *Hint*: See the differentiation of (2.6-7) in Section 2.6.

4.5-2 From problem 4.1-4(a) the least-squares weight $\hat{W}$ for estimating $X^*_{n,n}$ is given as

$$\hat{W} = \hat{W}(h = 0) = \begin{bmatrix} 1 & 0 \\ \dfrac{1}{T} & -\dfrac{1}{T} \end{bmatrix} \qquad \text{(P4.5-2a)}$$

for $m = 1$ and $L = 1$. Using (4.5-9), it follows that the one-step predictor least-squares weight is

$$\hat{W}(h = 1) = \Phi\hat{W}(h = 0) \qquad \text{(P4.5-2b)}$$

Using (P4.5-2a) and (P4.5-2b), obtain $\hat{W}(h = 1)$. Determine the components of $X^*_{2,1} = \Phi X^*_{1,1}$ using (P4.5-2b) and the $X^*_{1,1}$ of the solution of problem 4.1-4.

#4.5-3 Derive (4.5-2).

#4.5-4 Derive (4.5-5).

#5.4-1 Verify that (5.4-7) transitions X_n to X_{n+1} by substituting (5.4-4) to (5.4-6) into (5.4-1) and then multiplying by (5.4-7).

5.5-1 Verify that the 3×3 transition matrix (5.4-13) $\Phi(h)_z$ for the scaled state vector Z_n for $h = 1$ is obtained from (5.5-8) for the constant-accelerating target ($m = 2$) model.

5.5-2 Using (5.5-3a) to (5.5-8) obtain

$$\hat{W}(h = 1) = \begin{bmatrix} 2 & -1 \\ \dfrac{1}{T} & -\dfrac{1}{T} \end{bmatrix}$$

for the constant-velocity target ($m = 1$) with $L = 1$. This answer agrees with that of problem 4.5-2. In the process of obtaining $\hat{W}(h = 1)$, also obtain $\hat{W}(h = 0)$.

5.5-3 Using $\hat{W}(h = 0)$ of problem 5.5-2 and (5.6-3), verify (5.6-5) for $L = 1$.

5.5-4   Using (5.5-3a) to (5.3-9), verify that for $m = 1$ and $L = 4$ the one-step predictor least-squares weight is given by

$$W(h = 1) = \frac{1}{10} \begin{bmatrix} 8 & 5 & 2 & -1 & -4 \\ 2 & 1 & 0 & -1 & -2 \end{bmatrix}$$

#5.5-5   Using $\hat{W}(h)$ of problem 5.5-4 and (5.5-3), verify that (5.6-5) applies for $L = 4$.

5.6-1   Using (2.4-4f) and (2.4-1b), obtain (5.6-5) from (5.6-4) for $Q_n = 0$.

5.6-2   Using (5.6-7), (5.6-4), and (5.6-5), obtain the unscaled covariance matrices $S_{n,n}^*$ and $S_{n+1,n}^*$ for the least-squares filter for the constant-velocity ($m = 1$) target dynamics model.

#5.6-3   Using the $S_{n,n}^*$ obtained in problem 5.6-2, obtain $S_{n+1,n}^*$ using (2.4-4f) and (2.4-1b). How does $[S_{n+1,n}^*]_{0,0}$ vary with T?

#5.6-4   Verify that $[_sS_{n+1,n}^*]_{0,0}$ of (5.6-5) agrees with the variance of the one-step g–h expanding-memory polynomial given by (1.2-42) and with (1.2-19).

#5.6-5   Verify that $|_s S_{n+1,n}^*|_{1,1}$, $|_s S_{n,n}^*|_{0,0}$, and $|_s S_{n,n}^*|_{1,1}$ of (5.6-5) and (5.6-4) agree with (1.2-20) and (1.2-21).

#5.7-1   Find the memory needed for the first-degree ($m = 1$) fixed-memory smoothing filter that will result in $\sigma_{n+1,n}/\sigma_x = 3$. Compare the result to that of problem 1.2.7-2. What does this say the switching time is for a growing-memory track initiation filter for the g–h filter of problem 1.2.7-2?

#5.7-2   Using (5.7-3), find the T needed to make $[S_{n+1,n}^*]_{1,1} = (100\,\text{ft/sec})^2$ for the constant-velocity target, one-step predictor if $\sigma_x = 167\,\text{ft}$.

#5.8-1   Using Table 5.8-1, obtain $[S_{n+1,n}^*]_{0,0}$ and $[S_{n+1,n}^*]_{1,1}$ for the $m = 1$ one-step predictor filter for L large. Compare these results with those of (5.7-1) and (5.7-3). Do they agree?

#5.8-2   Using Table 5.8-2, obtain $[S_{n+1,h}^*]_{0,0}$ and $[S_{n+h,n}^*]_{1,1}$ for $h = -\frac{1}{2}L$, that is, for smoothing to the center of the data interval for the $m = 1$ one-step predictor filter for L large. Compare these results with those of problem 5.8-1.

5.8-3   (a) Show that the accuracy of the endpoint location obtained by predicting from the midpoint estimate is the same as the endpoint estimate of (5.8-4) for large L and $m = 1$.

(b) Show that the accuracy of the start-of-track location obtained by retrodiction from the endpoint estimate is the same as the accuracy of the end-of-track position given by (5.8-4) for L large and $m = 1$ target dynamics model.

6.5-1 Verify that for $m = 0$ degree growing-memory filter, the prediction output $x_{1,0}^*$ at time $n = 0$ is independent of the initial value $x_{0,-1}^* = x_0$ assumed.

6.5-2 Verify that for the $m = 1$ degree growing-memory filter, the prediction outputs $x_{2,1}^*$ and $\dot{x}_{2,1}^*$ are independent of the initial values $x_{0,-1}^* = x_0$ and $\dot{x}_{0,-1}^* = v_0$ assumed.

#6.5-3 Verify that for the $m = 2$ degree growing-memory filter, the prediction outputs $x_{3,2}^*$, $\dot{x}_{3,2}^*$, and $\ddot{x}_{3,2}^*$ are independent of the initial values $x_{0,-1}^*$, $\dot{x}_{0,-1}^*$, and $\ddot{x}_{0,-1}^*$.

7.2-1 (a) Put the degree 2 ($m = 2$) fading-memory filter of (1.3-2) and (1.3-3) into the form given in Table 7.2-2.

(b) Find, g, h, and k in terms of θ for the $m = 2$ fading-memory filter.

7.2-2 Find g, h, k for the growing-memory filter of Table 6.3-1 using the form for the g–h–k filter given in problem 7.2.1.

#7.2-3 Show that (7.2-5) leads to a recursive solution for $x_{n+1,n}^*$, specifically the feedback form of (P1.2.1-1a), when $i = 0$, $m = 1$, and $r = 1$, the one-stop prediction case.

#7.4-1 Verify that the variance of the one-step predictor for the critically damped g–h filter of (1.2-41) agrees with the results obtained from (7.4-2).

7.4-2 (a) Design a fading-memory $m = 2$ filter that has the same g of 0.87 as obtained for the Singer g–h–k filter of Section 2.10.

(b) Find $\sigma_{n+1,n}/\sigma_x$ for the above fading-memory filter. First use the expression for the VRF for the $m = 2$ fading-memory filter given in Table 7.4-1. Next use the expression for the VRF of the general g–h–k filter given by (1.3-4) and compare results. Compare these results with those of the Singer g–h–k filter of Section 2.10 having the same g.

(c) Obtain $\sigma_{n+1,n}/\sigma_x$ for the Singer g–h–k filter of Section 2.10 if the target has constant acceleration.

#7.5-1 Find the memory $L + 1$ of an $m = 1$ fixed-memory filter that has the same $\sigma_{n+1,n}/\sigma_x$ as obtained with the fading-memory g–h filter of problem 1.2.7-2(b) (for which $\theta = 0.1653$, $g = 0.973$, $h = 0.697$).

8.1-1 Obtain the A matrix of (8.1-10) for a constant-velocity target.

8.1-2 Using the results of problem 8.1-1 and (8.1-22), obtain the transition matrix for a constant-velocity target. Compare with (2.4-1b).

8.1-3 Derive the transition matrix given by (4.1-4) for a constant-accelerating target using (8.1-10a) and (8.1-22).

#8.1-4 Find the matrix A and transition matrices for a constant-jerk target, that is, a target for which $\dddot{x} = \text{const}$.

9.3-1 Derive (9.3-1) to (9.3-1d) from (9.2-1) to (9.2-1d) using the matrix inversion lemma given by (2.6-14).

#9.3-2 Verify that the Kalman filter equations of (9.3-1) to (9.3-1d) are identical to those of (2.4-4a) to (2.4-4j).

9.4-1 Derive (9.2-1) from (9.4-1). *Hint*: Add and subtract $M^T R_{n+1}^{-1} M \overset{\circ}{X}{}_{n+1,n}^{*}$ to respectively the first and second terms inside the bracket and then use (9.2-1b).

10.2-1 Show that (8.2-3) is the characteristic equation [given by (10.2-62)] for the general form of the matrix (8.2-6a). *Hint:* Use method of induction. Expand arbitrary $m' \times m'$ detriment $|A - \lambda I|$ using its first column cofactors [101].

#11.2-1 Find the Givens transformed matrix T of (11.2-1) for ε of 0.09 instead of 0.1.

†#11.2-2 Derive (11.2-3) to (11.2-10) using (11.2-14).

†#11.2-3 Derive (11.2-12) and (11.2-13).

#11.3.2-1 Determine the magnitude and phase of the complex number $-0.544 + j0.735$ using the CORDIC vectoring algorithm. Use an 11-bit representation, as done for the example of Table 11.3-2.

#11.3.2-2 Rotate the complex number $0.7881010754 + j0.615661475$ by $-142°$ using the CORDIC rotation algorithm. Use a 11-CORDIC rotation.

#11.3.2-3 Repeat problem 11.3.2-2 using the results of Table 11.3-2, that is, using the binary sequence of r_i's of Table 11.3-2.

#11.3.2-4 Repeat problem 11.3.2-3 for $+142°$ rotation.

#11.3.2-5 Derive CORDIC vectoring algorithm given by (11.3-16) to (11.3-25). *Hint:* Use Figure 11.3-6.

#11.3.2-6 Derive CORDIC rotation algorithms given by (11.3-31) to (11.3-37) plus (11.3-21), (11.3-22), (11.3-24), and (11.3-25).

#12.3-1 Find the two Householder transformations needed to transform T of (11.2-1) with $\varepsilon = 0.1$ into the upper triangular form of (11.2-2), the example used after (11.2-1) and also used in Figure 11.3-3 to demonstrate a systolic array implementation of the Givens procedure. Calculate the matrices obtained after each Householder transformation.

#13.2-1 Verify (13.2-2) and (13.2-3) and obtain $Q^T Q$ for both matrices.

#13.2-2 Calculate the orthonormal transformation matrix F for the matrix (13.2-1) using the Givens transformation procedure with computer round-off to four significant figures. How does it compare to $F = Q^T$ obtained using the MGS approach.

#13.2-3 Repeat problem 13.2-3 using the Householder procedure.

#13.2-4 Using the MGS procedure calculate the matrices R', Q', R, and Q for T given by (11.2-1) for $\varepsilon = 0.1$. Compare these results with those obtained using the Givens procedure [immediately after (11.2-1)] and the Householder procedure [problem (12.3-1)].

#14.1-1 Verify (14.1-5).

#14.1-2 Verify (14.1-6) to (14.1-9) using the MGS algorithm.

14.4-1 Show that the inverse of a square upper triangular matrix is upper triangular. *Hint*: Use the method of induction, first showing it is true for a 2×2 matrix and then showing it is true for an $r \times r$ matrix if true for an $(r - 1) \times (r - 1)$ matrix.

#16.2-1 Calculate $[M(\bar{X}_n)]_{ij}$ for $i, j = 1, 2, 3$ for the rectangular-to-spherical coordinate case in which the measurements are being made in spherical coordinates and the tracking is being done in rectangular coordinates.

#16.3-1 Derive (16.3-30).

#16.3-2 Extend (16.3-29) and (16.3-30) to the three-dimensional (x, y, z) coordinate case.

SYMBOLS AND ACRONYMS

Special Notation

$\|\cdot\|$

magnitude, also called Euclidean norm, of column matrix in s-dimensional hyperspace; see (4.2-40)

U^{-T}

$(U^T)^{-1}$ for matrix U; see (10.2-45) and the discussion immediately following

$\text{VAR}(\cdot)$

Variance of quantity in parentheses

Φ^{-i}

$(\Phi^{-1})^i$; see (4.1-8) and the discussion immediately following

$\int$

integral sign

$\sum$

Summation

$*$

Used as superscript to mean estimate; see (1.2-4) and related discussion; used in Section 4.4 and (11.3-39) as superscript to mean complex conjugate

Roman Letters

a_d

Target acceleration along radar line of sight; see discussion just prior to (3.5-14)

$\bar{a}_j$

$\bar{a}_k$ for $j = k$

$(\bar{a}_j)_n$

Estimate $\bar{a}_j$ of a_j based on measurements made up to and including time n; see (5.2-2) and the discussion following

a_k

Coefficient of kth term of accurate polynomial fit of degree d to trajectory; see (5.9-1)

$\bar{a}_k$

Coefficient of t^k term of polynomial approximation of degree m to data [see (4.1-44)]; $= \bar{a}_j$ for $k = j$

402

$(a_k)_n$	a_k for trajectory whose last time sample is at time n; see (5.10-1)
a_n	Random target acceleration occurring between time n and $n+1$ for Asquith–Friedland dynamics model; see Section 2.8
$a_1, a_2, \ldots, a_i$	ith time sample for auxiliary channel output sum voltage of sidelobe canceler; see (4.4-4) to (4.4-6) and Figure 4.4-1
A	Constant-coefficient matrix for linear state vector differential equation; see (8.1-10)
	Column matrix of auxiliary channel output sum voltages for sidelobe canceler; see (4.4-4) and (4.4-5) and Figure 4.4-1
$A_{(t)}$	Time-varying coefficient matrix for linear time-varying state vector differential equation; see (15.2-1)
$A_{\max}$	Maximum target random acceleration for Singer dynamic model; see Section 2.9
A_n	Trajectory accurate state vector representation in terms of its first d derivatives at time n; see (5.9-2)
$b^*(t)$	Systematic error for trajectory position estimate; see (5.9-3a) and discussion following
b^*	Steady-state prediction bias error of g–h filter when tracking constant acceleration target; see Section 1.2.4.3
$b^*(r) = b^*(L)$	Systematic error at respectively time r and L; see (5.10-2) and (7.7-1)
$B^*_{n,n}$	Systematic error in estimate $X^*_{n,n}$; see (5.9-3)
B_s	Signal bandwidth; see Section 3.5.1
$c(j, L)$	See (5.3-5a)
$c(j, \theta)$	See (7.2-1c)
c_i	Cosine element of ith Givens transformation matrix G_i; see (11.1-6), (11.1-6a), (11.1-13), and (11.1-13a)
c_j	See (5.3-5) and (7.2-1b), where c_j has different meanings for respectively the Legendre the Laguerre polynomials
c_D	Atmospheric dimensionless drag coefficient; see (2.4-9) and (16.3-19) and related discussions
D	Derivative with respect to time [see (5.3-12)]; $= D^m$ for $m = 1$; $= D_0$ without $\| q'_{m+1} \|$ term [see (13.1-40)]
D^d	dth derivation with respect to time; see (4.1-2a) and (8.1-6)
D^2	Statistical distance used for nearest-neighbor data association; see Section 3.3.1
	Second derivative of function; see (4.1-2a)

$D_{x^*_{n+1,n}}$	Transient error of one-step predictor; see Section 1.2.5
D_0	Diagonal matrix of elements $\parallel q'_1 \parallel$ to $\parallel q'_{m+1} \parallel$; see (13.1-24)
$e(X_n)$	Another form of e_n, the sum of squares of deviations between estimates and measurements; see (4.2-9) and related discussion
e_D	Total discounted weighted sum of squares of deviations between data points and straight-line fit to data; see (1.2-34)
e_n	sum of squares of deviations between estimates and measurements; see (4.1-37) and (4.1-38)
e_T	Total sum of squares of deviations between data points and straight-line estimate fit to date; see (1.2-33)
E	Total weighted sum of squared errors for one-dimensional case where only range is measured; see (2.5-12)
	Column matrix of deviations e_n, $n = 1, 2, \ldots$; see (4.3-22a) and (4.3-18), Figure 4.2-1, and Section 10.2
f_c	Radar carrier frequency; see (3.5-4)
f_d	Target Doppler shift, see (3.5-3)
$f_1, f_2 \ldots$	ith row of F; see (10.2-25); also (4.3-15)
F	Orthonormal transformation matrix; see (4.3-14) and related discussion and Chapters 10 to 14
$F[X(t), t]$	Vector of nonlinear function of elements of $X(t)$ used to define nonlinear dynamic model; see (16.3-1)
F_1	Part of transformation matrix F (first m' rows) that projects $Y_{(n)}$ onto T_p space and transforms T to matrix U; see discussion in Section 4.3 starting with paragraph containing (4.3-39) and Section 10.2; for 3-dimensional example of (4.2-1) to (4.2-4), T_p is the 2-dimensional plane of Figure 4.2-1
F_2	Part of transformation matrix F ($m' + 1$st row) that projects $Y_{(n)}$ onto coordinate perpendicular to T_p space, specifically the coordinate defined by unit vector $q_{m'+1}$ in direction $Y_{(n)} - Y'_1$; see discussion in Section 4.3 starting with paragraph containing (4.3-39); for 3-dimensional example of (4.2-1) to (4.2-4), $q_{m'+1} = q_3$ coordinate direction of Figure 4.3-1 perpendicular to T_p plane
F_3	Part of transformation matrix F (rows $m' + 2$ to s) that projects $Y_{(n)}$ onto space perpendicular to $(m' + 1)$-dimensional column space of T_0; see (4.3-58c) and (4.3-59) and related discussions

g	Weighting constant in g–h (and g–h–k) filter equations associated with position update; see Sections 1.2.1 and 1.3
g_i	ith row unit vector of G_1; see discussion in paragraph containing (11.1-18) and the four paragraphs immediately after
$(g_i)_2$	ith row vector of G_2; see discussion in third and fourth paragraph after one containing (11.1-24)
g_n	g at update time n
G_1'	2×2 matrix that represents rotation part of first Givens transformation G_1; see (11.1-7)
$G_1, G_2, \ldots, G_i$	ith Givens transformation; see Section 11.1
h	Weighting constant in g–h (and g–h–k) filter equation associated with velocity update; see Sections 1.2.1 and 1.3
h_n	h at update time n
$H(\omega)$	Transfer function of filter that produces Singer model correlated acceleration for white-noise input; see Section 2.9
H_n	Matrix giving tracking-filter constants; see (2.4-5) and (2.4-4e); also (9.2-1) to (9.2-1d) and (9.3-1) to (9.3-1d)
$H_1, H_2, \ldots, H_i$	ith Householder reflection transformations; see Chapter 12
i	Unit vector along x axis for x, y, z orthogonal coordinate system; see discussion just before (4.2-38) Time index as in (4.1-20)
I	Identity matrix, i.e., matrix whose diagonal elements are unity and whose off-diagonal elements are zero; see (4.2-29) and (4.5-6)
I_s	$s \times s$ identify matrix I; see (10.2-54)
j	Unit vector along y axis for x, y, z orthogonal coordinate system; see discussion just before (4.2-38)
J	Total weighted sum of square errors for multidimensional case; see Section 2.6 Time index representing data endpoint; see Section 2.10
k	Weighting constant in g–h–k filter equation associated with acceleration update; see Section 1.3 Unit vector along z axis for x, y, z orthogonal coordinate system; see discussion just before (4.2-38)
k_n	k at update time n
k_1, k_2	Weighting constants for combined linear estimate; see (2.5-1)

L	$L+1$ equals number of measurements made in fixed-memory filter (which is filter memory); see (5.2-1) and Figure 5.2-1 and related discussions
m	Target mass; see (2.4-9) and (16.3-20)
	Degree of approximate polynomial fit to data; see (4.1-44)
	Degree of tracking filter, i.e., degree of polynomial fit used by tracking filter; see (1.1-1) (where $m = 1$), (1.3-1) (where $m = 2$), and (4.1-44) and Tables 6.3-1 and 7.2-2 and relative discussions
m'	Number of states of state vector X_n; see paragraph after (4.1-2a)
m_{pq}	p,q element of true spatial covariance matrix of auxiliary elements of sidelobe canceler; see (4.4-17)
$\hat{m}_{pq}$	p,q element of $\hat{M}$; see (4.4-16) and following discussion
M	Observation matrix; see (2.4-3a) and Section 4.1
	Integer number, used to establish number of detections needed to establish track or drop track; see Section 3.2
$\hat{M}$	Estimate of spatial covariance matrix of auxiliary voltages of sidelobe canceler; see (4.4-16) and related discussions
n	Time or scan index
	Last time at which an observation was made for the expanding-memory filter; see (1.2-38a) and (1.2-38b) and (6.2-1)
$n_a(t)$	White noise; used for Singer dynamics model; see Section 2.9
N_n	Observation (measurement) error matrix at time n; see (2.4-3b) and Sections 2.4 and 4.1
$N_{(n)}$	Observation error matrix for times $n, n - 1, \ldots, n - L$; see (4.1-11a) and (4.1-10)
$p(r, j, L)$	$p_j(r)$; see (5.3-3) and (5.3-4)
$p(r; j, \theta)$	See (7.2-1d)
$p(t)$	Accurate polynomial fit of degree d to trajectory; see (5.9-1)
p^*	Abbreviated form of $[p^*(r)]_n$; see (5.2-3)
$p^*(r)$	Polynomial fit to data as a function of integer time index r; see (4.1-45) and discussion immediately after it; see also (5.2-3) and following discussion
$[p^*(r)]_n$	$p^*(r)$ with the subscript n indicating the last time a measurement was made for estimating $p^*(r)$; see (5.2-3) and (5.3-1)

$p^*(t)$	Polynomial fit to data as a function of time variable t; see (4.1-44)
$p_j(r)$	Unnormalized discrete-time orthogonal Legendre polynomial of degree j; see (5.3-1a), (5.3-3), and (5.3-4)
	Unnormalized discrete-time orthogonal Laguerre polynomial; see (7.2-1d) and related discussion
$p_j(t)$	Polynomial coefficient for sum of products of polynomials and expontential model for $x(t)$; see (8.2-1)
p_t	Projection of $Y_{(n)}$ onto T_p when T_p is a one-dimensional space, i.e., p_t is p_T^* when T_p is a line instead of a plane; see (4.2-32) and relative discussion
p_T	Vector formed by linear combination of columns of matrix T; hence a vector in column space of T; see (4.2-8) and related discussion; see also Figure 4.2-1 for special case where p_T is a two-dimensional vector in T_p plane
p_T^*	Projection of data measurements given by $Y_{(3)}$ onto plane T_p; this vector provided the optimum least-squares estimate; see (4.2-24) and related discussion; in general p_T^* is projection if $Y_{(n)}$ onto hyperspace formed by column space of T; see Chapter 10
p_1, p_2	Dimensionless Fitzgerald design parameters used for Singer g–h–k filter design curves; see Section 2.10
$p_1, p_2, \ldots, p_i$	p_i equals vector component of column vector t_1 along unit row vector f_i of $F(f_i = q_i^T)$; see (10.2-27)
$p_1, p_2, \ldots, p_s$	Components of p_T vector; see (4.2-8)
p_3	Fitzgerald normalized parameter used for determining performance of Singer g–h–k Kalman steady-state filter when chirp waveform is used by radar for tracking; see (3.5-16) and Section 3.5-2
P	Projection matrix, projects $Y_{(3)}$ onto plane T_p; see (4.2-25) and related discussion; in general, P projects $Y_{(n)}$ onto column space of T
$P_{\max}$	Probability of target having maximum acceleration $A_{\max}$ for Singer dynamics model; see Section 2.9
P_0	Probability of target having no acceleration for Singer dynamics model; see Section 2.9
q	Backward-shifting operator; see (7.2-5a)
q_{ij}	ith coordinate of unit vector q_j; see (4.3-5)
q_1, q_2, q_3	Orthonormal unit vectors for new coordinate system replacing x, y, z coordinate system; see Figure 4.3-1 and related discussion in Section 4.3; when T_p is an m-dimensional space, q_1, q_2 become $q_1, q_2, \ldots, q_{m'}$ and

	q_1, q_2, q_3 become $q_1, q_2, \ldots, q_s$; see the discussion in paragraphs following (4.3-52) and Chapter 13
$q_1, \ldots, q_{m'}$	Orthonormal unit vectors spanning m'-dimensional space T_p; see discussion following (4.3-52) and Chapter 13
$q_1, \ldots, q_s$	Orthonormal unit vectors spanning s-dimensional $Y_{(n)}$ space of the coordinates of $Y_{(n)}$; see discussions following (4.3-52) and Chapter 13; orthonormal vectors $q_1, q_2, \ldots, q_s$ formed from $q'_1, q'_2, \ldots, q'_s$ by making the latter unitary; see (13.1-22)
$q'_1, q'_2, \ldots, q'_s$	Orthogonal vectors spanning s-dimensional space of coordinates of $Y_{(n)}$; formed by applying Gram–Schmidt orthogonalization to columns of T_0; see Chapter 13
Q	Matrix of unit vectors $q_1, q_2, \ldots$; see (4.3-7), (13.1-23), and (13.1-34) and related discussion
Q'	Matrix of orthogonal vectors $q'_1, q'_2, \ldots, q'_{m'+1}$; see (13.1-17)
Q_n	Covariance of dynamic model driving noise vector U_n; see (2.4-4g)
r	$r + 1$ equals number of dimensions in which target is tracked; see (2.4-7) and (4.1-1a) and related discussions
r_{ij}	Physically the same as r'_{ij} except that it is computed using $t_j^{(i)}$ instead of t_j, where $t_j^{(i)}$ is t_j minus its vector components along $q_1, q_2, \ldots, q_{i-i}$; see discussion in Chapter 13 following (13.2-3)
r'_{ij}	Physically, magnitude of projection of t_j onto unit vector q_i normalized by $\| q'_i \|$; see (13.1-11); alternately, dot product of t_j and q'_i normalized by $\| q'_i \|^2$; see (13.1-11), r'_{ij} times q'_i equals vector component of t_j along q_i; here the projection r'_{ij} is computed using full vector t_j, i.e., using the CGS procedure; see discussion in Chapter 13 following (13.2-3)
r''_{ij}	Physically, magnitude of vector component of t_j along q_i; see (13.1-28a) and (13.1-30)
r_{12}	r'_{12}; see discussion in Chapter 13 following (13.2-3)
r'_{12}	Physically, magnitude of projection of t_2 onto unit vector q_1 normalized by $\| q'_1 \|$; see (13.1-5); alternately, dot product of t_2 and q'_i normalized by $\| q'_i \|^2$; r'_{12} times q'_1 equals t_{2c}, the vector components of t_2 along q_1; see (13.1-6) and the discussion just before it
R	Range to target; see Sections 1.5 and 1.1

	$s \times (m' + 1)$ matrix representing T_0 transformed by Q^T; see (13.1-35) and related discussion; example R for $s = 3$, $m' = 2$ given by (13.1-36).
R'	Matrix of r'_{ij} projection terms; see (13.1-21) and (13.1-39)
R_c	Range of closest approach for a target flyby trajectory; see Section 1.5 and Figure 1.5-1
	Moving-target ambiguous range indicated by radar using chirp waveform, also called extrapolated range; see Section 3.5.1.4 and (3.5-6) and (3.5-7)
R_n	Covariance of observation error matrix N_n or equivalently Y_n; see (2.4-4i) and Chapter 9
$R_{(n)}$	Covariance of measurement error matrix $N_{(n)}$ or equivalently $Y_{(n)}$; see (4.5-3)
R_1, R_2	Range for ith ($i = 1, 2$) target; see Section 1.1
s	Dimension of $Y_{(n)}$; see (4.1-10) to (4.1-11a) and (4.1-30) and related discussion; see also (4.3-57) and (10.2-7) and related discussion
s'	Number of Givens transformations needed to form orthonormal transformation F as in (10.2-8); see discussion in paragraph containing (11.3-2)
s_i	Sine element of ith Givens transformation matrix G_i; see (11.1-6), (11.1-6b), (11.1-13), and (11.1-13b)
$s_1, s_2, \ldots, s_i$	s_i is ith time sample for main antenna signal of sidelobe canceler; see (4.4-1) and Figure 4.4-1
S	Covariance matrix of state vector X; see (1.4-2) and discussion just before this equation
	Main antenna signal vector for sidelobe canceler; see (4.1-1) and Figure 4.4-1
$\mathring{S}^*_{n,n}$	Covariance of minimum-variance estimate $\mathring{X}^*_{n,n}$; see first paragraph of Section 4.5
$S^*_{n,n-1}$	Covariance of estimate state vector $X^*_{n,n-1}$; see (2.4-4f) and (2.4-4h)
$S^*_{n,n}$	Covariance of state vector $X^*_{n,n}$; see (2.4-4j)
$S_{n+h,n}$	Covariance of prediction state vector estimate $X^*_{n+h,n}$; see (5.6-6) and (2.4-4h) for $h = 1$
$[S^*_{n+h,n}]_{ij}$	i, j element of $S^*_{n+h,n}$; see (5.6-7)
$\mathring{S}_{n+h,n}$	Covariance of minimum-variance estimate $\mathring{X}_{n+h,n}$; see (4.5-10)
$_s\mathring{S}^*_{n+h,n}$	Covariance matrix of scaled least-squares estimate $Z^*_{n+h,n}$; see (5.6-2) and related discussions; see also Table 5.6-1
$[_sS_{n+h,n}]_{ij}$	i, j element of $_sS_{n+h,n}$; see (5.6-7)

S'_1 Projection of S onto space spanned by F_1, which in turn is the space spanned by V for sidelobe canceler; see (4.4-12c) and related discussions in paragraph containing this equation and following paragraph

S'_2 Projection of S onto space spanned by F_2 of sidelobe canceler; see (4.4-12d) and related discussion in paragraph containing this equation and the following paragraph

t Time variable; see Figure 1.1-7 and Sections 1.1 and 1.2.4.2

Matrix T when it is a column matrix; see (4.2-19) and (4.2-21d) and related discussion

t_i ith column of matrix T or T_0; see (4.2-2)

$\hat{t}_i$ t_i when it is a unit vector; see discussion before (4.2-28)

t_{ij} i,j element of transition–observation matrix T or its augmented form T_0; see (4.1-11b), (4.2-1) and (4.3-54), and (11.1-1), (11.1-25), and (11.1-29) and related discussions

$(t_{ij})_k$ i, j element of transition–observation matrix T or its augmented form T_0 after application of k Givens transformations; see beginning of Chapter 11

$t_j^{(i)}$ t_j minus its vector components along $q_1, q_2, \ldots, q_{i-1}$; see discussion in Chapter 13 following paragraph (13.2-3)

$(t_j)_{2c}$ Vector component of t_j along q_2 when computed using CGS method, subscript c indicating that CGS method is being used; see discussion in paragraph containing (13.2-6) and remaining part of this section

t_1, t_2 Time at scan $n - 1$ and n; see Section 1.1 and Figure 1.1-7

$\bar{t}_0$ Vector formed by first two coordinates (x,y coordinates) of vector t_1 (formed by first column of T or T_0); see (11.1-11) and Figure 11.1-1

$\bar{t}_1$ Vector $\bar{t}_0$ after rotation by G_1, i.e., after it is rotated into x axis; see (11.1-12) and Figure 11.1-1

t_{1F} Representation of vector t_1 as sum of vector components of t_1 along new transformed orthonormal axis directions defined by rows of transformation matrix F; see (10.2-28) and (10.2-29) and discussion 2 following

t_{2c} Vector component of t_2 along t_1 see (13.1-4) and discussion just before it

$\bar{t}_2$ Vector formed by first three coordinates (the x, y, z coordinates) of vector t_1 (formed by first column of T or T_0); see (11.1-16) and Figure 11.1-2

$\bar{t}_3$ Vector $\bar{t}_2$ after rotation by successively G_1 and G_2, i.e., after rotation onto x axis; see (11.1-17) and Figure 11.1-2

T	Scan-to-scan period or time between measurements; see Section 1.1
	When used as superscript indicates matrix transpose; see Section 1.4 and (1.4-1)
	Transition–observation matrix; see (4.1-11b), examples given after this equation, and discussion immediately after (4.1-24); see also discussion in paragraph immediately before that containing (4.1-12)
T'	T transformed by F or equivalently columns of T expressed in new $q_1, q_2 \ldots$ orthonormal coordinate system; see (10.2-8)
T_f	Fixed-memory filter smoothing time; see (5.8-3)
T_0	Augmented transition–observation matrix T; see (4.3-54) and (11.1-25) and related discussion
T_0'	T_0 transformed by F or equivalently columns of T_0 expressed in new $q_1, q_2, \ldots$ orthonormal coordinate system; see (12.2-6)
T_p	Column space of matrix T; see Section 4.2; see Figure 4.2-1 for special case where T_p is a plane
T_U	Uncompressed pulse width of chirp waveform; see Section 3.5-1
u_{ij}	i,j element of upper triangular transformed matrix U; see (4.3-29) and (4.3-29a) and (10.2-8) and (10.2-9); physically, u_{ij} is ith coordinate of column matrix vector t_j expressed in new orthonormal coordinate space $q_1, q_2, \ldots$; see paragraph containing (4.3-24) and Section 10.2-2
u_n	Random part of target velocity; change in target velocity from time n to $n + 1$; see Chapter 2 and (2.1-1b)
U	Upper triangular transformed matrix T; see (4.3-29) and (4.3-29a) and (10.2-8) and (10.2-9) and related discussions; see also Section 10.2.2 and definition of u_{ij} above
U_n	Dynamic model driving noise vector at time n; see (2.4-2)
v_d	Target Doppler velocity; see Section 3.5.1.4
v_{ij}	ith time sample for jth auxiliary antenna signal of sidelobe canceler; see (4.4-2) and Figure 4.4-1
v_0^*	Velocity estimate or equivalently slope of best-fitting line (constant-velocity trajectory) to data; see Section 1.2.6 and Figure 1.2-10
V	Matrix of auxiliary antenna voltage time samples for sidelobe canceler; see (4.4-2) and Figure 4.4-1

V_{im}	Row matrix of m' auxiliary antenna outputs at time i for sidelobe canceler; see (4.4-6a) and Figure 4.4-1
$\mathrm{VAR}(\cdot)$	Variance of quantity in parentheses; see Section 1.2.4.4
W	Linear estimate weight row matrix; see (4.1-30)
$\hat{W}$	Optimum least-squares weight matrix; see (4.1-32)
$\overset{\circ}{W}$	Optimum minimum-variance weight matrix; see (4.5-4)
W'_m	Column matrix of sidelobe canceler weights; see (4.4-3) and Figure 4.4-1
x	Target true range; see Sections 1.1 and 1.2.1
	x coordinate of x, y, z coordinate system; see Section 1.5 and Figures 1.5-2 and 4.3-1.
$\dot{x}$	Target true velocity; see Sections 1.1 and 1.2.1
$\ddot{x}$	Target true acceleration; see Section 1.2.4.2
$\dddot{x}$	Third derivative of x with respect to time; see (1.2-13) and (8.2-5) for example
$x^{(m)}$	See (5.3-4a)
x^*_c	Optimum combined linear estimate; see Section 2.5.1
x_n	True target range at scan n or time n; see Sections 1.1, 1.2.1, and 2.4
$\bar{x}_n$	$x^*_{n,n} =$ filtered estimate of x; see discussion just before (1.2-9a)
$\hat{x}_n$	$x^*_{n,n-1} =$ one-step prediction of x; see discussion just before (1.2-9a)
$\dot{x}_n$	True target velocity at scan n or time n; see Sections 1.1, 1.2.1, and 2.4
$\bar{\dot{x}}_n$	$\dot{x}^*_{n,n} =$ filtered estimate of $\dot{x}$; see discussion just before (1.2-9a)
$\hat{\dot{x}}_n$	$\dot{x}^*_{n,n-1} =$ one-step prediction of $\dot{x}$; see discussion just before (1.2-9a)
$\ddot{x}_n$	True target acceleration $\ddot{x}$ at scan n or time n; see Sections 1.2.4.3 and 1.3
$x^*_{n,n}$	Filtered estimate of x at time n (first subscript n) based on measurements made up to and including time n (second subscript n); see (1.2-7) and the discussion immediately before and after (1.2-4)
$X^*_{n+h,n}$	Predicted estimate of state vector X at time $n + h$ (first subscript $n + h$) based on measurements made up to and including time n (second subscript n); see (2.4-4a), where $h = 1$, and the discussion immediately before and after (1.2-4); this is h-step predictor if $h > 0$, it is retrodiction if

	$h < 0$, and it is filtered estimate if $h = 0$; see (4.5-9) and (5.5-1)
$x^*_{n+1,n}$	Predicted estimate of x at time $n + 1$ (first subscript $n + 1$) based on measurements made up to and including time n (second subscript n); see (1.2-10b) and the discussion immediately before and after (1.2-4); this is a one-step predictor
x^*_0	Intercept along y axis of best-fitting line to data; see Section 1.2.6 and Figure 1.2-10; physically x^*_0 is estimate of x at time $n = 0$
x^*_1, x^*_2	Two estimates of x; see Section 2.5.1
	First and second coordinates of $X^*_{n,n}$; see (10.2-17)
$x^*_1, x^*_2, x^*_3, x^*_4$	Components of state vector estimate $X^*_{n,n}$; see (10.2-17)
X_n	Target true state vector at time n; see (2.4-1a), (4.1-1b), and (4.1-3)
X'_n	X_n given by matrix of scaled jth–state derivatives $\left(z^*_i\right)_n$, hence it is scaled state vector; see (5.2-5) and (5.4-12)
$X^*_{n,n}$	Filtered estimate of target state vector X_n at time n (first subscript n) based on measurements made up to and including time n (second subscript n); see (2.4-4b) and discussion immediately before and after (1.2.1-4)
$\overset{\circ}{X}{}^*_{n,n}$	Minimum-variance estimate of $X^*_{n,n}$; see paragraph containing (4.5-9)
$\bar{X}_{n+h,n}$	Estimate of state vector at time $n + h$ given by its first $m + 1$ derivatives; last measurement on which estimate is based is made at time n; see (5.11-5a) and discussion immediately before and after (1.2-4)
$X^*_{n+1,n}$	Predicted estimate of state vector X_{n+1} at time $n + 1$ (first subscript $n + 1$) based on measurements made up to and including time n (second subscript n); see (2.4-4a), (2.4-4c), and (2.4-6)
y	y coordinate of x, y, z orthonormal coordinate system; see Section 1.5 and Figures 1.5-2 and 4.3-1
y'_i	Amplitude of ith coordinate of $Y_{(3)}$ expressed in q_1, q_2, q_3 coordinate system; see (4.3-4) and related discussions
$(y_i)_k$	ith component of column matrix $Y_{(n)}$ after k Givens transformations; see (11.1-30)
y_n	Target range measurement at time n; see Section 1.2.1; measurement of any general parameter at time n; see (5.2-1)
Y_n	Measurement matrix for time n; see (2.4-3), (2.4-7), and (2.4-8) and Section 4.1

$Y_{(n)}$ Measurement matrix for times $n, n-1, \ldots, n-L$; see (4.1-11) and (4.1-11a)

Y_1' For three-dimensional example of (4.2-1 to 4.2-4) [$s = 3$ for $Y_{(n)}$ of (4.1-11a)], it is projection of $Y_{(3)}$ onto plane T_p; Y_1' is expressed in q_1, q_2, q_3 coordinates; see (4.3-34a), (4.3-42), and (4.3-42a) and discussion leading up to this equation in Section 4.3; for the general case of arbitrary s, Y_1' is projection of $Y_{(n)}$ onto T_p column space of T; see discussion following (4.3-49) of Section 4.3

Y_2' For 3-dimensional example of (4.2-1) to (4.2-7) [$s = 3$ for $Y_{(n)}$ of (4.1-11a)], it is projection of $Y_{(3)}$ onto direction q_3 perpendicular to plane T_p; Y_2' is expressed in q_1, q_2, q_3 coordinates; see (4.3-42), (4.3-42b), and (4.3-50) and related discussion to these equation; for general case of arbitrary s, Y_2' is projection of $Y_{(n)}$ onto coordinate perpendicular to space T_p formed by column space of T, specifically coordinate defined by unit vector $q_{m'+1}$ in direction $Y_{(n)} - Y_1'$; see (4.3-50) and discussion just preceding it; physically $(Y_2')^2$ is minimum least-squares error; see (12.2-7) [Note: Because Y_2' is a one element matrix we have taken the liberty of writing $\|Y_2'\|^2$ as $(Y_2')^2$.

$Y_{(3)}'$ $Y_{(3)}$ expressed in q_1, q_2, q_3 coordinate system; see (4.3-13) and related discussion

z z coordinate of x, y, z orthonormal coordinate system; see Section 1.5 and Figures 1.5-2 and 4.3-1

Z_n Scaled state vector X_n; see (5.4-12)

$Z_{n+h,n}^*$ Scaled estimate state vector $X_{n+h,n}^*$; see (5.5-2)

Greek Letters

α Weighting constant in α–β (and α–β–δ) filter equation associated with position update; see Sections 1.2.2 and 1.3

β Weighting constant in α–β (and α–β–δ) filter equation associated with velocity update; see Sections 1.2.2 and 1.3

          Atmospheric ballistic coefficient; see (2.4-9) and (16.3-20).

$(\beta_j)_n$ β_k with $k = j$ and time n, the last time at which a measurement was made; see (5.3-1) and (7.2-4) and discussion just prior to latter equation

β_k kth coefficient of least-squares discrete-time orthogonal Legendre or Laguerre polynomial fit to data; see (4.1-45), (5.3-1), and (7.2-4) and related discussions

γ	Weighting constant in $\alpha-\beta-\gamma$ filter equation associated with acceleration update; see Section 1.3
$\gamma_0, \gamma_1, \ldots$	Coefficients of linear, constant-coefficient differential vector equation given by (8.2-2)
$\delta p(t)$	Difference between accurate polynomial fit of degree d and estimated polynomial fit of degree m to trajectory; see (5.11-4) and related discussion
Δr	Radar range resolution; see Section 3.5.2 and (4.5-16)
ΔR	Apparent range displacement of target's observed position due to its Doppler velocity when chirp waveform is used; position target would have at future (past) time Δt if upchirp (downchirp) waveform is used and it had no acceleration; see Section 3.5.1.4
ΔR_a	Shift of target position due to target acceleration; see (3.5-15) and Section 3.5.1.4
Δt	Small (delta) change in time t; see (1.2-13)
	Time later that ambiguous range measurement is correct for target moving away when upchirp is used; see Section 3.5.1.4
δy_{n-i}	Deviation of y_{n-i} from estimate p^*_{n-i}; see (5.11-2)
$\delta Y_{(n)}$	Matrix of $L+1$ δy_{n-i} values; see (5.11-3)
$\Delta\tau$	Time later at which a longer range target echo arrives; see Section 3.5.1.4 and (3.5-5)
ε'_i	ith coordinate of deviation column matrix E expressed in $q_1, q_2, \ldots$ coordinates, i.e., magnitude of component of E along q_i direction; see (4.3-30) to (4.3-32)
$\varepsilon_1, \varepsilon_2, \varepsilon_n, \ldots$	Deviation between measurement y_n at time n and estimate x^*_n; see Figure 1.2-10 and (1.2-32), (4.3-22a), and (4.3-23)
	Difference voltages at output of sidelobe canceler at time n; see (4.4-7) to (4.4-9) and Figure 4.4-1.
ς	Small change in time; see (8.1-11)
θ	Discounting factor used for discounted least-squares filter; see Section 1.2.6 and (7.1-2), (7.1-2a), and related discussions
	Azimuth angle of target; see Sections 1.4 and 1.1 and Figure 1.5-2
	Azimuth between vectors; see for example (4.2-36).
θ_i	Rotation performed by ith Givens transformation; see (11.1-6a), (11.1-6b), (11.1-13a), and (11.1-13b)
θ_3	3-dB radar beamwidth; see Section 2.10

λ	Weighting factor; see (1.2-26)
	Radar wavelength; see Section 3.5.1.4
λ_{ij}	Constant in equation for covariance of least-squares estimate $X^*_{n+h,n}$; see (5.8-1) and (7.4-1) and Tables 5.8-1, 5.8-2, and 7.4-2 and related discussions
λ_j	Root of equation given by (8.2-3) which is the characteristic equation for the matrix A of (8.2-6a); see also (10.2-62) and related discussion
	Eigenvalues of matrix; see (10.2-62) and related discussion
$\hat{\rho}$	Estimate of cross correlation between main antenna and auxiliary antenna outputs of sidelobe canceler; see (4.4-16) and related discussion
ρ_j	jth cross correlation between main antenna and jth auxiliary channel output of sidelobe canceler; see (4.4-19)
$\hat{\rho}_j$	jth element of $\hat{\rho}$, i.e., estimate of ρ_j; see (4.4-20)
σ	rms of sum of range prediction and measurement; see Section 1.1
$\sigma(x^*_{n+1,n})$	$\sigma_{n+1,n}$; see Section 1.2.4.5
$\dot{\sigma}$	Abbreviated form for rms of steady-state Singer g–h–k Kalman filter filtered and one-step prediction velocity estimates; see Figure 2.10-6
$\ddot{\sigma}$	Abbreviated form for rms of steady-state Singer g–h–k Kalman filter filtered and one-step prediction acceleration estimates; see Figure 2.10-7
σ_a	rms of a_n; see (2.8-2)
	rms of target random acceleration for Singer model; see Section 2.9
σ_c	rms of optimum combined linear estimate x^*_c; see Section 2.5
σ_{cx}	rms of cross-range measurement; see Section 2.10 and (2.10-6)
σ_i	rms of ith measurement; see (4.5-11) and (4.5-15) and related discussion
$\sigma_{n+1,n}$	rms of prediction estimate $x^*_{n+1,n}$; see Section 1.2.4.5
σ_u	rms of u_n; see Chapter 2, specifically discussion just after (2.1-1b)
σ_x	σ_ν; see Section 1.2.4.4
σ_θ	rms of radar angle measurement; see Section 2.10
σ_ν	rms error of range measurement y_n; rms of ν_n; see Section 1.2.4.4

σ_1, σ_2	rms of estimates x_1^* and x_2^*; see Section 2.5
τ	Correlation time of random target acceleration for Singer model; see Section 2.9
τ_c	Compressed pulse width for chirp waveform; see Section 3.5.1
ϕ	Target elevation angle; see Section 1.5 and Figure 1.5-2
$\phi_j(r)$	Discrete-time orthonormal Legendre polynomial of degree j; see (5.3-1a)
	Discrete-time orthonormal Laguerre polynomial; see (7.2-1).
Φ	Sate transition matrix; see Section 2.4 and Chapter 4
$\Phi(T, \tau)$	Transition matrix for Singer dynamics model; see Section 2.9
Φ_i	Transition matrix for transitioning state vector X_{n-i} from time $n - i$ to n; see (4.1-6)
Φ_z	Transition matrix for scaled state vector Z_n; see (5.4-13) and (5.5-8) for example
Ψ	Inverse of Φ; see (8.1-29)
ω	Radian frequency $= 2\pi f$; see Section 2.9, specifically (2.9-3) and related discussion

Acronyms

APL	Applied Physical Laboratory
ASR	Airport Surveillance Radar; see Sections 1.1 and 1.5
BCGS	Basic classical Gram–Schmidt (circuit); see Figure 13.1-4
BMEWS	Ballistic Missile Early Warning System; see Figure 1.1-11
BMGS	Basic modified Gram–Schmidt (circuit); see Figure 13.2-1
CFAR	Constant false-alarm rate; see Section 3.1.4
CGS	Classical Gram–Schmidt; see first paragraph of Chapter 13 and discussion after paragraph containing (13.1-43)
CORDIC	Coordinate Rotation Digital Computer; see last paragraph of Section 11.3.1 and Section 11.3.2
CPI	Coherent processing interval; see Section 3.1.2.1
DCS	Dual coordinate system; see Section 1.5
DOLP	Discrete–time orthogonal Legendre polynomials; see Sections 5.3 and 14.4
FAA	Federal Aviation Administration
HiPAR	High Performance Precision Approach Radar; see Section 1.1 and Figure 1.1-8

IADT	Integrated Automatic Detection and Tracking; see Section 1.5
ICBM	Intercontinental ballistic missile; see Sections 1.1 and 1.5
IF	Intermediate frequency of radar receiver; see Section 3.5.1
JPDA	Joint probabilistic data association; see Section 3.3.2
LFM	Linear frequency modulation; see Section 3.5.1
MGS	Modified Gram–Schmidt; see discussion after paragraph containing (13.1-43)
MTD	moving-target detector; see Section 3.1.2.1
NAFEC	National Aviation Facilities Engineering Center; see Section 3.1.2.1
PAR	Pulse Acquisition Radar; see Section 1.1 and Figure 1.1-5
PC	Pulse compression ratio for chirp waveform; see (3.5-1) and related discussion
ROTHR	Relocatable Over-the-Horizon Radar; see Section 1.1 and Figure 1.1-17
SLC	Sidelobe canceler; see Figure 4.4-1
SNR	Signal-to-noise ratio; see Section 2.2
TWS	Track-while-scan; see Sections 1.1 and 3.4 and Figure 1.1-2 to 1.1-6
VAR	Variance; see Section 1.2.4.4
VHSIC	Very high speed integrated circuit; see Section 3.3.2
VLSI	Very large scale integrated circuitry; see Section 3.3.2
VRF	Variance reduction factor, the normalized variance of an estimate; see Section 1.2.4.4 and (5.8-2)

SOLUTION TO SELECTED PROBLEMS

1.2.1-1 From (1.2-11a) and (1.2-11b)

$$\dot{x}^*_{n+1,n} = \dot{x}^*_{n,n-1} + \frac{h}{T}(y_n - x^*_{n,n-1})$$

$$x^*_{n+1,n} = x^*_{n,n-1} + T\dot{x}^*_{n+1,n} + g(y_n - x^*_{n,n-1})$$

The corresponding prediction equations for n are given by

$$\dot{x}^*_{n,n-1} = \dot{x}^*_{n-1,n-2} + \frac{h}{T}(y_{n-1} - x^*_{n-1,n-2}) \qquad \text{(P1.2.1-1b)}$$

$$x^*_{n,n-1} = x^*_{n-1,n-2} + T\dot{x}^*_{n,n-1} + g(y_{n-1} - x^*_{n-1,n-2}) \qquad \text{(P1.2.1-1c)}$$

Substituting (1.2-11a) into (1.2-11b) for $\dot{x}^*_{n=1,n}$ yields

$$x^*_{n+1,n} = x^*_{n,n-1} + T\dot{x}^*_{n,n-1} + h(y_n - x^*_{n,n-1}) + g(y_n - x^*_{n,n-1})$$
$$= x^*_{n,n-1}(1 - g - h) + (g + h)y_n + T\dot{x}^*_{n,n-1} \qquad \text{(P1.2.1-1d)}$$

Solving for $T\dot{x}^*_{n,n-1}$ in (P1.2.1-1c) and substituting into (P1.2-1d) yield

$$x^*_{n+1,n} = (1 - g - h)x^*_{n,n-1} + (g + h)y_n + x^*_{n,n-1} - x^*_{n-1,n-2} - g(y_{n-1} - x^*_{n-1,n-2})$$
$$= (2 - g - h)x^*_{n,n-1} + (g - 1)x^*_{n-1,n-2} + (g + h)y_n - gy_{n-1}$$
$$\text{(P1.2.1-1e)}$$

which is the same as (P1.2.1-1a).

Recursive expressions similar to (P1.2.1-1a) for $x^*_{n+1,n}$ can also be obtained for the filtered position estimate $x^*_{n,n}$ and the velocity estimate $\dot{x}^*_{n,n}$.

1.2.4.3-1 For $x^*_{n,n-1}$ and $x^*_{n-1,n-2}$ we get

$$x^*_{n,n-1} = \tfrac{1}{2}\ddot{x}(n)^2 T^2 + b^* \qquad \text{(P1.2.4.3-1c)}$$

$$x^*_{n-1,n-2} = \tfrac{1}{2}x(n-1)^2 T^2 + b^* \qquad \text{(P1.2.4.3-1d)}$$

From (P1.2.4.3-1a)

$$y_{n-1} = \tfrac{1}{2}\ddot{x}(n-1)^2 T^2 \qquad \text{(P1.2.4.3-1e)}$$

Substituting (P1.2.4.3-1a) to (P1.2.4.3-1e) into (P1.2.1-1a) and solving for b^* yield (1.2-15).

1.2.4.4-1 The variance of $x^*_{n+1,n}$ will be independent of the constant-velocity target trajectory assumed. Hence for ease in the analysis we will assume a zero-velocity target at range zero. Thus y_n will consist only of the measurement noise ν_n. Hence the expected value of $x^{*2}_{n+1,n}$ is the variance of $x^*_{n+1,n}$. Thus squaring (P1.2.1-1a) and obtaining its expected value yield

$$E\left(x^{*2}_{n+1,n}\right) = \sigma^*_{n+1,n}$$
$$= (g+h)^2\sigma_x^2 + g^2\sigma_x^2 + (2-g-h)^2\sigma^2_{n+1,n} + (g-1)^2\sigma^2_{n+1,n}$$
$$2g(2-g-h)E[y_{n-1}x^*_{n,n-1}] + 2(2-g-h)(g-1)E[x^*_{n,n-1}x^*_{n-1,n-2}]$$
$$\text{(P1.2.4.4-1a)}$$

where use was made of

$$E\left[y_n x^*_{n,n-1}\right] = E\left[y_n x^*_{n-1,n-2}\right] = E\left[y_{n,n-1} x^*_{n-1,n-2}\right] = 0$$

To obtain $E\left[y_{n-1}x^*_{n,n-1}\right]$ in (P1.2.4.4-1a) above, rewrite (P1.2.1-1a) for $x^*_{n,n-1}$, multiply by y_{n-1}, and obtain the expected value of the resulting equation to yield

$$E\left[y_{n-1}x^*_{n,n-1}\right] = (g+h)\sigma_x^2 \qquad \text{(P1.2.4.4-1b)}$$

To obtain $E\left[x^*_{n,n-1}x^*_{n-1,n-2}\right]$, multiply (P1.2.1-1a) rewritten for $x^*_{n,n-1}$ by $x^*_{n-1,n-2}$ and obtain the expected value, yielding

$$(2-g)E\left[x^*_{n,n-1}x^*_{n-1,n-2}\right] = -g(g+h)\sigma_x^2 + (2-g-h)\sigma^2_{n+1,n} \quad \text{(P1.2.4.4-1c)}$$

Substituting (P1.2.4.4-1b) and (P1.2.4.4-c) into (P1.2.4.4-1a) yields (1.2-19) after much manipulation, as we desired to show.

1.2.5-1 (a) In Figure 1.2-7, $b^* = 3\sigma_{n+1,n}$ for $A_N = 3.0$ when the $A_N = 3.0$ curve crosses the $3\delta = 3\sigma_{n+1,n}/\sigma_x$ curve. At this point, $b_N^* = 3\delta = 4.7$. Hence $E_{TN} = 9.4$. Also at this point $g = 0.86$ so that, from (1.2-36), $h = 0.649$.

(b) From Figure 1.2-9, for $A_N = 3.0$ the minimum E_{TN} is 9.3, obtained, for $g = 0.91$, so that $h = 0.760$. For this minimum design $b_N^* = 4.15$, $3\delta = 5.15$.

(c) For the minimum design E_{TN} is 1% smaller. Although $b_N^* \neq 3\delta$, they are almost equal with $3\delta/b_N^* = 1.24$.

1.2.5-2 (a) For $A_N = 0.001$, $b_N^* = 3\delta = 0.62$ for $g = 0.048$. Hence $E_{TN} = 1.24$ and $h = 0.0012$.

(b) For $A_N = 0.001$ the minimum E_{TN} is 1.1. At this point $b_N^* = 0.2$, $3\delta = 0.8$, and $g = 0.10$. Hence $h = 0.0098$.

(c) For the minimum design E_{TN} is 11% smaller. At this minimum point $3\delta/b_N^* = 4.0$ instead of 1.0. We can conclude from this that the minimum point is very broad.

1.2.6-1 The z-transform of $x(n) = x_n$ is defined by [132, 133]

$$F[x(n)] = X(z) = \sum_{n=-\infty}^{\infty} x(n)z^{-n} = \sum_{n=-\infty}^{\infty} x_n z^{-n} \qquad (P1.2.6\text{-}1a)$$

$$F[x(n-m)] = X(z)z^{-m} \qquad (P1.2.6\text{-}1b)$$

Applying the above to (P1.2.1-1a) yields

$$X_{n+1,n}^* = (g+h)Y_n - gY_n z^{-1} + (2-g-h)X_{n+1,n}^* z^{-1} + (g-1)X_{n+1,n}^* z^{-2}$$

and in turn

$$\frac{X_{n+1,n}^*}{Y_n} = H_p(z) = H_p = \frac{g+h-gz^{-1}}{1-(2-g-h)z^{-1}-(g-1)z^{-2}}$$

$$H_p = \frac{z[(g+h)z - g]}{z^2 + (g+h-2)z + (1-g)} \qquad (P1.2.6\text{-}1c)$$

(b) The denominator is a quadratic function in z, and its roots, which are the poles of H_p, are readily found to be

$$z_{1,2} = \tfrac{1}{2}[2-g-h] \pm \sqrt{(2-g-h)^2 + 4(g-1)]} \qquad (P1.2.6\text{-}1d)$$

(c) For critical damped conditions the poles must be real and equal; hence from (P1.2.6-1d) it is necessary that

$$(2-g-h)^2 + 4(g-1) = 0$$

which leads to (1.2-36), as we desired to show.

1.2.6-2 The input to the g–h filter consists of the sequence $y_n, n = -\infty, \ldots, n$, where the y_n's consists of the actual target range x_n plus the measurement noise ν_n given by (1.2-17). The output of the filter will consist of the deterministic part plus a noise part due to the input measurement noise ν_n. Let the output noise part be designated as $\nu_0(n)$ at time n. The variance of this output, in steady state, will be $\sigma^2_{n+1,n}$. The output due to the input noise sequence ν_n can be readily determined from the inpulse response of the filter. Let $h_n, n = 0, \ldots, \infty$, be the impulse response of the filter to an impulse applied at time zero. Then

$$\nu_0(n) = \sum_{i=n}^{\infty} h_i \nu_{n-i} \tag{P1.2.6-2a}$$

The impulse response can be obtained from the inverse of $H_p(z)$ given by (P1.2.6-1c). This inverse response is the sum of the residues of the poles of $H_p(z)$ inside the unit circle in the z-plane [132]. The variance of $\nu_0(n)$ is obtained by squaring $\nu_0(n)$ and obtaining its expected value to yield

$$E[\nu_0^2(n)] = \sigma^2_{n+1,n} = \sigma^2_x \sum_{i=0}^{\infty} h_i^2 \tag{P1.2.6-2b}$$

use being made of the independence of ν_i from ν_j for $i \neq j$. Thus

$$\text{VRF}(x^*_{n+1,n}) = \sum_{i=0}^{\infty} h_i^2 \tag{P1.2.6-2c}$$

But [130, 131]

$$\sum_{i=0}^{\infty} h_i^2 = \frac{1}{2\pi i} \oint H_p(z) H_p(z^{-1}) z^{-1} dz \tag{P1.2.6-2d}$$

in which the path of integration is the unit circle in the z-plane. Thus $\text{VRF}(x^*_{n+1,n})$ is the sum of the residues of the integrand of (P1.2.6-2d) inside the unit circle. Let the integrand of (P1.2.6-2d) be represented by $F(z)$; then $2\pi i$ times the residue of $F(z)$ at the simple first-order pole z_i is given by [132].

$$R_i = (z - z_i)F(z)\Big|_{z=z_i} \tag{P1.2.6-2e}$$

or

$$R_i = (z - z_i)H_p(z)H_p(z^{-1})z^{-1}\Big|_{z=z_i} \tag{P1.2.6-2f}$$

The poles of $F(z)$ inside the unit circles are given by the poles of $H_p(z)$, which were determined to be given by (P1.2.6-1d). Thus substituting z_1 and z_2 of (P1.2.6-1d) into (P1.2.6-2f) to obtain R_1 and R_2 and adding these give (P1.2.6-2d), which is (P1.2.6-2c) and hence (1.2-19).

Alternately, one can use the following [131, p. 420] relationships to evaluate (P1.2.6-2d): If $H(z)$ is the z-transform of h_n and

$$H(z) = \frac{b_0 z^2 + b_1 z + b_2}{a_0 z^2 + a_1 z + a_2} \qquad \text{(P1.2.6-2g)}$$

then

$$\sum_{i=0}^{\infty} h_i^2 = \frac{a_0 e_1 B_0 - a_0 a_1 B_1 + (a_1^2 - a_2 e_1)B_2}{a_0[(a_0^2 - a_2^2)e_1 - (a_0 a_1 - a_1 a_2)a_1]} \qquad \text{(P1.2.6-2h)}$$

where

$$B_0 = b_0^2 + b_1^2 + b_2^2 \qquad B_1 = 2(b_0 b_1 + b_1 b_2) \qquad \text{(P1.2.6-2i)}$$
$$B_2 = 2 b_0 b_2 \qquad e_1 = a_0 + a_2 \qquad \text{(P1.2.6-2j)}$$

1.2.6-3   A unity noiseless jump in velocity at time $n = 0$ is given by

$$y_n = nT \qquad n = 0, \ldots, n \qquad \text{(P1.2.6-3a)}$$

Its z-transform is [130]

$$Y(z) = Y_n = \frac{T z^{-1}}{(1 - z^{-1})^2} \qquad \text{(P1.2.6-3b)}$$

The filter for $x^*_{n+1,n}$ gives the value for time $n + 1$ at time n. Hence ideally it produces y_{n+1} at time n. Thus the desired filter output sequence is

$$s_n = (n + 1)T \qquad n = 0, 1, \ldots, n \qquad \text{(P1.2.6-3c)}$$

From (P1.2.6-1a) we see that the coefficient of z^{-n} for the z-transform of y_n gives the amplitude of the nth time sample of y_n. To make this time sample occur at time $n - 1$, we multiply the z-transform of y_n by z so that the amplitude of z^{-n+1} is now y_n with the result that y_n occurs at time $n - 1$. Thus the z-transform of (P1.2.6-3c), designate as $S_n(z)$, is obtained by multiplying (P1.2.6-3b) by z to obtain

$$S_n(z) = \frac{T}{(1 - z^{-1})^2} \qquad \text{(P1.2.6-3d)}$$

The output of the prediction filter to y_n is obtained by multiplying (P1.2.6-1c) by (P1.2.6-3b). Subtracting from this (P1.2.6-3d) gives the z-transform, $E(z)$, of the error transient $\varepsilon_n, n = 0, 1, 2, \ldots, n$, given by

$$E(z) = X^*_{n+1,n} - S_n \tag{P1.2.6-3e}$$

or

$$E(z) = \frac{-T}{1 - (2 - g - h)z^{-1} - (g - 1)z^{-2}} \tag{P1.2.6-3f}$$

From (P1.2.6-2d) it follows that

$$\sum_{i=0}^{\infty} \varepsilon_i^2 = \frac{1}{2\pi i} \oint E(z)E(z^{-1})z^{-1}dz \tag{P1.2.6-3g}$$

in which again the path of intergration is the unit circle in the z-plane. Thus the transient error is the sum of the residues of the integrand of (P1.2.6-3g) inside the unit circle; see solution to problem 1.2.6-2. Evaluating the sum of these residues gives (1.2-28) for a unity step in velocity, that is, for $\Delta v = 1$. In a similar manner (1.2-29) and (1.2-30) can be derived.

1.2.6-4 The z-transform of the noiseless measurements of a constant-accelerating target as given by (P1.2.4.3-1a) is [130]

$$y_n(z) = Y_n = \ddot{x}\frac{T^2z^{-1}(1 + z^{-1})}{2(1 - z^{-1})^3} \tag{P1.2.6-4b}$$

The desired output is given by

$$s_n = y_{n+1} \tag{P1.2.6-4c}$$

Hence the z-transform of s_n, designated as s_n or $S_n(z)$, is z times (P1.2.6-4b), or

$$S_n(z) = S_n = \ddot{x}\frac{T^2(1 + z^{-1})}{2(1 - z^{-1})^3} \tag{P1.2.6-4d}$$

The z-transform of $x^*_{n+1,n}$, designated as $X^*_{n+1,n}(z) = X^*_{n+1,n}$ when constant-accelerating target noiseless measurements are inputs, is given by

$$X^*_{n+1,n} = Y_nH_p \tag{P1.2.6-4e}$$

where H_p is given by (P1.2.6-1c) and Y_n by (P1.2.6-4b). The z-transform of the error $\varepsilon_n, n = 0, 1, \ldots, n$, is then given by

$$E(Z) = X^*_{n+1,n} - S_n \tag{P1.2.6-4f}$$

Applying the final value theorem to (P1.2.6-4f) yields (1.2-15)

1.2.6-5 The z-transforms of (1.2-8a), (1.2-8b), (1.2-10a), and (1.2-10b) are

$$\dot{X}^*_{n,n-1} = z^{-1}\dot{X}^*_{n-1,n-1} \tag{P1.2.6-5a}$$

$$X^*_{n,n-1} = z^{-1}X^*_{n-1,n-1} + T\dot{X}^*_{n,n-1} \tag{P1.2.6-5b}$$

$$\dot{X}^*_{n,n} = \dot{X}^*_{n,n-1} + \frac{h}{T}[Y_n - X^*_{n,n-1}] \tag{P1.2.6-5c}$$

$$X^*_{n,n} = X^*_{n,n-1} + g[Y_n - X^*_{n,n-1}] \tag{P1.2.6-5d}$$

where (1.2-10a) and (1.2-10b) were written for prediction to time n instead of $n + 1$. Dividing by Y_n yields

$$H'_{pd} = z^{-1}H_{fd} \tag{P1.2.6-5e}$$

$$H'_p = z^{-1}H_f + TH'_{pd} \tag{P1.2.6-5f}$$

$$H_{fp} = H'_{pd} + \frac{h}{T}\left[1 - H'_p\right] \tag{P1.2.6-5g}$$

$$H_f = H'_p + g\left[1 - H'_p\right] \tag{P1.2.6-5h}$$

where

$$H'_{pd} = \frac{\dot{X}^*_{n,n-1}}{Y_n} \tag{P1.2.6-5i}$$

$$H'_p = \frac{X^*_{n,n-1}}{Y_n} \tag{P1.2.6-5j}$$

$$H_f = \frac{X^*_{n,n}}{Y_n} \tag{P1.2.6-5k}$$

$$H_{fd} = \frac{\dot{X}^*_{n,n}}{Y_n} \tag{P1.2.6-5l}$$

where the subscript f stands for filtered, p for predicted, and d for derivative.

The set of equations (P1.2.6-5e) to (P1.2.6-5h) represents four equations with four unknown transfer functions which can be solved for. Here, H'_{pd} is given by (P1.2.6-5e) and can hence be eliminated from (P1.2.6-5f) to (P1.2.6-5h), leaving three equations with three unknowns. Solving yields [131]

$$H'_p = \frac{z(g+h) - g}{z^2 + (g+h-2)z + 1 - g} \tag{P1.2.6-5m}$$

$$H_f = \frac{z(gz + h - g)}{z^2 + (g+h-2)z + 1 - g} \tag{P1.2.6-5n}$$

$$H_{fd} = \frac{h}{T} \frac{z(z-1)}{z^2 + (g+h-2)z + 1 - g} \tag{P1.2.6-5o}$$

$$H'_{pd} = z^{-1} H_{fd} \tag{P1.2.6-5p}$$

What we really want instead of H'_p is

$$H_p = \frac{X^*_{n+1,n}}{Y_n} \tag{P1.2.6-5q}$$

see solution to problem 1.2.6-1. Because

$$X^*_{n,n-1} = z^{-1} X^*_{n+1,n} \tag{P1.2.6-5r}$$

from (P1.2.6-5j)

$$H'_p = \frac{z^{-1} X^*_{n+1,n}}{Y_n} = z^{-1} H_p \tag{P1.2.6-5s}$$

and

$$H_p = zH'_p \tag{P1.2.6-5t}$$

$$H_p = \frac{z[(g+h)z - g]}{z^2 + (g+h-2)z + (1-g)} \tag{P1.2.6-5u}$$

which is identical to (P1.2.6-1c). Using (P1.2.6-5q) for H_p in (P1.2.6-5u), multiplying out the denominator, and obtaining the inverse z-transform immediately lead to (P1.2.1-1a). In a similar manner, the recursive equations for $x^*_{n,n}$ and $\dot{x}^*_{n,n}$ can be obtained.

1.2.6-6 (a) For $A_N = 3.0$, Figure 1.2-13 indicates that $b^*_N = 3\delta = 5.0$ for $g = 0.95$. Hence $E_{TN} = 10.0$, and from (1.2-35a) and (1.2-35b), $\theta = 0.224$ and $h = 0.603$.

(b) For $A_N = 3.0$, from Figure 1.2-15 the minimum E_{TN} is 9.4 with $b^*_N = 3.7$ and $3\delta = 5.7$. At this point $g = 0.98$ so that $\theta = 0.141$ and $h = 0.737$.

(c) For the minimum design, E_{TN} is 6% smaller. At this minimum point, b^*_N almost equals 3δ; specifically, $3\delta/b^*_N = 1.54$.

1.2.6-7 (a) For $A_N = 0.001$, $b_N^* = 3\delta = 0.66$ for $g = 0.08$. Hence $E_{TN} = 1.32$, $\theta = 0.959$, and $h = 0.00167$.

(b) For $A_N = 0.001$ from Figure 1.2-15 the minimum E_{TN} is 1.15 with $b_N^* = 0.20$, $3\delta = 0.95$, and $g = 0.135$ so that $\theta = 0.930$ and $h = 0.0049$.

(c) For the minimum design E_{TN} is 13% smaller. At this point $3\delta/b_N^* = 4.75$.

1.2.6-8 (a) For $A_N = 0.003$, $b_N^* = 3\delta = 0.83$ for $g = 0.115$. Hence $E_{TN} = 1.66$, $\theta = 0.941$, and $h = 0.0035$.

(b) For $A_N = 0.003$, from Figure 1.2.6-15 the minimum E_{TN} is 1.4 with $b_N^* = 0.25$, $3\delta = 1.15$, and $g = 0.19$ so that $\theta = 0.90$ and $h = 0.01$.

(c) For the minimum design E_{TN} is 16% smaler. At this point $3\delta/b_N^* = 4.6$.

1.2.7-2 (b) Here $\sigma_{n+1,n}/\sigma_x = 3$. Hence from (1.2-41), $\theta = 0.1653$. From (1.2-35a) and (1.2-35b), $g = 0.973$ and $h = 0.697$. Using (1.2-23) and (1.2-15) yields $T = 3.13$ sec; this update time is possibly a little fast for some track-while-scan systems.

(c) Because $\sigma_{n+1,n}/\sigma_x = 3$, again $\theta = 0.1653$, $g = 0.973$, and $h = 0.697$. But now $T = 5.72$ sec, a reasonable update time for a track-while-scan system. To require large update times, it is necessary that $\sigma_{n+1,n}$ be large and $\ddot{x}_{max}$ be low. Note that the g–h filter obtained has a short memory, θ being small.

1.2.7-3

$$VRF\left(x_{n+1,n}^*\right) = \left(\frac{608'}{500'}\right)^2 = (1.216)^2 = 1.479$$

From (1.2-19)

$$VRF(x_{n=1,n}^*) = \frac{2g^2 + 2h + gh}{g(4 - 2g - h)} = 1.479 \tag{1}$$

and from (1.2-27)

$$h = \frac{g^2}{2 - g} \tag{2}$$

Solving (1) and (2) yields $g = 0.739$ and $h = 0.434$. Alternatively Table 1.2-4 could be used or Figure 1.2-7. Then $b^* = -\ddot{x}T^2/h = -(0.2g)10^2/0.434 = -1475$ ft, and

$$\frac{D_{x_{n+1,n}}^*}{\Delta v^2} = \frac{T^2(2 - g)}{gh(4 - 2g - h)} = 188.3 \text{ sec}^2$$

Note that $b^*/3\sigma_{n+1,n} = 0.809$ here.

1.2.7-5   From (1.2.41), $\theta = 0.400$.

From (1.2-16a), and (1.2-16b) $g = 0.840$, $h = 0.360$.

From (1.2-28), $Dx^*_{n+1,n}/\Delta v^2 = 195.9 \, \text{sec}^2$, larger than the value of $188.3 \, \text{sec}^2$ obtained for the Benedict–Bordner filter design of problem 1.2.7-3.

From (1.2-15), $b^* = -1778 \, \text{ft}$ versus $-1475 \, \text{ft}$ for problem 1.2.7-3.

Note that $b^*/3\sigma_{n+1,n} = 0.975$. One could reduce b^* to $-1445 \, \text{ft}$ by decreasing T to 9 sec, in which case $b^*/3\sigma_{n+1,n} = 0.790$.

1.2.7-8

$$g = 0.628 \qquad h = 0.287 \qquad T = 1.421 \, \text{sec}$$

$$\sigma_{n+1,n} = 150 \, \text{ft} \qquad \sigma_{n,n} = 109.5 \, \text{ft} \qquad \dot{\sigma}_{n+1,n} = 34.5 \, \text{ft/sec}$$

1.2.9-1   For the g–h to be stable, the poles of (P1.2.6-1c) in the z-plane must be within the unit circle centered about the origin. Alternatively the poles must be in the left hand of u given by the transformation of (P1.2.9-1a). To apply (P1.2.9-1a), we rewrite it as [132]

$$z = \frac{1+u}{1-u} \qquad\qquad \text{(P1.2.9-1b)}$$

and substitute into the denominator of (P1.2.6-1c) and set the result equal to zero to obtain

$$u^2(4 - 2g - h) + u(2g) + h = 0 \qquad\qquad \text{(P1.2.9-1c)}$$

The Routh–Hurwitz criterion [133] states that for a second-degree polynomial the roots are in the left-hand plane if all the coefficients are greater than zero, which leads immediately to (1.2-37a) to (1.2-37b) as we desired to show.

1.2.10-1   (a) The expanding-memory polynomial g–h filter used for track initiation is

$$h_n = \frac{6}{(n+2)(n+1)} \qquad\qquad \text{(1.2-38a)}$$

$$g_n \frac{2(2n+1)}{(n+2)(n+1)} \qquad\qquad \text{(1.2-38b)}$$

(1) Assume the target first range measurement is made at $n = 0$ and is given by y_0. We do not know the target velocity yet, so logically the prediction of range at time $n = 1$ is chosen to be $x^*_{1,0} = y_0$. The growing-memory g–h filter also gives this result. For $n = 0$, $g_0 = 1$ from (1.2.-38b). Hence from (1.2-8b)

$$x^*_{0,0} = x^*_{0,-1} + (y_0 - x^*_{0,-1})$$

$$= y_0$$

Because we do not know a priori the target velocity at time $n = 0$, in (1.2-10b) we set $\dot{x}^*_{0,1} = 0$ so that (1.2-10b) yields

$$x^*_{1,0} = x^*_{0,0} = y_0$$

(2) When the second range measurement y_1 at time $n = 1$ is obtained, the target velocity at time $n = 1$ can be estimated. It is logically given by

$$\dot{x}^*_{1,1} = \frac{y_1 - y_0}{T}$$

This is the same estimate obtained using the g–h growing-memory filter. At $n = 1$, from (1.2-38a), $h = 1$. Using (1.2-8a), we have

$$\dot{x}^*_{1,1} = 0 + \frac{y_1 - x^*_{1,0}}{T} = 0 + \frac{y_1 - y_0}{T}$$

the same as that obtained above.

From (1.2-38), $g_1 = 1$. Hence from (1.2-8b) we have

$$\begin{aligned}
x^*_{1,1} &= x^*_{1,0} + (y_1 - x^*_{1,0}) \\
&= y_0 + (y_1 - y_0) \\
&= y_1
\end{aligned}$$

Physically, this also follows because the least-squares best-fitting line to the first two measurements is a line going through these first two measurements.

From (1.2-38a), $\dot{x}^*_{2,1} = \dot{x}^*_{1,1}$, and from (1.2-10b)

$$\begin{aligned}
x^*_{2,1} &= y_1 + T\frac{y_1 - y_0}{T} = y_1 + y_1 - y_0 \\
&= 2y_1 - y_0
\end{aligned}$$

(b) Switchover to Benedict–Bordner filter occur when

$$\begin{aligned}
&\text{VAR}(x^*_{n+1,n}) \text{ for expanding-memory polynomial filter} \\
&= \text{VAR}(x^*_{n+1,n}) \text{ for Benedict–Bordner filter}
\end{aligned} \qquad (1.2\text{-}40)$$

Substituting (1.2-19) and (1.2-42) yields

$$\frac{2(2n + 3)}{(n + 1)n} = \frac{2g^2 + 2h + gh}{g(4 - 2g - h)} = \left(\frac{608'}{500'}\right)^2 = 1.479$$

Solving for n yields $n = 3.04 = 3$; hence the switchover occurs at $n = 3$.

1.2.10-2 (b) 0.26%
  (c) 4.56%.

1.2.10-3 The track initiation filter is exactly the same as obtained for the Benedict–Bordner filter, the switch over occurring at the same time.

1.2.10-4 From (1.2-19)

$$\text{VRF}\left(x^{*}_{n+1,n}\right) = \frac{2g^2 + 2h + gh}{g(4 - 2g - h)}$$

Substituting (1.2-35a) (1.2-35b) yields

$$
\begin{aligned}
\text{VRF}(x^{*}_{n+1,n}) &= \frac{2(1-\theta^2)^2 + 2(1-\theta)^2 + (1-\theta^2)(1-\theta)^2}{(1-\theta^2)[4 - 2(1-\theta^2) - (1-\theta)^2]} \\
&= \frac{2(1-\theta)^2(1+\theta)^2 + 2(1-\theta)^2 + (1-\theta^2)(1-\theta)^2}{(1-\theta)(1+\theta)[4 - 2 + 2\theta^2 - 1 + 2\theta - \theta^2]} \\
&= \frac{2(1-\theta)(1+\theta)^2 + 2(1-\theta) + (1-\theta^2)(1-\theta)}{(1+\theta)[1 + 2\theta + \theta^2]} \\
&= \frac{(1-\theta)[2(1-\theta)^2 + 2 + (1-\theta^2)]}{(1+\theta)[1 + 2\theta + \theta^2]} \\
&= \frac{(1-\theta)[2 + 4\theta + 2\theta^2 + 2 + 1 - \theta^2]}{(1+\theta)^3} \\
&= \frac{(1-\theta)(5 + 4\theta + \theta^2)}{(1+\theta)^3}
\end{aligned}
$$

1.3-1 Substituting (1.3-10c) into (1.3-9) yields

$$2h - g\left(g + h + \frac{h^2}{4g}\right) = 0 \qquad (P1.3-1)$$

Multiplying out (P1.3-1) and transposing terms yield

$$2h = g^2 + gh + \frac{1}{4}h^2$$

or

$$g^2 + gh + \left(\tfrac{1}{4}h^2 - 2h\right) = 0$$

Solving for g yields (1.3-10a).

2.4-1 From (2.4-4e) we have, in steady state,

$$H_n = S^*_{n,n-1}M^T[R_n + MS^*_{n,n-1}M^T]^{-1}$$

Substituting (2.4-4j) into (2.4-4f), we have

$$S^*_{n,n-1} = \Phi[1 - H_{n-1}M]S^*_{n-1,n-2}\Phi^T + Q_n \qquad (P.4\text{-}1a)$$

Dropping the subscript n for steady state and replacing the predictor covariance matrix $S^*_{n,n-1}$ by $\hat{S}$, the above equations become

$$H = \hat{S}M^T[R + M\hat{S}M^T]^{-1} \qquad (P2.4\text{-}1b)$$

$$\hat{S} = \Phi[1 - HM]\hat{S}\Phi^T + Q \qquad (P2.4\text{-}1c)$$

The matrix H is given by (2.4-5), but for simplicity we shall write it as

$$H = \begin{bmatrix} w_1 \\ w_2 \end{bmatrix} \qquad (P2.4\text{-}1d)$$

w_1 and w_2 being the Kalman filter weights. The terms Φ and M are given by respectively (2.4-1b) and (2.4-3a). Let $\hat{S}$ be given by

$$\hat{S} = \begin{bmatrix} s_{11} & s_{12} \\ s_{12} & s_{22} \end{bmatrix} \qquad (P2.4\text{-}1e)$$

The matrix R from (2.4-4i) and (2.4-3b) and the fact that $E[v_n^2] = \sigma_x^2$ is the 1×1 matrix

$$R = [\sigma_x^2] \qquad (P2.4\text{-}1f)$$

We will write σ_x^2 as r for simplicity. Substituting into (P2.4-1b) yields

$$\begin{bmatrix} w_1 \\ w_2 \end{bmatrix} = \begin{bmatrix} s_{11} & s_{12} \\ s_{12} & s_{22} \end{bmatrix}\begin{bmatrix} 1 \\ 0 \end{bmatrix}\left[r + \begin{bmatrix} 1 & 0 \end{bmatrix}\begin{bmatrix} s_{11} & s_{12} \\ s_{12} & s_{22} \end{bmatrix}\begin{bmatrix} 1 \\ 0 \end{bmatrix}\right]^{-1} \qquad (P2.4\text{-}1g)$$

which readily yields

$$w_1 = \frac{s_{11}}{r + s_{11}} \qquad (P2.4\text{-}1h)$$

$$w_2 = \frac{s_{12}}{r + s_{11}} \qquad (P2.4\text{-}1i)$$

Proceeding in a similar way with (P2.4-1c), after substitution and setting identical terms equal, we obtain

$$s_{11} = [(1 - w_1) - Tw_2]s_{11} + Ts_{12} + [(1 - w_1) - Tw_2]Ts_{12} + T^2s_{22} \quad \text{(P2.4-1j)}$$

$$s_{12} = [(1 - w_1) - Tw_2]s_{12} + Ts_{22} \quad \text{(P2.4-1k)}$$

$$s_{22} = -w_2s_{12} + s_{22} + u \quad \text{(P2.4-1l)}$$

where u is used in place of σ_u^2 for simplicity.

Solve (P2.4-1k) for T^2s_{22} and substitute into (P2.4-1j). Next divide (P2.4-1i) by (P2.4-1h) to obtain

$$s_{12} = \frac{w_2}{w_1}s_{11} \quad \text{(P2.4-1m)}$$

Substitute this into the new (P2.4-1j) and get

$$Tw_2 = \frac{w_1^2}{2 - w_1} \quad \text{(P2.4-1n)}$$

From (2.4-5) we know that

$$w_1 = g \quad \text{(P2.4-1o)}$$

$$w_2 = \frac{h}{T} \quad \text{(P2.4-1p)}$$

Applying the above to (P2.4-1n) yields (2.1-4), which we desired to get.

Using (P2.4-1h), we solve for s_{11} to obtain

$$s_{11} = \frac{w_1 r}{1 - w_1} \quad \text{(P2.4-1q)}$$

From (P2.4-1l) we get

$$s_{12} = \frac{u}{w_2} \quad \text{(P2.4-1r)}$$

Substituting (P2.4-1q) and (P2.4-1r) into (P2.4-1i) yields

$$\frac{u}{r} = \frac{w_2^2}{1 - w_1} \quad \text{(P2.4-1s)}$$

Using (P2.4-1n) yields

$$T^2 \frac{u}{r} = \frac{w_1^4}{(2 - w_1)^2(1 - w_1)} \quad \text{(P2.4-1t)}$$

Using (P2.4-1o) yields (2.1-5), as we desired to show.

2.4-2 Let $\hat{s}_{ij}$ and $\bar{s}_{ij}$ be the ij components of respectively $\hat{S}$ and $\bar{S}$; then

$$\hat{s}_{11} = \frac{g\sigma_x^2}{1-g} \tag{P2.4-2a}$$

$$\hat{s}_{12} = \frac{\sigma_u^2}{h}T \tag{P2.4-2b}$$

$$\hat{s}_{22} = \frac{\sigma_u^2}{h}(g+h) = \bar{s}_{22} + \sigma_u^2 \tag{P2.4-2c}$$

$$\bar{s}_{11} = (1-g)\hat{s}_{11} = g\sigma_x^2 \tag{P2.4-2d}$$

$$\bar{s}_{12} = (1-g)\hat{s}_{12} \tag{P2.4-2e}$$

$$\bar{s}_{22} = \sigma_u^2\frac{g}{h} \tag{P2.4-2f}$$

2.4-3 (a) $g = 0.742$, $h = 0.438$, $\sigma_{n+1,n} = 255$ ft, $\sigma_{n,n} = 129.2$ ft, $\dot{\sigma}_{n+1,n} = 149.2$ ft/sec, and $\dot{\sigma}_{n,n} = 118.3$ ft/sec.
(b) $\sigma_{n+1,n} = 183.4$ ft, $\sigma_{n,n} = 121.0$ ft, $\dot{\sigma}_{n+1,n} = 52.7$ ft/sec and $b^* = 295$ ft
($= 1.608\sigma_{n+1,n}$).

$$\hat{S} = \begin{bmatrix} \hat{s}_{11} & \hat{s}_{12} \\ \hat{s}_{12} & \hat{s}_{22} \end{bmatrix} = \begin{bmatrix} \bar{s}_{11} + 2T\bar{s}_{12} & +T^2\bar{s}_{22} & \bar{s}_{12} + T\bar{s}_{22} \\ \bar{s}_{12} + T\bar{s}_{22} & & \bar{s}_{22} \end{bmatrix} \tag{2.4-3a}$$

2.10-1 (a) From Figure 2.10-4, for $\sigma_{n+1,n}/\sigma_x = 2$, $p_2 \leq 1$ if the solution is at the peak of a $p_2 = $ contstant curve:

$$p_2 = T^2\frac{\sigma_a}{\sigma_x} = \frac{32\,\text{ft/sec}^2}{170\,\text{ft}}T^2 = 0.1882\,T^2$$

Hence $0.1882T^2 \leq 1$ or $T \leq 2.31$ sec. Choose $T = 2.3$ sec.
Then from (2.10-1), $p_1 = \tau/T = 3/2.3 = 1.304$, which is consistent with the value in Figure 2.10-4 for $p_2 = 1$ and no iterations are needed, just as was the case for the example in Section 2.10. From Figures 2.10-1 to 2.10-3 and 2.10-5, for $p_1 = 1.30$ and $p_2 = 1$, $g = 0.8$, $h = 0.56$, $k = 0.066$, and $\sigma_{n,n}/\sigma_x = 0.88$. Hence $\sigma_{n,n} = 150$ ft. It is worth noting that this example and that of Section 2.10 required no iteration primarily because we are working in a region of Figure 2.10-4 where the $p_2 = $ const curves are very flat. We will find this to be almost the case for problems 2.10-2 through 2.10-4. The region where the peak of the $p_2 = $ const curves are less flat (more peaked require large $p_2(> 10)$ and large $\sigma_{n+1,n}/\sigma_x(> 8)$). Typically our designs do not have $\sigma_{n+n,1}/\sigma_x > 8$. Alternatively, a region away from the peak where the curves are not flat requires that p_2 be large and p_1 be very small or large for $\sigma_{n+1,n}/\sigma_x$ to have a practical value.
(b) From Figures 2.10-6 and 2.10-7 $T\dot{\sigma}_{n+1,n}/\sigma_x = 1.6$, $T\dot{\sigma}_{n,n}/\sigma_x = 1.0$, and $T^2\ddot{\sigma}_{n+1,n}/\sigma_x \doteq T^2\ddot{\sigma}_{n,n}/\sigma_x = 1.0$. Thus $\dot{\sigma}_{n+1,n} = 118$ ft/sec, $\dot{\sigma}_{n,n} = 73.9$ ft/sec,

and $\ddot{\sigma}_{n,n} = \ddot{\sigma}_{n+1,n} = 32.1$ ft/sec^2. From Figures 2.10-8 to 2.10-10 $\sigma_{m,n}/\sigma_{n,n} =$ 0.64, $\dot{\sigma}_{m,n}/\dot{\sigma}_{n,n} = 0.42$, and $\ddot{\sigma}_{m,n}/\ddot{\sigma}_{n,n} = 0.70$. Thus $\sigma_{m,n} = 0.64$ (150 ft) $= 96$ ft, $\dot{\sigma}_{m,n} = 0.42(73.9\,\text{ft/sec}) = 31.0$ ft/sec, and $\ddot{\sigma}_{m,n} = 0.70(32.1) = 22.5$ ft/sec^2.

2.10-2   Based on the solution for problem 2.10-1, again in Figure 2.10-4 first try $p_2 \leq 1$ and again $p_2 = 0.1882T^2 \leq 1$. Hence $T = 2.3$ sec and $p_1 = \tau/T = 20/2.3 = 8.7$. Checking with Figure 2.10-4 yields, for $p_1 = 8.7$ and $\sigma_{n+1,n}/\sigma_x = 2$, $p_2 = 1.05$, close to the value of $p_2 = 1.0$. Choosing to do an iteration (one may argue that an interation really is not necessary), we now choose $p_2 = 1.05$ for the iteration. Then $p_2 = 0.1882T^2 = 1.05$ and $T = 2.36$ sec. Thus $p_1 = 20/2.36 = 8.5$, which is consistent with $\sigma_{n+1,n}/\sigma_x = 2$ and $p_2 = 1.05$ in Figure 2.10-4. From Figures 2.10-1 to 2.10-3 and 2.10-5, for $p_1 = 8.5$ and $p_2 = 1.05$, $g = 0.8$, $h = 0.54$, $k = 0.093$, and $\sigma_{n,n}/\sigma_x = 0.89$. Hence $\sigma_{n,n} = 151$ ft.

2.10-3   Again, based on the solution for problem 2.10-1, from Figure 2.10-4, first try $p_2 \leq 1$, and from (2.10-2)

$$p_2 = T^2 \frac{\sigma_a}{\sigma_x} = T^2 \frac{5(32)}{30} = 5.33T^2$$

Hence $p_2 = 5.33T^2 \leq 1$ so that $T \leq 0.433$ sec. Choose $T = 0.433$ sec; then from (2.10-1) $p_1 = \tau/T = 5/0.433 = 11.55$. Checking with Figure 2.10-4 yields for $p_1 = 11.55$ and $\sigma_{n+1,n}/\sigma_x = 2$, $p_2 = 1.05$, again close to the value of $p_1 = 1.0$. Choosing to do an iteration, we now choose $p_2 = 1.05$. Then $T = 0.444$ sec and $p_1 = 11.3$, which is consistent with $\sigma_{n+1,n}/\sigma_x = 2$ and $p_2 = 1.05$ in Figure 2.10-4. From Figures 2.10-1 to 2.10-3 and 2.10-5, for $p_1 = 11.3$ and $p_2 = 1.05$, $g = 0.78$, $h = 0.52$, $k = 0.087$, and $\sigma_{n,n}/\sigma_x = 0.88$. Hence $\sigma_{n,n} = 26.4$ ft.

2.10-4   From Figure 2.10-4 for $\sigma_{n+1,n}/\sigma_x = 3.33$, $p_2 \leq 2.5$ for the solution to be at the peak of a $p_2 =$const curve, as was the case for the example of Section 2.10. Also from problem 2.10-1 again $p_2 = 0.1882T^2$. For $p_2 = 0.1882T^2 \leq 2.5$, $T \leq 3.64$ sec. Choose $T = 3.64$ sec. From (2.10-1), $p_1 = \tau/T = 20/3.64 = 5.49$. Checking with Figure 2.10-4 yields, for $p_1 = 5.49$ and $\sigma_{n+1,n}/\sigma_x = 3.33$, $p_2 = 2.8$, which is not consistent with the value of $p_2 = 2.5$. For the second iteration, we use $p_2 = 2.8$. Then $T \leq 3.86$ sec. Use $T = 3.86$ sec. Now $p_1 = 5.2$, which is consistent with $p_2 = 2.8$ and $\sigma_{n+1,n}/\sigma_x = 3.33$ in Figure 2.10-4. From Figure 2.10-1 to 2.10-3 and 2.10-5, for $p_1 = 5.2$ and $p_2 = 2.8$, $g = 0.87$, $h = 0.83$, $k = 0.17$, and $\sigma_{n,n}/\sigma_x = 0.93$. Hence $\sigma_{n,n} = 158$ ft.

It is worth comparing this filter with the g–h critically damped filter design of problem 1.2.7-2(c) for which $\ddot{x}_{\max}$, $\sigma_{n+1,n}$, and σ_x are almost the same. The g–h–k filter obtained here has a longer memory, $g = 0.87$ and $h = 0.83$ here and $g = 0.973$ and $h = 0.697$ for problem 1.2.7-2(c). From (1.2-35a) the g of 0.87 is

Table P2.10-1. Summary of Singer g–h–k Filter Designs

Text Reference	σ_x (ft)	$\sigma_{n+1,n}/\sigma_x$	τ (sec)	$\ddot{x}$ (ft/sec²)	p_1	p_2	T (sec)	$\sigma_{n,n}/\sigma_x$	g	h	k
Section 2.10	164	3.33	3	98.4	1.50	2.4	2	0.93	0.87	0.81	0.14
Problem 2.10-1	170	2	3	32	1.30	1.0	2.3	0.88	0.8	0.56	0.066
Problem 2.10-2	170	2	20	32	8.5	1.05	2.36	0.89	0.8	0.54	0.093
Problem 2.10-3	30	2	5	160	11.3	1.05	0.444	0.88	0.78	0.52	0.087
Problem 2.10-4	170	3.33	20	32	5.2	2.8	3.86	0.93	0.87	0.83	0.17
Problem 2.10-6	164	2.03	—[a]	98.4	—	—	2	0.905	0.87	0.82	0.13
Problem 2.10-7	164	2.57	—[a]	98.4	—	—	2	0.897	0.931	1.1	0.20
Problem 2.10-8	164	1.93	—[a]	98.4	—	—	2	0.779	0.83	0.69	0.14
Problem 2.10-9	164	2.27	—[a]	98.4	—	—	2	0.862	0.90	0.93	0.25

[a] Assumed target acceleration to be independent between measurements.

435

equivalent to a θ of 0.36 for a critically damped filter, whereas $\theta = 0.1653$ for the critically damped g–h filter of problem 1.2.7-2(c). (From problem 7.2-2 we find that for a g–h–k critically damped filter $g = 1 - \theta^3$ so that $\theta = 0.507$ actually.) Also note that a shorter update is needed with the g–h–k ($T = 3.86\,\text{sec}$) than the g–h critically damped filter ($T = 5.72\,\text{sec}$), by about a factor of about 1.5.

Table P2.10-1 summarizes the Singer g–h–k filter designs given in Section 2.10 and problems 2.10-1 to 2.10-4.

2.10-5   Enough parameters are given here to define p_1 and p_2 directly from (2.10-1) and (2.10-2), without any iteration, yielding $p_1 = 5$ and $p_2 = 3$. From Figures 2.10-4 and 2.10-5, it then follows that $\sigma_{n+1,n}/\sigma_x = 3.7$ and $\sigma_{n,n}/\sigma_x = 0.94$; Hence $\sigma_{n+1,n} = 630\,\text{ft}$ and $\sigma_{n,n} = 160\,\text{ft}$. Because p_1 and p_2 are close to the value of problem 2.10-4, the values of g, h, and k as obtained from Figures 2.10-1 to 2.10-3 will be about the same as for problem 2.10-4; see Table P2.10-1.

2.10-6   For the example of Section 2.10, $\sigma_a = 30\,\text{m/sec}^2$. Hence $\sigma_w = \sigma_a = 30\,\text{m/sec}^2$, and from (2.4-13), $\ddot{x} = 10\,\text{m/sec}^3$. From Figure 1.3-1, $J_N = T^3 \ddot{x}/\sigma_{cx} = (2\,\text{sec})^3\,(10\,\text{m/sec}^3)/50\,\text{m} = 1.6$, where we used the value $T = 2$ sec of the example. From Figure 1.3-1 it follows that $b_N^* = 3\delta$ for $3\delta = 6.1 = b_N^*$. Then $\delta = 2.03$ and $E_{TN} = 12.2$. Also from Figure 1.3-1, $h = 0.82$. Because $b_N^* = J_N/2k$ (see Figure 1.3-1), it follows that $k = 0.13$. From (1.3-8c) it follows in turn that $\theta = 0.362$. Hence from (1.3-8a), $g = 0.870$. Using (1.3-6) yields $\sigma_{n,n}/\sigma_x = 0.905$. Note that these values for g, h, k, δ, and $\sigma_{n,n}/\sigma_x$ are very similar to those obtained for the g–h–k filter of Section 2.10; see Table P2.10-1. As a check, we used (1.3-8b) with $\theta = 0.362$ to yield $h = 0.83$, which is in good agreement with the values obtained above. Using (1.3-4) yields $\delta = 1.92$, which is also in good agreement with the value obtained above. The differences are due to the inaccuracies in reading the curves.

2.10-7   From Figure 1.3-3, for $J_N = 1.6$ the minimum E_{TN} design yields $E_{TN} = 11.7$, $h = 1.1$, $3\delta = 7.7$, $b_N^* = 4.0$, and $\delta = 2.57$. Because $b_N^* = J_N/2k$, $k = 0.20$. From (1.3-8c), $\theta = 0.263$. Thus from (1.3-8a), $g = 0.931$. Using (1.3-6) yields $\sigma_{n,n}/\sigma_x = 0.897$. Note that these values for g, h, k differ somewhat from the values obtained for problem 2.10-6 and the example of Section 2.10. The values for δ and $\sigma_{n,n}/\sigma_x$ are similar; see Table P2.10-1.

2.10-8   As in problem 2.10-6, $\sigma_w = \sigma_a = 30\,\text{m/sec}^2$, $\ddot{x} = 10\,\text{m/sec}^3$, and $J_N = 1.6$. From Figure 1.3-4 it follows that $b_N^* = 3\delta$ for $3\delta = 5.8 = b_N^*$. Then $E_{TN} = 11.6$ and $\delta = 1.93$. Also from Figure 1.3-4, $h = 0..69$. Because $b_N^* = J_N/2k$ (see Figure 1.3-4), it follows that $k = 0.14$. Using (1.3-10a) yields $g = 0.83$. Using (1.3-6) yields $\sigma_{n,n}/\sigma_x = 0.779$. These values for g, h, k, δ, and $\sigma_{n,n}/\sigma_x$ are somewhat similar to those of the g–h–k filter of the examples of Section 2.10 and of problem 2.10-6; see Table P2.10-1.

2.10-9   Again as in problem 2.10-6, $\sigma_w = \sigma_a = 30\,\text{m/sec}^2$, $\ddot{x} = 10\,\text{m/sec}^3$, and $J_N = 1.6$. From Figure 1.3-6, for $J_N = 1.6$ the minimum E_{TN} design yields $E_{TN} = 10.0$, $h = 0.93$, $3\delta = 6.8$, $b_N^* = 3.2$, and $\delta = 2.27$. Because $b_N^* = J_N/2k$, $k = 0.25$. Using (1.3-10a) yields $g = 0.90$. Using (1.3-6) yields $\sigma_{n,n}/\sigma_x = 0.862$. These values for g, h, k, δ, and $\sigma_{n,n}/\sigma_x$ are in reasonable agreement with those of the example of Section 2.10; see Table P2.10-1.

2.10-17

	Steady-State Singer g–h–k Filter	Steady-State g–h Kalman Filter
g	0.71	0.742
h	0.4	0.438
k	0.03	0.000
$\sigma_{n+1,n}$ (ft)	237	255
$\sigma_{n,n}$ (ft)	126	129.2
$\dot{\sigma}_{n+1,n}$ (ft/sec)	137	149.2
$\sigma_{n,n}$ (ft/sec)	73.9	118.3

We see on comparing the values of the steady-state g–h–k Singer filter with those of the steady-state g–h Kalman that they are indeed approximately equivalent, as we desired to show.

3.5.1.4-1   (a) From (3.5-9), $\Delta t = 2.47\,\text{sec}$. Thus $\Delta R = 37,000\,\text{ft}$; $\Delta r = c/2B_s = 492\,\text{ft}$. Hence $\Delta R/\Delta r = 75.3$.

(b) From (3.5-9), now $\Delta t = 6.83\,\text{msec}$. Thus $\Delta R = 95.6\,\text{ft}$. Now $\Delta r = 2.46\,\text{ft}$, and $\Delta R/\Delta r = 38.9$.

(c) From (3.5-9), $\Delta t = 0.0370\,\text{sec}$, and now $\Delta R = 555\,\text{ft}$, $\Delta r = 98.4\,\text{ft}$, and $\Delta R/\Delta r = 5.64$, one-seventh of the $\Delta R/\Delta r$ in (b) because T_U is one-seventh as large.

(d) From (3.5-9), $\Delta t = 0.370\,\text{sec}$ and now $\Delta R = 5550\,\text{ft}$, $\Delta r = 98.4\,\text{ft}$, and $\Delta R/\Delta r = 56.4$, 10 times the $\Delta R/\Delta r$ in (c) because T_U is 10 times larger.

3.5.1.4-2   Now ΔR_a is 100 times bigger so that, for example 1, $\Delta R_a = 1600\,\text{ft}$ and $\Delta R = 2106\,\text{ft}$, instead of the 506 ft for $a_d = 0$; for example 2, $\Delta R_a = 4\,\text{ft}$, so that $\Delta R = 28.7\,\text{ft}$ instead of the 24.7 ft for $a_d = 0$.

3.5.2-1   (a) From the solution to problem 2.10-5, $p_1 = 5$, and $p_2 = 3$. From (3.5-9) and (3.5-16), $p_3 = 10$. Hence from Figure 3.5-7, $\sigma_{n,n}/\sigma_x = 0.30$, so that $\sigma_{n,n} = 51\,\text{ft}$ when we use an upchirp. In problem 2.10-5, $\sigma_{n,n} = 160\,\text{ft}$ when a nonchirped waveform was used. Thus using an upchirp waveform improves the accuracy of $\sigma_{n,n}$ by $160/51 = 3.1$, about a factor of 3. For this example we assumed that the range measurement accuracy was about one-third of the waveform resolution of $\Delta r = c/2B_s = 492\,\text{ft}$ to yield $\sigma_x = 170\,\text{ft}$.

(b) For a downchirp, $p_3 = -10$, and from Figure 3.5-7, we see that $\sigma_{n,n}/\sigma_x \gg 10$, so that $\sigma_{n,n} \gg 1700$ ft, way over 10 times worse than with a nonchirped waveform and way over 30 times worse than with the downchirp waveform.

4.1-1   From (4.1-15) and (2.4-3a)

$$\Phi^{-1} = \begin{bmatrix} 1 & -T \\ 0 & 1 \end{bmatrix}$$

$$M = \begin{bmatrix} 1 & 0 \end{bmatrix}$$

Thus, $M\Phi^{-1}$ of (4.1-13) becomes

$$M\Phi^{-1} = \begin{bmatrix} 1 & 0 \end{bmatrix} \begin{bmatrix} 1 & -T \\ 0 & 1 \end{bmatrix} = \begin{bmatrix} 1 & -T \end{bmatrix}$$

Substituting the above and (2.4-3a) into (4.1-13) yields (4.1-17).

4.1-4   (a) From (4.1-17)

$$T = \begin{bmatrix} 1 & 0 \\ 1 & -T \end{bmatrix}$$

Hence

$$T^T = \begin{bmatrix} 1 & 1 \\ 0 & -T \end{bmatrix}$$

and

$$T^T T = \begin{bmatrix} 1 & 1 \\ 0 & -T \end{bmatrix} \begin{bmatrix} 1 & 0 \\ 1 & -T \end{bmatrix} = \begin{bmatrix} 2 & -T \\ -T & T^2 \end{bmatrix}$$

From matrix theory the α_{ij} element of the inverse A^{-1} of a matrix A is given by

$$\alpha_{ij} = -\frac{A_{ji}}{|A|} \tag{P4.1-4a}$$

where A_{ji} is the ji cofactor of A. Specifically, A_{ji} is $(-1)^{i+j}$ times the dererminant of the matrix A with its jth row and ith column deleted and $|A|$ is the determinant of A. Hence

$$[T^T T]^{-1} = \begin{bmatrix} 1 & \dfrac{1}{T} \\ \dfrac{1}{T} & \dfrac{2}{T^2} \end{bmatrix}$$

Check

$$
\begin{bmatrix} 1 & \dfrac{1}{T} \\ 1 & \dfrac{2}{T^2} \end{bmatrix} \begin{bmatrix} 2 & -T \\ -T & T^2 \end{bmatrix} = \begin{bmatrix} 1 & 0 \\ 0 & 1 \end{bmatrix}
$$

Therefore

$$
\hat{W} = [T^T T]^{-1} T^T = \begin{bmatrix} 1 & \dfrac{1}{T} \\ 1 & \dfrac{2}{T^2} \end{bmatrix} \begin{bmatrix} 1 & 1 \\ 0 & -T \end{bmatrix} = \begin{bmatrix} 1 & 0 \\ \dfrac{1}{T} & -\dfrac{1}{T} \end{bmatrix} \qquad \text{(P4.1-4b)}
$$

$$
\hat{W}T = \begin{bmatrix} 1 & 0 \\ \dfrac{1}{T} & -\dfrac{1}{T} \end{bmatrix} \begin{bmatrix} 1 & 0 \\ 1 & -T \end{bmatrix} = \begin{bmatrix} 1 & 0 \\ 0 & 1 \end{bmatrix}
$$

Hence

$$
\begin{bmatrix} 1 & 0 \\ 1 & -T \end{bmatrix}^{-1} = \begin{bmatrix} 1 & 0 \\ \dfrac{1}{T} & -\dfrac{1}{T} \end{bmatrix}
$$

Above $\hat{W}$ is the inverse of T. This follows because we are dealing with a case where there are as many equations in (4.1-11) as there are unknowns when we try to solve for $X_n = X_{n,n}^*$, the $N_{(n)}$ term being dropped. In this case T is nonsingular and (4.1-11) can be solved for $X_{n,n}^*$ by inverting T to give

$$
X_{n,n}^* = T^{-1} Y_{(n)} \qquad \text{(P4.1-4c)}
$$

This is called the deterministic case. For this case our best least-squares fitting line of Figure 1.2-10 fits through the data points; there are only two points and a straight line fits exactly through these two data points. The error e_n of (4.1-31) is zero for this case. Here

$$
X_{1,1}^* = \begin{bmatrix} x_{1,1}^* \\ \dot{x}_{1,1}^* \end{bmatrix} = T^{-1} Y_1 = \begin{bmatrix} 1 & 0 \\ 1 & -T \end{bmatrix}^{-1} \begin{bmatrix} y_1 \\ y_0 \end{bmatrix} = \begin{bmatrix} 1 & 0 \\ \dfrac{1}{T} & -\dfrac{1}{T} \end{bmatrix} \begin{bmatrix} y_1 \\ y_0 \end{bmatrix}
$$

Therefore

$$
x_{1,1}^* = y_1 \qquad \dot{x}_{1,1}^* = \frac{y_1 - y_0}{T}
$$

as one expects for the straight-line fit to the two data points.

4.5-1 From (2.6-10) we see that the derivative of (4.5-13) is

$$
2(Y_{(n)} - TX_{n,n}^*)^T R_{(n)}^{-1}(-T) = 0
$$

Hence

$$(Y_{(n)}^T - X_{n,n}^{*T} T^T)R_{(n)}^{-1} T = 0$$

Expanding yields

$$Y_{(n)}^T R_{(n)}^{-1} T = X_{n,n}^T T^T R_{(n)}^{-1} T$$

Taking the transpose of both sides yields

$$T^T R_{(n)}^{-1} Y_{(n)} = T^T R_{(n)}^{-1} T X_{n,n}$$

Thus

$$X_{n,n}^* = \underbrace{(T^T R_{(n)}^{-1} T)^{-1} T^T R_{(n)}^{-1}}_{\overset{\circ}{W}} Y_{(n)}$$

The coefficient of Y_n above is $\overset{\circ}{W}$, which agrees with (4.5-4).

4.5-2 Substituting (P4.5-2a) and (2.4-1b) into (P4.5-2b) yields

$$\hat{W}(h = 1) = \Phi\hat{W}(h = 0) = \begin{bmatrix} 1 & T \\ 0 & 1 \end{bmatrix} \begin{bmatrix} 1 & 0 \\ \dfrac{1}{T} & -\dfrac{1}{T} \end{bmatrix} = \begin{bmatrix} 2 & -1 \\ \dfrac{1}{T} & -\dfrac{1}{T} \end{bmatrix} \quad \text{(P4.5-2c)}$$

Therefore extending the solution to problem 4.1-4a, we get

$$X_{2,1}^* = \begin{bmatrix} x_{2,1}^* \\ \dot{x}_{2,1}^* \end{bmatrix} = \begin{bmatrix} 2 & -1 \\ \dfrac{1}{T} & -\dfrac{1}{T} \end{bmatrix} \begin{bmatrix} y_1 \\ y_0 \end{bmatrix}$$

$$x_{2,1}^* = 2y_1 - y_0 = y_1 + (y_1 - y_0)$$

$$\dot{x}_{2,1}^* = \frac{y_1 - y_0}{T}$$

5.5-1 In (5.5-8)

$$\binom{j}{i} = \frac{j!}{i!(j-i)!}$$

Hence from (5.5-8), for $h = 1$

$$\Phi(1)_{00} = 1$$

$$\Phi(1)_{01} = \binom{1}{0} 1^1 = 1$$

$$\Phi(1)_{02} = \binom{2}{0} 1^2 = \frac{2!}{0!2!} = 1$$

$$\Phi(1)_{10} = \binom{0}{1} 1^{-1} = \frac{0!}{1!(-1)!} = \frac{1}{\infty} = 0$$

because $(-n)! = \infty$ when n is a positive integer. Also

$$\Phi(1)_{11} = \binom{1}{1}1^0 = \frac{1!}{1!(0!)} = 1$$

$$\Phi(1)_{12} = \binom{2}{1}1^1 = \frac{2!}{1!1!} = 2$$

$$\Phi(1)_{20} = \binom{0}{2}1^{-2} = \frac{0!}{2!(-2!)} = \frac{1}{\infty} = 0$$

$$\Phi(1)_{21} = \binom{1}{2}1^{-1} = \frac{1!}{2!(-1)!} = \frac{1}{\infty} = 0$$

$$\Phi(1)_{22} = \binom{2}{2}1^0 = \frac{2!}{2!0!} = 1$$

Thus (5.4-13) follows, as we desired to show.

5.5-2   From (5.5-3a)

$$W(h)_z = \Phi(h)_z SGCB$$

Thus

$$W(h = 1)_z = \Phi(1)_z SGCB \qquad\qquad\text{(P5.5-2a)}$$

and

$$W(h = 0)_z = SGCB \qquad\qquad\text{(P.5.5-2b)}$$

and

$$W(h = 1)_z = \Phi(1)_z W(h = 0) \qquad\qquad\text{(P5.5-2c)}$$

From the solution to problem 5.5-1 it follows that $\Phi(1)_z$ of (5.5-3a) is given by

$$\Phi(1)_z = \begin{bmatrix} 1 & 1 \\ 0 & 1 \end{bmatrix} \qquad\qquad\text{(P5.5-2d)}$$

From (5.5-4), for $m = 1$, S is a 2×2 matrix and $[S]_{00} = 1$, $[S]_{01} = 0 = [S]_{10}$, $[S]_{11} = [S]_{00} + 0 \cdot 0 = 1$. Thus

$$S = \begin{bmatrix} 1 & 0 \\ 0 & 1 \end{bmatrix} \qquad\qquad\text{(P5.5-2e)}$$

From (5.5-5), G is also a 2×2 matrix for $m = 1$, and from (5.5-5) and (5.3-4a)

$$[G]_{00} = (1)(1)(1)\frac{1}{1} = 1$$

$$[G]_{01} = (-1)\binom{1}{0}\binom{1}{0}\frac{1}{1} = -1$$

$$[G]_{10} = (-1)^0 \binom{0}{1}\binom{1}{1}\frac{1}{L^{(1)}} = \frac{0!}{1!(-1)!L} = 0 \qquad \text{(P5.5-2f)}$$

$$[G]_{11} = (-1)^1 \binom{1}{1}\binom{2}{1}\frac{1}{L^{(1)}} = (-1)\left(\frac{1!}{1!0!}\right)\left(\frac{2!}{1!1!}\right)\frac{1}{L} = -\frac{2}{L}$$

Therefore for $L = 1$

$$G = \begin{bmatrix} 1 & -1 \\ 0 & -2 \end{bmatrix}$$

Here, C is also a 2×2 matrix for $m = 1$, and from (5.3-5) and (5.3-5a), C is diagonal, with

$$[C]_{jj} = \frac{1}{c_j^2} \quad \text{for } 0 \le j \le m \qquad (5.5\text{-}7)$$

where

$$c_j^2 = \frac{(L+j+1)^{(j+1)}}{(2j+1)L^{(j)}} \qquad (5.3\text{-}5a)$$

and

$$c_0^2 = \frac{L+1}{(1)(1)} = L+1 = 2$$

$$c_1^2 = \frac{(L+2)^{(2)}}{(3)L^{(1)}} = \frac{(L+2)(L+1)}{3L} = 2$$

Thus for $L = 1$

$$[C]_{00} = \tfrac{1}{2} \qquad [C]_{11} = \tfrac{1}{2}$$

and

$$C = \begin{bmatrix} \tfrac{1}{2} & 0 \\ 0 & \tfrac{1}{2} \end{bmatrix} \qquad \text{(P5.5-2g)}$$

For $m = 1$, $L = 1$, B is also a 2×2 matrix, and from (5.5-6), (5.3-3), and (5.3-4)

$$[B]_{ij} = p_i(L - j) = p(L - j; i, L)$$

$$= \sum_{\nu=0}^{i} (-1)^{\nu} \binom{i}{\nu} \binom{i+\nu}{\nu} \frac{(L-j)^{(\nu)}}{L^{(\nu)}}$$

$$[B]_{00} = p_0(L - 0) = (-1)^0 \binom{0}{0} \binom{0}{0} \frac{(L-0)^{(0)}}{L^{(0)}} = +1$$

$$[B]_{01} = p_0(L - 1) = (-1)^0 \binom{0}{0} \binom{0}{0} \frac{(L-1)^{(0)}}{L^{(0)}} = +1$$

$$[B]_{10} = p_1(L - 0) = \left[(-1)^0 \binom{1}{0} \binom{1+0}{0} \frac{(L-0)^0}{L^{(0)}} = 1 \right]$$

$$+ \left[(-1)^1 \binom{1}{1} \binom{1+1}{1} \frac{(L-0)^{(1)}}{L^{(1)}} = -2 \right]$$

$$= -1$$

$$[B]_{11} = p_1(L - 1) = \left[(-1)^0 \binom{1}{0} \binom{1}{0} \frac{(L-1)^{(0)}}{L^{(0)}} = 1 \right]$$

$$+ \left[(-1)^1 \binom{1}{1} \binom{2}{1} \frac{(L-1)^1}{L^{(1)}} = -2 \frac{L-1}{L} = 0 \right]$$

$$= +1$$

Therefore

$$B = \begin{bmatrix} 1 & 1 \\ -1 & 1 \end{bmatrix} \tag{P5.5-2h}$$

Substituting (P5.5-2e) to (P5.5-2h) into (P5.5-2b) yields

$$W(h = 0)_z = \begin{bmatrix} 1 & 0 \\ 0 & 1 \end{bmatrix} \begin{bmatrix} 1 & -1 \\ 0 & -2 \end{bmatrix} \begin{bmatrix} \frac{1}{2} & 0 \\ 0 & \frac{1}{2} \end{bmatrix} \begin{bmatrix} 1 & 1 \\ -1 & 1 \end{bmatrix}$$

$$= \begin{bmatrix} 1 & -1 \\ 0 & -2 \end{bmatrix} \begin{bmatrix} 1 & 1 \\ -1 & 1 \end{bmatrix} \begin{pmatrix} 1 \\ 2 \end{pmatrix}$$

$$= \begin{bmatrix} 2 & 0 \\ 2 & -2 \end{bmatrix} \frac{1}{2}$$

$$= \begin{bmatrix} 1 & 0 \\ 1 & -1 \end{bmatrix} \tag{P5.5-2i}$$

From (P5.5-2c) and (P5.5-2d)

$$W(h = 1)_z = \begin{bmatrix} 1 & 1 \\ 0 & 1 \end{bmatrix} \begin{bmatrix} 1 & 0 \\ 1 & -1 \end{bmatrix} = \begin{bmatrix} 2 & -1 \\ 1 & -1 \end{bmatrix} \qquad \text{(P5.5-2j)}$$

The above is $W(h = 1)$ for $Z^*_{n+1,n}$. On comparing (5.4-1) for X_n with (5.4-12) for Z_n, we see that, for $m = 1$,

$$Z_n = \begin{bmatrix} 1 & 0 \\ 0 & T \end{bmatrix} X_n \qquad \text{(P5.5-2k)}$$

or

$$\begin{bmatrix} x \\ TDx \end{bmatrix} = \begin{bmatrix} 1 & 0 \\ 0 & T \end{bmatrix} \begin{bmatrix} x \\ Dx \end{bmatrix}$$

Because

$$\begin{bmatrix} 1 & 0 \\ 0 & T \end{bmatrix}^{-1} = \begin{bmatrix} 1 & 0 \\ 0 & \dfrac{1}{T} \end{bmatrix}$$

it follows that

$$X_n = \begin{bmatrix} 1 & 0 \\ 0 & \dfrac{1}{T} \end{bmatrix} Z_n \qquad \text{(P5.5-2l)}$$

Thus $W(h = 1)$ for $X^*_{n+1,n}$ is given by

$$W(h = 1) = \begin{bmatrix} 1 & 0 \\ 0 & \dfrac{1}{T} \end{bmatrix} W(h = 1)_z = \begin{bmatrix} 1 & 0 \\ 0 & \dfrac{1}{T} \end{bmatrix} \begin{bmatrix} 2 & -1 \\ 1 & -1 \end{bmatrix}$$

$$= \begin{bmatrix} 2 & -1 \\ \dfrac{1}{T} & -\dfrac{1}{T} \end{bmatrix}$$

as we desired to show. Also from (P5.5-2l) and (P5.5-2i)

$$W(h = 0) = \begin{bmatrix} 1 & 0 \\ 0 & \dfrac{1}{T} \end{bmatrix} W(h = 0)_z = \begin{bmatrix} 1 & 0 \\ 0 & \dfrac{1}{T} \end{bmatrix} \begin{bmatrix} 1 & 0 \\ 1 & -1 \end{bmatrix}$$

$$= \begin{bmatrix} 1 & 0 \\ \dfrac{1}{T} & -\dfrac{1}{T} \end{bmatrix}$$

which agrees with (P4.5-2a) and (P4.1-4b).

5.5-3 From (P5.5-2j)

$$\hat{W}(h = 1)_z = \begin{bmatrix} 2 & -1 \\ 1 & -1 \end{bmatrix}$$

From (5.6-3)

$$
{}_sS^*_{n+h,n} = \sigma_x^2 W(h)_z W(h)_z^T
$$
$$
= \sigma_x^2 \begin{bmatrix} 2 & -1 \\ 1 & -1 \end{bmatrix} \begin{bmatrix} 2 & 1 \\ -1 & -1 \end{bmatrix} = \sigma_x^2 \begin{bmatrix} 5 & 3 \\ 3 & 2 \end{bmatrix} \tag{P5.5-3a}
$$

for $h = 1$. From (5.6-5) for $L = 1$

$$
{}_sS^*_{n+1,n} = \sigma_x^2 \begin{bmatrix} 5 & 3 \\ 3 & 2 \end{bmatrix}
$$

5.5-4 As in problem 5.5-2, all the matrices except B of (5.5-3a) have dimension 2×2 with the matrices S and $\Phi(h = 1)_z$ the same as in problem 5.5-2. The matrix B has dimension $(L + 1) \times 2$; see (5.5-6). Now from (5.5-5) for $L = 4$

$$
[G]_{00} = (-1)^0 (1)(1)\frac{1}{1} = 1
$$
$$
[G]_{01} = (-1)^0 \binom{1}{0}\binom{1}{0}(1)\frac{1}{1} = -1
$$
$$
[G]_{10} = (-1)^0 \binom{0}{1}\binom{1}{1}\frac{1}{4^{(1)}} = \frac{0!}{1!(-1)!}\frac{1}{4} = 0
$$
$$
[G]_{11} = (-1)\binom{1}{1}\binom{2}{1}\frac{1}{4^{(1)}} = -\frac{1}{2}
$$

Therefore for $L = 4$

$$
G = \begin{bmatrix} 1 & -1 \\ 0 & -\frac{1}{2} \end{bmatrix} \tag{5.5-4a}
$$

From (5.5-7) and (5.3-5a), C is diagonal, with

$$
[C]_{jj} = \frac{1}{c_j^2}
$$
$$
c_j^2 = \frac{(L+j+1)^{(j+1)}}{(2j+1)L^{(j)}}
$$

Hence (see solution to problem 5.5-2)

$$
c_0^2 = L + 1
$$
$$
c_1^2 = \frac{(L+2)(L+1)}{3L}
$$

and for $L = 4$, $c_0^2 = 5$ and $c_1^2 = \frac{5}{2}$. Thus

$$C = \begin{bmatrix} \frac{1}{5} & 0 \\ 0 & \frac{2}{5} \end{bmatrix} \tag{P5.5-4b}$$

We could determine the components of the matrix B using (5.5-6), (5.3-3), and (5.3-4), as was done in problem 5.5-2. Instead we shall now make use of Table 5.3-1 with its simple expressions of the orthogonal discrete Legendre polynomials. From Table 5.3-1

$$p(x; 0, L) = 1 \tag{P5.5-4c}$$

$$p(x; 1, L) = 1 - 2\frac{x}{L} \tag{P5.5-4d}$$

From (5.5-6)

$$[B]_{ij} = p_i(r = L - j) = p_i(L - j) = p(L - j; i, L) \tag{P5.5-4e}$$

Thus from (P5.5-4b) to (P5.5-4d)

$$[B]_{00} = p(L; 0, L) = 1$$
$$[B]_{01} = p(L - 1; 0, L) = 1$$
$$[B]_{0j} = p(L - j; 0, L) = 1 \quad \text{for} \quad j = 2, 3, 4$$
$$[B]_{1j} = p(L - j; 1, L) = 1 - 2\,\frac{L - j}{L} = \frac{2j - L}{L} \quad \text{for} \quad j = 0, 1, 2, 3, 4$$

Thus for $L = 4$

$$[B]_{10} = -1 \qquad [B]_{11} = -\tfrac{1}{2} \qquad [B]_{21} = 0 \qquad [B]_{31} = \tfrac{1}{2} \qquad [B]_{41} = 1$$

Therefore

$$B = \begin{bmatrix} 1 & 1 & 1 & 1 & 1 \\ -1 & -\frac{1}{2} & 0 & \frac{1}{2} & 1 \end{bmatrix} \tag{P5.5-4f}$$

Substituting (P5.5-2d), (P5.5-2e), (P.5.5-4a), (P5.5-4b), (P5.5-4f) into (5.5-3a) yields the sought-after $W(h = 1)$.

5.6-1

$$\Phi = \begin{bmatrix} 1 & T \\ 0 & 1 \end{bmatrix} \tag{2.4-1b}$$

$$S_{n+1,n}^* = \Phi S_{n,n}^* \Phi^T + Q_n \tag{2.4-4f}$$

$$_sS^*_{n,n} = \sigma^2_x \begin{bmatrix} \dfrac{2(2L+1)}{(L+2)(L+1)} & \dfrac{6}{(L+2)(L+1)} \\ \dfrac{6}{(L+2)(L+1)} & \dfrac{12}{(L+2)(L+1)L} \end{bmatrix} \tag{5.6-4}$$

$$\Phi_z = \begin{bmatrix} 1 & 1 \\ 0 & 1 \end{bmatrix} \quad \therefore \Phi^T_z = \begin{bmatrix} 1 & 0 \\ 1 & 1 \end{bmatrix}$$

$$\Phi_z \, _sS^*_{n,n} = \sigma^2_x \begin{bmatrix} \dfrac{4(L+2)}{(L+2)(L+1)} & \dfrac{6}{(L+1)L} \\ \dfrac{6}{(L+2)(L+1)} & \dfrac{12}{(L+2)(L+1)L} \end{bmatrix}$$

$$\Phi_z \, _sS^*_{n,n} \Phi^T_z = \begin{bmatrix} \dfrac{2(2L+3)}{(L+1)L} & \dfrac{6}{(L+1)L} \\ \dfrac{6(L+2)}{(L+2)(L+1)L} & \dfrac{12}{(L+2)(L+1)L} \end{bmatrix}$$

5.6-2

$$\left[S^*_{n+h,n} \right]_{ij} = \underbrace{\dfrac{i!\,j!}{T^{i+j}}}_{a_{ij}} \left[_sS^*_{n+h,n} \right]_{i,j} \tag{5.6-7}$$

$$a_{00} = 1 \qquad a_{10} = \dfrac{1}{T} \qquad a_{01} = \dfrac{1}{T} \qquad a_{11} = \dfrac{1}{T^2}$$

Thus

$$S^*_{n+1,n} = \begin{bmatrix} \dfrac{2(2L+3)}{(L+1)L} & \dfrac{6}{T(L+1)L} \\ \dfrac{6}{T(L+1)L} & \dfrac{12}{T^2(L+2)(L+1)L} \end{bmatrix}$$

$$S^*_{n,n} = \sigma^2_x \begin{bmatrix} \dfrac{2(2L+1)}{(L+2)(L+1)} & \dfrac{6}{T(L+2)(L+1)} \\ \dfrac{6}{T(L+2)(L+1)} & \dfrac{12}{T^2(L+2)(L+1)L} \end{bmatrix}$$

5.8-3 (a) The endpoint location estimate in terms of the midpoint location estimate is, constant velocity target model ($m = 1$),

$$x^*_{n,n} = x^*_{m,n} + \dfrac{1}{2} T_f \dot{x}^*_{m,n}$$

where m in the subscript is the midpoint time index, that is, $m = \frac{1}{2}n$. Because the estimate is unbiased, the variance of $x_{n,n}^*$ is given by

$$\text{VAR}(x_{n,n}^*) = \overline{(x_{n,n}^*)^2} = \overline{(x_{m,n}^*)^2} + \overline{T_f x_{m,n}^* \dot{x}_{m,n}^*} + \tfrac{1}{4} T_f^2 \overline{(\dot{x}_{m,n}^*)^2}$$

where the overbar signifies ensemble average here. The cross-correlation term $\overline{x_{m,n}^* \dot{x}_{m,n}^*}$ is zero for the midterm estimate; hence, using (5.8-6) and (5.8-7), we get

$$\frac{\text{VAR}(x_{n,n}^*)}{\sigma_x^2} = \frac{1}{L} + \frac{T_f^2}{4} \frac{12}{T_f^2 L} = \frac{4}{L}$$

which agrees with (5.8-4), as we desired to show.

(b) The start-of-track estimate in terms of the end-of-track estimate is given by

$$x_{0,n}^* = x_{n,n}^* - T_f \dot{x}_{n,n}^*$$

The variance of $x_{0,n}^*$ is given by

$$\text{VAR}(x_{0,n}^*) = \overline{(x_{0,n}^*)^2} = \overline{(x_{n,n}^*)^2} - 2T_f \overline{x_{n,n}^* \dot{x}_{n,n}^*} + T_f^2 \overline{(\dot{x}_{n,n}^*)^2}$$

Here the cross-selection term $\overline{x_{n,n}^* \dot{x}_{n,n}^*}$ is not zero. It is given by the 0,1 term of (5.6-4), which becomes $6/L^2$ for large L. Thus from (5.8-4) and (5.8-5) and using (5.6-7)

$$\text{VAR}(x_{0,n}^*) = \frac{4}{L} - 2T_f \frac{6}{TL^2} + T_f^2 \frac{12}{T_f^2 L} = \frac{4}{L}$$

which again agrees with (5.8-4), as we desired to show.

6.5-1 From Table 6.3-1 for $m = 0$

$$x_{n+1,n}^* = x_{n,n-1}^* + \frac{1}{n+1}(y_n - x_{n,n-1}^*)$$

For $n = 0$ we get

$$x_{1,0}^* = x_{0,-1}^* + (y_n - x_{0,-1}^*)$$

or

$$x_{1,0}^* = y_n$$

as we desired to prove.

6.5-2   From Table 6.3-1 and (1.2-11a) and (1.2-11b), for the growing-memory filter when $m = 1$, we get

$$x^*_{n+1,n} = \dot{x}^*_{n,n-1} + \frac{h_n}{T}\varepsilon_n \tag{P6.5-2a}$$

$$\dot{x}^*_{n+1,n} = x^*_{n,n-1} + T\dot{x}^*_{n+1,n} + g_n\varepsilon_n \tag{P6.5-2b}$$

where

$$g_n = \frac{2(2n+1)}{(n+2)(n+1)} \tag{P6.5-2c}$$

$$h_n = \frac{6}{(n+2)(n+1)} \tag{P6.5-2d}$$

$$\varepsilon_n = y_n - x^*_{n,n-1} \tag{P6.5-2e}$$

[Note that the equations for g_n and h_n given above agree with (1.2-38a) and (1.2-38b).] Thus $g_0 = 1$, $h_0 = 3$, $g_1 = 1$ and $h_1 = 1$. We have assumed initial position and velocity values x_0 and v_0 for $n = 0$. Thus for $n = 0$ we get

$$\dot{x}^*_{1,0} = v_0 + \frac{h_0}{T}(y_0 - x_0) = v_0 + \frac{y_0}{T} - \frac{x_0}{T}$$

$$\dot{x}^*_{1,0} = x_0 + T\dot{x}^*_{1,0} + g_0(y_0 - x_0)$$

$$= x_0 + Tv_0 + h_0(y_0 - x_0) + g_0(y_0 - x_0)$$

$$= x_0 + Tv_0 + 3y_0 - 3x_0 + y_0 - x_0$$

$$= Tv_0 + 4y_0 - 3x_0$$

For $n = 1$ we get

$$\dot{x}^*_{2,1} = \dot{x}^*_{1,0} + \frac{h_1}{T}(y_1 - x^*_{1,0})$$

$$= v_0 + \frac{h_0}{T}(y_0 - x_0) + \frac{h_1}{T}(y_1 - Tv_0 - 4y_0 + 3x_0)$$

$$= \frac{3y_0}{T} + \frac{y_1}{T} - \frac{4y_0}{T}$$

or

$$\dot{x}^*_{2,1} = \frac{y_1 - y_0}{T} \tag{P6.5-2f}$$

which is independent of the initial values x_0 and v_0.

For $n = 1$ we get for

$$x^*_{2,1} = x^*_{1,0} + T\dot{x}^*_{2,1} + g_1(y_1 - x^*_{1,0})$$

$$= (y_1 - y_0) + y_1$$

or

$$x_{2,1}^* = 2y_1 - y_0 \qquad \text{(P6.5-2g)}$$

which is also independent of x_0 and v_0, as we desired to show. Finally note that the above values for $x_{2,1}^*$ and $\dot{x}_{2,1}^*$ given by (P6.5-2f) and (P6.5-2g) agree with the least-squares fixed-memory filter solution of problem 4.5-2

7.2-1 (a) After a little manipulation (1.3-2) and (1.3-3) can be put in the form

$$\ddot{x}_{n+1,n}^* = \ddot{x}_{n,n-1}^* + \frac{2k}{T^2}\varepsilon_n \qquad \text{(P7.2-1a)}$$

$$\dot{x}_{n+1,n}^* = \dot{x}_{n,n-1}^* + \ddot{x}_{n+1,n}^* T + \frac{h}{T}\varepsilon_n \qquad \text{(P7.2-1b)}$$

$$x_{n+1,n}^* = x_{n,n-1}^* + \dot{x}_{n+1,n}^* T - \ddot{x}_{n+1,n}^* \frac{T^2}{2} + g\varepsilon_n \qquad \text{(P7.2-1c)}$$

where

$$\varepsilon_n = y_n - x_{n,n-1}^* \qquad \text{(P7.2-1d)}$$

Next making use of (5.4-12) yields

$$(z_0^*)_{n+1,n} = z_{n+1,n}^* = x_{n+1,n}^* \qquad \text{(P7.2-1e)}$$

$$(z_1^*)_{n+1,n} = \dot{z}_{n+1,n}^* = T\dot{x}_{n+1,n}^* \qquad \text{(P7.2-1f)}$$

$$(z_2^*)_{n+1,n} = \ddot{z}_{n+1,n}^* = \frac{T^2}{2!}\ddot{x}_{n+1,n}^* \qquad \text{(P7.2-1g)}$$

Applying (P7.2-1e) to (P7.2-1g) to (P7.2-1a) to (P7.2-1c) yields, finally,

$$\ddot{z}_{n+1,n}^* = \ddot{z}_{n,n-1}^* + k\varepsilon_n \qquad \text{(P7.2-1h)}$$

$$\dot{z}_{n+1,n}^* = \dot{z}_{n,n-1}^* + 2\ddot{z}_{n+1,n}^* + h\varepsilon_n \qquad \text{(P7.2-1i)}$$

$$z_{n+1,n}^* = z_{n,n-1}^* + \dot{z}_{n+1,n}^* - \ddot{z}_{n+1,n}^* + g\varepsilon_n \qquad \text{(P7.2-1j)}$$

(b) Comparing (P7.2-1h) to (P7.2-1j) with the corresponding $m = 2$ equations of Table 7.2-2 yields

$$g = 1 - \theta^3 \qquad \text{(P7.2-1k)}$$

$$h = \tfrac{3}{2}(1 - \theta)^2(1 + \theta) = \tfrac{3}{2}(1 - \theta^2)(1 - \theta) \qquad \text{(P7.2-1l)}$$

$$k = \tfrac{1}{2}(1 - \theta)^3 \qquad \text{(P7.2-1m)}$$

7.2-2 Comparing (P7.2-1a) to (P7.2-1c) with the results of Table 6.3-1 for $m = 2$ yields, that for the growing-memory g–h–k filter,

$$g_n = \frac{3(3n^2 + 3n + 2)}{(n+3)(n+2)(n+1)} \tag{P7.2-2a}$$

$$h_n = \frac{18(2n+1)}{(n+3)(n+2)(n+1)} \tag{P7.2-2b}$$

$$k_n = \frac{30}{(n+3)(n+2)(n+1)} \tag{P7.2-2c}$$

7.4-2 (a) From the solution to problem 7.2-1

$$g = 1 - \theta^3 \tag{P7.2-2k}$$

Hence for $g = 0.87$, $\theta = 0.507$. From (P7.2-1l) and (P7.2-1m), $h = 0.550$ and $k = 0.0601$, which are much smaller than the values of $g = 0.81$, $h = 0.81$, and $k = 0.14$ for the Singer filter of Section 2.10; see Table P2.10-1.

(b) Form Table 7.4-1

$$\left(\frac{\sigma_{n+1,n}}{\sigma_x}\right)^2 = \frac{1-\theta}{(1+\theta)^5}(19 + 24\theta + 16\theta^2 + 6\theta^3 + \theta^4) \tag{P7.4-2a}$$

Thus $\sigma_{n+1,n}/\sigma_x = 1.51$. Using (1.3-4) gives the same result. This result for $\sigma_{n+1,n}/\sigma_x$ is somewhat smaller than the value of 3.33 obtained with the Singer filters for the target having the dynamics defined by Singer; see Table P2.10-1.

(c) For this Singer filter $g = 0.87$, $h = 0.81$, and $k = 0.14$, and using (1.3-4), we obtain $\sigma_{n+1,n}/\sigma_x = 1.9$, which is much smaller than the value of 3.33 of Table P2.10-1. This we attribute to the smaller tracking and prediction error resulting when the target has a constant acceleration rather than the Singer dynamics model. The critically damped filter has a still smaller $\sigma_{n+1,n}/\sigma_x$ of 1.5. This we attribute to the smaller h and k of the critically damped filter than for the Singer filter. A smaller h and k implies more memory in the filter.

8.1-1

$$Dx(t) = \dot{x}(t) \qquad D^2x(t) = 0$$

Therefore

$$\begin{bmatrix} Dx(t) \\ D^2x(t) \end{bmatrix} = \begin{bmatrix} 0 & 1 \\ 0 & 0 \end{bmatrix} \begin{bmatrix} x(t) \\ Dx(t) \end{bmatrix} \tag{P8.1-1a}$$

$$A = \begin{bmatrix} 0 & 1 \\ 0 & 0 \end{bmatrix} \tag{P8.1-1b}$$

8.1-2 Using (8.1-21),

$$X(t_n + \zeta) = \Phi(\zeta)X(t_n)$$

yields

$$X(t_n + T) = \Phi(T)X(t_n) \tag{P8.1-2a}$$

But from (8.1-22)

$$\Phi(T) = I + TA + \frac{T^2}{2!}A^2 + \frac{T^3}{3!}A^3$$

Using

$$A = \begin{bmatrix} 0 & 1 \\ 0 & 0 \end{bmatrix} \tag{P8.1-1b}$$

we get

$$A^2 = \begin{bmatrix} 0 & 1 \\ 0 & 0 \end{bmatrix}\begin{bmatrix} 0 & 1 \\ 0 & 0 \end{bmatrix} = \begin{bmatrix} 0 & 0 \\ 0 & 0 \end{bmatrix}$$

Therefore

$$\Phi(T) = \begin{bmatrix} 1 & 0 \\ 0 & 1 \end{bmatrix} + T\begin{bmatrix} 0 & 1 \\ 0 & 0 \end{bmatrix} \tag{P8.1-2b}$$

or

$$\Phi(T) = \begin{bmatrix} 1 & T \\ 0 & 1 \end{bmatrix} \tag{P8.1-2c}$$

which agrees with (2.4-1b).

8.1-3 Using

$$A = \begin{bmatrix} 0 & 1 & 0 \\ 0 & 0 & 1 \\ 0 & 0 & 0 \end{bmatrix} \tag{8.1-10a}$$

we get

$$A^2 = \begin{bmatrix} 0 & 1 & 0 \\ 0 & 0 & 1 \\ 0 & 0 & 0 \end{bmatrix}\begin{bmatrix} 0 & 1 & 0 \\ 0 & 0 & 1 \\ 0 & 0 & 0 \end{bmatrix} = \begin{bmatrix} 0 & 0 & 1 \\ 0 & 0 & 0 \\ 0 & 0 & 0 \end{bmatrix}$$

$$A^3 = \begin{bmatrix} 0 & 0 & 1 \\ 0 & 0 & 0 \\ 0 & 0 & 0 \end{bmatrix} \begin{bmatrix} 0 & 1 & 0 \\ 0 & 0 & 1 \\ 0 & 0 & 0 \end{bmatrix} = \begin{bmatrix} 0 & 0 & 0 \\ 0 & 0 & 0 \\ 0 & 0 & 0 \end{bmatrix}$$

Substituting into (8.1-22) yields

$$\Phi(T) = \begin{bmatrix} 1 & 0 & 0 \\ 0 & 1 & 0 \\ 0 & 0 & 1 \end{bmatrix} + \begin{bmatrix} 0 & T & 0 \\ 0 & 0 & T \\ 0 & 0 & 0 \end{bmatrix} + \begin{bmatrix} 0 & 0 & \frac{1}{2}T^2 \\ 0 & 0 & 0 \\ 0 & 0 & 0 \end{bmatrix}$$

or

$$\Phi(T) = \begin{bmatrix} 1 & T & \frac{1}{2}T^2 \\ 0 & 1 & T \\ 0 & 0 & 1 \end{bmatrix}$$

which agrees with (4.1-4).

9.3-1　Applying (2.6-14) to the right-hand side of (9.2-1b) yields

$$(\overset{\circ}{S}{}^{*-1}_{n+1,n} + M^T R_{n+1}^{-1} M)^{-1} = \overset{\circ}{S}{}^{*}_{n+1,n} - \overset{\circ}{S}{}^{*}_{n+1,n} M^T (R_{n+1} + M\overset{\circ}{S}{}^{*}_{n+1,n} M^T)^{-1} M \overset{\circ}{S}{}^{*}_{n+1,n} \tag{1}$$

Postmultiplying both sides of the above by $M^T R_{n+1}^{-1}$ yields

$$\begin{aligned}
(\overset{\circ}{S}{}^{*-1}_{n+1,n} &+ M^T R_{n+1}^{-1} M)^{-1} M^T R_{n+1}^{-1} \\
&= \overset{\circ}{S}{}^{*}_{n+1,n} M^T R_{n+1}^{-1} - \overset{\circ}{S}{}^{*}_{n+1,n} M^T \\
&\quad \times (R_{n+1} + M\overset{\circ}{S}{}^{*}_{n+1,n} M^T)^{-1} M \overset{\circ}{S}{}^{*}_{n+1,n} M^T R_{n+1}^{-1} \\
&= \overset{\circ}{S}{}^{*}_{n+1,n} M^T (R_{n+1} + M\overset{\circ}{S}{}^{*}_{n+1,n} M^T)^{-1} \\
&\quad \times [(R_{n+1} + M\overset{\circ}{S}{}^{*}_{n+1,n} M^T) R_{n+1}^{-1} - M\overset{\circ}{S}{}^{*}_{n+1,n} M^T R_{n+1}^{-1}] \\
&= \overset{\circ}{S}{}^{*}_{n+1,n} M^T (R_{n+1} + M\overset{\circ}{S}{}^{*}_{n+1,n} M^T)^{-1}
\end{aligned} \tag{2}$$

Substituting the left-hand side of (9.2-1b) for the inverse on the left-hand side of (2) yields

$$\overset{\circ}{S}{}^{*}_{n+1,n+1} M^T R_n^{-1} = \overset{\circ}{S}{}^{*}_{n+1,n} M^T (R_{n+1} + M\overset{\circ}{S}{}^{*}_{n+1,n} M^T)^{-1} \tag{3}$$

But from (9.2-1a) the left-hand side of (3) equals $\overset{\circ}{H}_{n+1}$. This alternate form for $\overset{\circ}{H}_{n+1}$ is the Kalman filter form of (9.3-1a). Using this alternate form for $\overset{\circ}{H}_{n+1}$ on the right-hand side of (1) yields

$$(\mathring{S}^{*-1}_{n+1,n} + M^T R^{-1}_{n+1} M)^{-1} = \mathring{S}^{*}_{n+1,n} - \mathring{H}_{n+1} M \mathring{S}^{*}_{n+1,n}$$
$$= (1 - \mathring{H}_{n+1} M) \mathring{S}^{*}_{n+1,n} \tag{4}$$

The left-hand side of (4) is, from (9.2-1b), equal to $\mathring{S}^{*}_{n+1,n+1}$. Making this substitution gives (9.3-1b). This completes our derivation of the Kalman filter equations.

9.4-1 From (9.4-1)

$$\mathring{X}^{*}_{n+1,n+1} = \mathring{S}^{*}_{n+1,n+1} \left[\mathring{S}^{*-1}_{n+1,n} \mathring{X}^{*}_{n+1,n} + M^T R^{-1}_{n+1} Y_{n+1} \right]$$
$$+ \mathring{S}^{*}_{n+1,n+1} M^T R^{-1}_{n+1} M \mathring{X}^{*}_{n+1,n} - \mathring{S}^{*}_{n+1,n+1} M^T R^{-1}_{n+1} M \mathring{X}^{*}_{n+1,n}$$
$$= \mathring{S}^{*}_{n+1,n+1} (\mathring{S}^{*-1}_{n+1,n} + M^T R^{-1}_{n+1} M) \mathring{X}^{*}_{n+1,n}$$
$$+ \mathring{S}^{*}_{n+1,n+1} M^T R^{-1}_{n+1} (Y_{n+1} - M \mathring{X}^{*}_{n+1,n})$$

Using (9.2-1b) lets us replace the term in the first bracket above by $S^{*-1}_{n+1,n+1}$ to yield

$$\mathring{X}^{*}_{n+1,n+1} = \mathring{X}^{*}_{n+1,n} + \mathring{S}^{*}_{n+1,n+1} M^T R^{-1}_{n+1} [(Y_{n+1}) - M \mathring{X}^{*}_{n+1,n}]$$

Finally using (9.2-1a) yields (9.2-1).

14.4-1 Assume a 2×2 upper triangle matrix

$$B_2 = \begin{bmatrix} 1 & \alpha \\ 0 & 1 \end{bmatrix}$$

Let its inverse be

$$B_2^{-1} = \begin{bmatrix} a & b \\ c & d \end{bmatrix}$$

Then

$$\begin{bmatrix} 1 & \alpha \\ 0 & 1 \end{bmatrix} \begin{bmatrix} a & b \\ c & d \end{bmatrix} = \begin{bmatrix} a + \alpha c & b + \alpha d \\ c & d \end{bmatrix} = \begin{bmatrix} 1 & 0 \\ 0 & 1 \end{bmatrix}$$

Hence

$$c = 0 \qquad d = 1 \qquad a = 1 \qquad b = -\alpha d = -\alpha$$

and the inverse of B_2 is

$$B_2^{-1} = \begin{bmatrix} 1 & -\alpha \\ 0 & 1 \end{bmatrix}$$

which is upper triangular.

We now show that if the inverse of a square upper triangular matrix of dimension $(r-1) \times (r-1)$ is upper triangular, then so is the inverse of the square upper triangular matrix of dimension $r \times r$ upper triangular, thus completing our proof. Let B_r be an upper triangular matrix of dimension $r \times r$. Let it be partitioned as shown below:

$$B_r = \begin{bmatrix} B_{r-1} & | & C \\ -- & -- & -- \\ D & | & E \end{bmatrix} \begin{matrix} \}r-1 \\ \\ \}1 \end{matrix}$$
$$\underbrace{}_{r-1} \quad \underbrace{}_{1}$$

where B_{r-1} is an upper triangular $(r-1) \times (r-1)$ matrix; E is a 1×1 matrix, which we shall now call e; and D is a $1 \times (r-1)$ matrix of zeroes. From [134] the inverse of B_r can be written as

$$B_r^{-1} = \begin{bmatrix} X & | & Y \\ -- & -- & -- \\ Z & | & U \end{bmatrix} \begin{matrix} \}r-1 \\ \\ \}1 \end{matrix}$$
$$\underbrace{}_{r-1} \quad \underbrace{}_{1}$$

where

$$X = \left[B_{r-1} - Ce^{-1}D \right]^{-1}$$
$$U = \left[e - DB_{r-1}^{-1}C \right]^{-1}$$
$$Y = -B_{r-1}^{-1}CU$$
$$Z = -e^{-1}DX$$

Because the elements of D are composed of zeroes, it follows that

$$X = B_{r-1}^{-1} \qquad U = \frac{1}{e} \qquad Z = 0$$

Thus B_r^{-1} becomes

$$B_r^{-1} = \begin{bmatrix} B_{r-1}^{-1} & | & Y \\ -- & -- & -- \\ 0 & | & \frac{1}{e} \end{bmatrix} \begin{matrix} \}r-1 \\ \\ \}1 \end{matrix}$$
$$\underbrace{}_{r-1} \quad \underbrace{}_{1}$$

which is upper triangular if B_{r-1}^{-1} is upper triangular. This completes our proof.

REFERENCES

1. Brookner, E. (Ed.), *Aspects of Modern Radar*, LexBook, 282 Marrett Rd., Lexington, MA (formerly published by Artech House, Norwood, MA), 1988.
2. Brookner, E., "Phased-Array Radars," *Scientific American,* Vol. 252, No. 2, February 1985, pp. 94–102.
3. Brookner, E., *Radar Technology*, LexBook, 282 Marrett Rd., Lexington, MA (formerly published by Artech House, Norwood, MA), 1977.
4. Mailloux, R. J., "Limited Scan Arrays, Part 1," Lecture 9, in *Practical Phased Array Antenna Systems,* E. Brookner (Ed.), LexBook, 282 Marrett Rd., Lexington MA (formerly published by Artech House, Norwood, MA), 1991.
5. Morrison, N., *Introduction to Sequential Smoothing and Prediction,* McGraw-Hill, New York, 1969.
6. Farina, A. and F. A. Studer, *Radar Data Processing,* Vol. 1, *Introduction and Tracking* (1985), and Vol. 2, *Advanced Topics and Applications* (1986), Wiley and Research Studies Press Ltd., Letchworth, Hertfordshire, England.
7. Gelb, A. (Ed.), written by the Technical Staff of the Analytical Sciences Corporation, *Applied Optimal Estimation*, MIT Press, Cambridge, MA, 1979.
8. Blackman, S. S., *Multiple-Target Tracking with Radar Applications,* Artech House, Norwood, MA, 1986.
9. Bar-Shalom, Y. and T. E. Fortmann, *Tracking the Data Association,* Academic, New York, 1988.
10. Benedict, R. T. and G. W. Bordner, "Synthesis of an Optimal Set of Radar Track-While-Scan Smoothing Equations," *IRE Transactions on Automatic Control*, July 1962, pp. 27–32.
11. Asquith, C. F. and A. B. Woods, "Total Error Minimization in First and Second Order Prediction Filters," Report No. RE-TR-70-17, U.S. Army Missile Command, Redstone Arsenal, AL, November 24, 1970.

12. Asquith, C. F., "Weight Selection in First Order Linear Filters," Report No. RG-TR-69-12, U.S. Army Missile Command, Redstone Arsenal, AL, July 23, 1969.

13. Simpson, H. R., "Performance Measures and Optimization Condition for a Third-Order Sampled-Data Tracker," *IEEE Transactions on Automatic Control*, Vol. AC-8, April 1963 (correspondence), pp. 182–183.

14. Wood, A. B., "Comparison of Truncation Errors of Optimum Steady-State and Critically Damped Second-Order Filters," Report No. RG-TR-69-11, U.S. Army Missile Command, Redstone Arsenal, AL, July 23, 1969.

15. Kalata, P. R., "The Tracking Index: A Generalised Parameter for α–β and α–β–γ Target Trackers," *IEEE Transactions on Aerospace and Electronic Systems*, Vol. AES-20, No. 2, March 1984, pp. 174–182.

16. Anderson, J. R. and D. Karp, "Evaluation of the MTD in a High-Clutter Environment," *Proceedings of 1980 IEEE International Radar Conference*, pp. 219–224. Published by the IEEE New York, 1980.

17. Bath, W. G., F. R. Castella, and S. F. Haase, "Techniques for Filtering Range and Angle Measurements from Colocated Surveillance Radars," *Proceedings of 1980 IEEE International Radar Conference*, pp. 355–360. Published by the IEEE, New York, 1980.

18. Daum, F. E. and R. J. Fitzgerald, "Decoupled Kalman Filters for Phased Array Radar Tracking," *IEEE Transactions on Automatic Control*, Vol. AC-28, No. 3, March 1983, pp. 269–283.

19. Kalman, R. E., "A New Approach to Linear Filtering and Prediction Problems," *Journal of Basic Engineering*, Trans. of ASME, Ser. D, Vol. 82, No. 1, March 1960, pp. 35–45.

20. Kalman, R. E. and R. S. Bucy, "New Results in Linear Filtering and Prediction Theory," *Transactions of the ASME, Journal of Basic Engineering*, Vol. 83, No. 3, December 1961, pp. 95–107.

21. Sittler, R., "Tracking for Terminal Air Traffice Control," Boston IEEE Aerospace and Electronics Systems Group Meeting, May 6, 1976.

22. Asquith, C. F., "A Discrete Kalman-Derived First Order Filter," Report No. RG-TR-70-16, U.S. Army Missile Command, Redstone Arsenal, AL, November 25, 1970.

23. Friedland, B., "Optimum Steady-State Position and Velocity Estimation Using Noisy Sampled Position Data," *IEEE Transactions on Aerospace and Electronic Systems,* Vol. AES-9, November 1973, pp. 907–911; corrections Vol. AES-11, July 1975, p. 675.

24. Singer, R. A., "Estimating Optimal Tracking Filter Performance of Manned Maneuvering Tragets," *IEEE Transactions on Aerospace and Electronic Systems*, Vol. AES-6(4), 1970, pp. 473–484.

25. Fitzgerald, R. J., "Simple Tracking Filters: Steady-State Filtering and Smoothing Performance," *IEEE Transactions on Aerospace and Electronic Systems*, Vol. AES-16, No. 6, November 1980, pp. 860–864.

26. Singer, R. A. and K. W. Behnke, "Real-Time Tracking Filter Evaluation and Selection for Tactical Applications," *IEEE Transactions on Aerospace and Electronic Systems*, Vol. AES-7, No.1, January 1971, pp. 100–110.

27. Mirkin, M. I., C. E. Schwartz, and S. Spoerri, "Automated Tracking with Netted Ground Surveillance Radars," *Proceedings of 1980 IEEE International Radar Conference*, pp. 371–377. Published by IEEE, New York, 1980.

28. McAulay, R. J. and E. Denlinger, "A Decision -Directed Adaptive Tracker," *IEEE Transactions on Aerospace and Electronic Systems*, Vol. AES-9, No. 2, March 1973, pp. 229–236.

29. Bogler, P. L., *Radar Principles with Applications to Tracking Systems*, Wiley, New York, 1990.

30. Bar-Shalom, Y. and X. Li, *Multitarget-Multisensor Tracking: Principles and Techniques*, YBS, Storrs, CT, 1995.

31. Daum, F., private communication, Raytheon Comp., 1995.

32. Prengaman, R. J., R. E. Thurber, and W. B. Bath, "A Retrospective Detection Algorithm for Extraction of Weak Targets in Clutter and Interference Environments," paper presented at the IEE 1982 International Radar Conference, Radar-82, London, 1982, pp. 341–345.

33. Bath, W. B., L. A. Biddison, and R. E. Thurber, "Noncoherent Subclutter Visibility through High-Speed Contact Sorting," *Proceedings of Military Microwave Conference*, October 24–26, 1984, pp. 621–624, Microwave Exhibitions and Pub. Ltd., Kent, England, 1982.

34. Ewell, G. W., M. T. Tuley, and W. F. Horne, "Temporal and Spatial Behavior of High Resolution Sea Clutter 'Spikes'," *Proceedings of 1984 IEEE National Radar Conference*, March 13–14, 1984. Published by IEEE, New York, 1984.

35. Olin, I. E., "Amplitude and Temporal Statistics of Sea Spike Clutter," *IEEE International Radar Conference*, 1983, pp. 198–202. Published by IEEE, New York, 1983.

36. Nathanson, F., *Radar Design Principles*, McGraw-Hill, New York, 1969.

37. Brookner, E., "Radar of the '80s and Beyond," IEEE Electro-84, Session 4, Boston, MA, May 15–17, 1984.

38. Brookner, E., "Array Radars: An Update," *Microwave Journal*, Pt. 1, Vol. 30, No. 2, February 1987, pp. 117–138; Pt. 2, Vol. 30, No. 3, March 1987, pp. 167–1974.

39. Brookner, E., "Radar of the '80s and Beyond—An Update," IEEE Electro-86, May 13–15, 1986.

40. Brookner, E., "Radar Trends to the Year 2000," *Interavia*, May 1987, pp. 481–486.

41. Rabinowitz, S. J., C. H. Gager, E. Brookner, C. E. Muehe, and C. M. Johnson, "Applications of Digital Technology to Radar," *Proceedings of the IEEE*, Vol. 73, No. 2, February 1985, pp. 325–339.

42. Muehe, C. E., L. Cartledge, W. H. Drury, E. M. Hofstetter, M. Labitt, P. B. McCorison, and V. J. Sferring, "New Techniques Applied to Air Traffic Control Radars," *Proceedings of the IEEE*, Vol. 62, No. 6, June 1974, pp. 716–723.

43. Morrison, N., "Tracking and Smoothing," Chapter 25, and Sheats, L., "Tracking and Smoothing," Chapter 26, in *Radar Technology*, E. Brookner (Ed.), LexBook, 282 Marrett Rd., Lexington, MA (formerly Published by Artech House, Norwood, MA), 1977.

44. Castella, F. R. and J. T. Miller, Jr., "Moving Target Detector Data Utilization Study," IEE Radar-77, London, October 25–28, 1977, pp. 182–185.

45. Cartledge, L. and R. M. O'Donnell, "Description and Performance Evaluation of the Moving Target Detector," Report No. FAA-RD-76-190, Lincoln Laboratory Project Report ATC-69, March 8, 1977.

46. O'Donnell, R. M. and L. Cartledge, "Comparison of the Performance of the Moving Target Detector and the Radar Video Digitizer," Report No. FAA-RD-76-191, Lincoln Laboratory Project Report ATC-70, April 26, 1977.

47. Skolnik, M. I., *Introduction to Radar Systems*, 2nd ed., McGraw-Hill, New York, 1980.

48. Skolnik, M. I. (Ed.), *Radar Handbook*, McGraw-Hill, New York, 1970.

49. Proceedings of Chinese International Conferences on Radar, 1986 and 1991. Chinese Academic Pub., Beijing China, 1986, by Int. Academic Pub., Beijing, 1991.

50. Schleher, D. C., *MTI Radar*, Artech House, Norwood, MA, 1978.

51. Schleher, D. C., *Automatic Detection and Radar Data Processing*, Artech House, Norwood, MA, 1980.

52. Brookner, E. and T. F. Mahoney, "Derivation of a Satellite Radar Architecture for Air Surveillance," IEEE EASCON-83 Conference Rec., September, 19–21, 1983, pp. 465–475

53. Proceedings of IEEE International Radar Conferences, 1975, 1980, 1985, 1990, 1995, Pub. by IEEE, New York.

54. IEEE National Radar Conference Proceedings, Radar-84, -86, -87, -91, -93, -94, -96, Pub. by IEEE, New York.

55. Proceedings of the International Radar Symposium, India-83, Bangalore, October 9–12, 1983. The Inst. of Electronic and Telecommunication Engineers, Bangalore, India, 1983.

56. Bath, W. G., L. A. Biddison, S. F. Haase, and E. C. Wetzlar, "False Alaram Control in Automated Radar Surveillance Systems," paper presented at the IEE 1982 International Radar Conference, Radar—82, London, 1982, pp. 71–75.

57. Klauder, J. R., A. C. Price, S. Darlington, and W. J. Albersheim, "The Theory and Design of Chirp Radars," *Bell System Technical Journal*, Vol. 39, July 1960, pp. 745–808.

58. Military Microwave Conference Proceedings, October 25–27, 1978, October 20–22, 1982, October 24–26, 1984, June 24–26, 1986, July 5–7, 1988, October 14–16, 1992, Microwave Exhibitions and Pub. Ltd. Kent, England, 1992.

59. Cook, C. E. and M. Bernfeld, *Radar Signals: An Introduction to Theory and Applications*, Academic, New York, 1967.

60. Proceedings of International IEE Radar Conferences, Radar-73, -78, -82, -87, -92, England. Pub. by IEE, London.

61. Rihaczek, A. W., *Principles of High-Resolution Radar,* McGraw-Hill, New York, 1969; reprinted by Artech House, Norwood, MA.

62. Trunk, G. V., "Survey of Radar ADT," *Microwave Journal*, July 1983, pp. 77–88.

63. Trunk, G. V., "Automatic Detection, Tracking, and Sensor Integration," Naval Research Lab. Report, No. 9110, June 8, 1988. Washington, D.C.

64. Trunk, G. V., "Automatic Detection, Tracking, and Sensor Integration," In *Radar Handbook*, 2nd ed., M. Skolnik (Ed.), McGraw-Hill, New York, 1990.

65. Brookner, E., "Developments in Digital Radar Processing," (pp. 7–12), and "A Guide to Radar Systems and Processors," (pp. 13–23), in *Trends and Perspectives*

in Signal Processing, Vol. 2, V. Oppenheim (Ed.), MIT Press, Cambridge, MA, 1982.

66. Brookner, E., "Trends in Radar Signal Processing," *Microwave Journal,* Vol. 25, No. 10, October 1982, pp. 20–39.

67. Mevel, J., "Studies Relating to the Detailed Structure of Aircraft Echoes," paper presented at the International Conference on Radar, Paris, May 21–24, 1984, pp. 275–280. Comité Radar 84, Paris.

68. Maaloe, J., "Classification of Ships Using an Incoherent Marine Radar," paper presented at the IEE International Conference on Radar, Radar-82, October 18–20, 1982, pp. 274–277.

69. Cantrell, B. H., G. V. Trunk, and J. D. Wilson, "Tracking System for Two Asynchronously Scanning Radars," Naval Research Lab. Report 7841, December 1974. Washington, D.C.

70. Sinsky, A., "Waveform Selection and Processing" (Chapter 7), and Cook, C., "Large Time-Bandwidth Radar Signals" (Chapter 8), in *Radar Technology*, E. Brookner (Ed.), LexBook, 282 Marrett Rd., Lexington, MA (formerly published by Artech House, Norwood, MA), 1977.

71. Fitzgerald, R., "Effect of Range-Doppler Coupling on Chirp Radar Tracking Accuracy," *IEEE Transactions on Aerospace and Electronic Systems,* Vol. AES-10, No. 4, July, 1974, pp. 528–532.

72. Taenzer, E., "Some Properties of Critically Damged gh and ghk Filters," Raytheon Internal Memo No. ET-67-1, May 11, 1967, Bedford, MA.

73. Patton, W. T., "Compact, Constrained Feed Phased Array for AN/SPY-1," Lecture 8, in *Practical Phased Array Antenna Systems*, E. Brookner (Ed.), LexBook, 282 Marrett Rd., Lexington MA (formerly published by Artech House, Norwood, MA), 1991.

74. Sklansky, J., "Optimizing the Dynamic Parameters of a Track-While-Scan System," *RCS Review*, Vol. 18, June 1957, pp. 163–185.

75. Scheffé, H., *The Analysis of Variance*, Wiley, New York, 1959.

76. Strang, G., *Linear Algebra and Its Applications,* Academic, New York, 1980.

77. Shanmugan, K. S. and A. M. Breipohl, *Random Signals: Detection, Estimation and Data Analysis,* Wiley, New York, 1988.

78. Giordano, A. A. and F. M. Hsu, *Least Square Estimation with Applications to Digital Signal Processing*, Wiley, New York, 1985.

79. Bierman, G. J., *Factorization Methods for Discrete Sequential Estimation*, Academic, New York, 1977.

80. Golub, G. H. and C. F. VanLoan, *Matrix Computations*, John Hopkins University Press, Baltimore, MD, 1989.

81. Lawson, C. L. and R. J. Hanson, *Solving Least-Squares Problems,* Prentice-Hall, Englewood Cliffs, NJ, 1974.

82. Dahlquist, G. and A. Bjorck, *Numerical Methods*, translated by N. Anderson, Prentice-Hall, Englewood Cliffs, NJ, 1974.

83. Ward, C. R., P. J. Hargrave, and J. G. McWhirter, "A Novel Algorithm and Architecture for Adaptive Digital Beamforming," *IEEE Transactions on Antennas and Propagation*, Vol. AP-34, No. 3, 1986, pp. 338–346.

84. Monzingo, R. A. and T. W. Miller, *Introduction to Adaptive Arrays*, Wiley, New York, 1980.

85. Farina, A., *Antenna-Based Signal Processing Techniques for Radar Systems*, Artech House, Norwood, MA, 1992.

86. Nitzberg, R., *Adaptive Signal Processing*, Artech House, Norwood, MA, 1992.

87. Brookner, E., "Sidelobe Cancelling and Adaptive Array Processing," paper presented at the Tutorial Seminar on "Adaptive Radar Processing," sponsored by IEE, October 22–23, 1987. Pub. by IEE, London, 1987.

88. Brookner, E., "Antenna Array Fundamentals, Part 1," Lecture 2, in *Practical Phased Array Antenna Systems*, E. Brookner (Ed.), LexBook, 282 Marrett Rd., Lexington, MA (formerly published by Artech House, Norwood, MA), 1991.

89. McWhirter, J. G., "Recursive Least-Squares Minimization Using Systolic Array," in Proceedings of the SPIE, 1983, Vol. 431. Real-Time Signal Processing VI, 2983.

90. Rader, C. M., "Wafer-Scale Systolic Array for Adaptive Antenna Processing," IEEE, 1988 ICASSP, pp. 2069–2071.

91. Rader, C. M., D. L. Allen, D. B. Glasco, and C. E. Woodward, "MUSE—A Systolic Array for Adaptive Nulling with 64 Degrees of Freedom, Using Givens Transformations and Wafer Scale Integration," Technical Report 86, MIT Lincoln Laboratory, Lexington, MA, May 18, 1990.

92. Rader, C. M., "Wafer-Scale Integration of a Large Systolic Array for Adaptive Nulling," *Lincoln Laboratory Journal*, Vol. 4, No. 1, Spring 1991, pp. 3–30.

93. McWhirter, J. G., T. J. Shephard, C. R. Ward, R. A. Jones, and P. J. Hargave, "Application of Linear Constraints to an Adaptive Beamformer," *Proceedings of the IEEE International Conference on Acoustics, Speech and Signal Processing*, Tokyo, 1986.

94. McWhirter, J. G., "Adaptive Antenna Array Signal Processing," presented at the Tutorial Seminar on "Adaptive Radar Processing," sponsored by IEEE, October 22–23, 1987.

95. Archer, D., "Lens-Fed Multiple-Beam Arrays," *Microwave Journal*, pp. 37–42, October 1975.

96. Brookner, E. and J. M. Howell, "Adaptive-Adaptive Array Processing," Phased Arrays Symposium Proceedings, The Mitre Corp., Bedford, MA, October 15–18, 1985, pp. 133–146. See also: RADC Report No. RADC-TR-85-171, Electromagnetics Science Div., RADC, Hanscom AFB, Bedford, MA, Air Force Systems Command, August 1985; IEE International Conference on Radar, Radar-87, London, October 19–21, 1987; Proceedings of the IEEE, Vol. 74, No. 4, April 1986, pp. 602–604;

97. Aitken, A. C., "On Least-Squares and Linear Combinations of Observations," *Proceedings of the Royal Society of Edinb.* A, 55, 42-7, 1934.

98. Gauss, K. F., "Theory of the Motion of the Heavenly Bodies Moving About the Sun in Conic Section," 1809, English translation, Dover Publications, New York, 1963.

99. Barton, D. K., "Detection and Measurement," in *Radar Technology*, E. Brookner (Ed.), LexBook, 282 Marrett Rd., Lexington, MA (Artech House, Norwood MA), Chapter 2; see also Barton, D. K., *Modern Radar System Analysis*, Artech House, Norwood MA, 1998.

100. Lehmann, E. L., *Testing Statistical Hypothesis*, Wiley, New York, 1959.

101. Bronson, R., *Schaum's Outline of Theory and Problems of Matrix Operations*, McGraw-Hill, New York, 1989.

102. Golub, G. H., "Numerical Methods for Solving Linear Least-Squares Problems, *Numer. Math.*, Vol. 7, pp. 206–216, 1965.

103. Rader, C. M. and A. O. Steinhardt, "Hyperbolic Householder Transformations," *IEEE Transactions on Acoustics, Speech and Signal Processing*, Vol. ASSP-34, No. 6, December 1986, pp. 1589–1602.

104. Scheffé, M., formerly Raytheon Company, now at Lincoln Lab, MIT, Lexington, MA, private communication 1989.

105. Arbenz, K. and A. Wohlhauser, *Advanced Mathematics for Practicing Engineers*, Artech House, Norwood, MA, 1986.

106. Kung, H. T. and W. M. Gentleman, "Matrix Triangularization by Systolic Arrays," *Proceedings of the SPIE*, 1981, Vol. 298, Real-Time Signal Processing IV, pp. 19–26.

107. Kung, H. T. and C. L. Leiserson, "Systolic Arrays (for VLSI)," in *Sparse Matrix Proceedings 1978*, I. S. Duff and G. W. Stewart (Eds.), Society of Industrial and Applied Mathematics, Philadelphia, Pub., 1979, pp. 256–282. A slightly different version appears in *Introduction to VLSI Systems*, by C. A. Mead and L. A. Conway, Addison-Wesley, Reading, MA, 1980, Section 8.3.7.

108. Gentleman, W. M., "Least-Squares Computations by Givens Transformation without Square Roots," *J. Inst. Math. Appl.*, Vol. 2, 1973, pp. 329–336.

109. Hammarling, S., "A Note on Modifications to the Givens Plane Rotations." *J. Inst. Math. Appl.*, Vol. 13, 1974, pp. 215–218.

110. Volder, J. E., "The CORDIC Trigonometric Computing Technique," *IRE Transactions on Electronic Computers*, September 1959, pp. 330–334.

111. Cochran, D. S., "Algorithms and Accuracy in the HP-35," *Hewlett-Packard Journal*, June 1972, pp. 10–11.

112. Despain, A. M. "Fourier Transform Computers Using CORDIC Iterations," *IEEE Transactions on Computers*, Vol. C-23, No. 10, October 1974, pp. 993–1001.

113. Schreiber, R. and P. Kuekes, "Systolic Linear Algebra Machines in Digital Signal Processing," in *VLSI and Modern Signal Processing*, S. Y. Kung et al. (Eds.), Prentice-Hall, Englewood Cliffs, NJ, 1985, p. 401.

114. Steinhardt, A. O., "Householder Tranforms in Signal Processing," *IEEE ASSP Magazine*, July, 1988, pp. 4–12.

115. Kaminski, P. G., E. Bryson, Jr., and S. F. Schmidt, "Discrete Square Root Filtering: A Survey of Current Techniques," *IEEE Transactions on Automatic Control*, Vol. AC-16, No 6, December 1971, pp. 727–736.

116. Orfanidis, S. J., *Optimum Signal Processing*, 2nd ed., MacMillan New York, 1988.

117. Longley, J. W., "Least Squares Computations Using Orthogonalization Methods," in *Notes in Pure and Applied Mathematics*, Marcel Dekker, New York, 1984.

118. Bjorck, A., "Solving Linear Least Squares Problems by Gram-Schmidt Orthogonalization," *BIT*, Vol. 7, pp. 1–21, 1976.

119. Maybeck, P. S., *Stochastic Models, Estimation and Control*, Vol. 1, Academic New York, 1979.

120. Rice, J. R., "Experiments on Gram-Schmidt Orthogonalization," *Math. Comp.* Vol. 20, 1966, pp. 325–328.

121. Brown, R. G. and P. Y. C. Hwang, *Introduction to Random Signals and Applied Kalman Filtering.* 2nd ed., Wiley, New York, 1992.

122. Grewal, M. S., and A. P. Andrews, *Kalman Filtering Theory and Practice*, Prentice-Hall, Englewood Cliffs, NJ, 1993.

123. Swerling, P., "First Order Error Propagation in a Stagewise Smoothing Procedure for Statellite Observations," *Journal of Astronautical Sciences*, Vol. 6, Autumn 1959, pp. 46–52.

124. Morrison, N., "Tracking and Smoothing," Boston IEEE Modern Radar Technology, Lecture Series, November 20, 1973.

125. Kolmogorov, A. N., "Interpolation and Extrapolation of Stationary Random Sequences," translated by W. Doyle and J. Selin, Report No. RM-3090-PR, Rand Corp., Santa Monrica, CA, 1962 (first published in 1941 in the *USSR Science Academy Bulletin*).

126. Wiener, N., "The Extrapolation, Interpolation, and Smoothing of Stationary Time Series," OSRD 370, Report to the Services 19, Research Project DIC-6037, MIT, Cambridge, MA, February 1942.

127. Kailath, T., "A View of Three Decades of Linear Filtering Theory," IEEE Transactions on Information Theory, Vol. IT-20, 1974, pp. 146–181.

128. Lee, R. C. K., *Optimal Estimation, Identification and Control*, Research Monograph 28, MIT Press, Cambridge, MA, Chapter 3.

129. Rader, C., Lincoln Laboratory, MIT, Lexington, MA, private communication 1991

130. Ragazzini, J. R. and G. F. Franklin, *Sampled-Data Control Systems,* McGraw-Hill, New York, 1958.

131. Cadzow, T. A., *Discrete-Time Systems*, Prentice-Hall, Englewood Cliffs, NJ, 1973.

132. Papoulis, A., *Signal Analysis*, McGraw-Hill, New York, 1977.

133. Kuo, B. C., *Automatic Control Systems*, Prentice-Hall, Englewood Cliffs, NJ, 1962.

134. Selby, S. M., *Standard Mathematical Tables*, Chemical Rubber Co., Boca Raton, FL, 1973.

135. Proceedings of Paris International Radar Conferences, December 1978, April 1984, and May, 1994. Pub. by Comité Radar, Paris.

136. Sarcione, M., J. Mulcahey, D. Schmidt, K. Chang, M. Russell, R. Enzmann, P. Rawlinson, W. Guzak, R. Howard, and M. Mitchell, "The Design, Development and Testing of the THAAD (Theater High Altitude Area Defense) Solid State Phased Array (formerly Ground Based Radar)," paper presented at the 1996 IEEE International Symposium of Phased Array Systems and Technology, October 15–18, 1996, Boston, MA, pp. 260–265.

137. Brookner, E., "Major Advances in Phased Arrays: Part I," *Microwave Journal,* May 1997, pp. 288–294, "Major Advances in Phased Arrays: Part II," *Microwave Journal,* June 1997, pp. 84–92.

138. Steinhardt, A., "Orthogonal Transforms in Adaptive Nulling," Boston IEEE/AESS Lecture, July 6, 1989.

139. Bjorck, A., *Numerical Methods for Least Squares Problems*, Society for Industrial and Applied Mathematics (SIAM), Philadelphia, 1996.

140. McWhirter, J. G., "Signal Processing Architectures," Royal Signals and Radar Establishment, Malvern, England, October, 1987.

141. Kay, S. M. *Fundamentals of Statistical Signal Processing, Vol. 1, Estimation Theory,* Prentice-Hall, 1993.

142. Sorenson, H. W., "On the development of practical nonlinear filters," *Information Sciences,* Vol. 7, pp. 253–270, 1974.

143. Tanizaki, H., *Nonlinear Filters,* second edition, Sprinter-Verlag, 1996.

144. Daum, F. E., "New Exact Nonlinear Filters," Chapter 8 in *Bayesian Analysis of Time Series and Dynamic Models,* ed. by J. C. Spall, Marcel-Dekker Inc., 1988.

INDEX

465